WOOL & WATER

WOOL & WATER

THE GLOUCESTERSHIRE WOOLLEN INDUSTRY AND ITS MILLS

JENNIFER TANN

For
Roger,
Edmund, Oliver,
with love and gratitude

Cover illustrations. Front: Ebley Mill, a mid-nineteenth-century painting (Museum in the Park, Stroud). *Back, top*: Stanley Mill showing an arcaded cast-iron frame (Annie Blick); *below*: Gig at Longfords, 1963 (Jennifer Tann).

First published 2012

The History Press
The Mill, Brimscombe Port
Stroud, Gloucestershire, GL5 2QG
www.thehistorypress.co.uk

© Jennifer Tann, 2012

The right of Jennifer Tann to be identified as the Author of this work has been asserted in accordance with the Copyrights, Designs and Patents Act 1988.

All rights reserved. No part of this book may be reprinted or reproduced or utilised in any form or by any electronic, mechanical or other means, now known or hereafter invented, including photocopying and recording, or in any information storage or retrieval system, without the permission in writing from the Publishers.

British Library Cataloguing in Publication Data.
A catalogue record for this book is available from the British Library.

ISBN 978 0 7524 6215 8

Typesetting and origination by The History Press
Printed in Great Britain
Manufacturing managed by Jellyfish Print Solutions Ltd

CONTENTS

	Acknowledgements	6
Part I		8
Chapter One	The Medieval Woollen Industry	8
Chapter Two	The Gloucestershire Woollen Industry in Tudor Times	20
Chapter Three	Organisation and Location of the Woollen Industry from the Seventeenth Century to the Eve of the Industrial Revolution	31
Chapter Four	Woollen Trade Fluctuations and State Intervention in the Seventeenth and Eighteenth Centuries	54
Chapter Five	Industrial Revolution 1790–1835: the Advent of Machinery	63
Chapter Six	The Age of Machinery: Sources of Power	90
Chapter Seven	Industrial Revolution 1790–1835: Labour and its Displacement	103
Chapter Eight	Concentration and Decline 1836–80	117
Chapter Nine	Decline 1880–1914	138
Chapter Ten	Postscript	155
Appendix	Enquiries and Orders for Boulton & Watt Steam Engines in Gloucestershire Woollen Mills 1802–39	165
	Bibliography and Sources	168
Part II	**Mills and Associated Buildings by River Valley**	173
	Index	253

ACKNOWLEDGEMENTS

In the wave of enthusiasm for industrial archaeology I wrote *Gloucestershire Woollen Mills*. This is a different book in which I seek to tell the story of the woollen industry of Gloucestershire while correcting and updating, in Part II, a record of fulling and cloth mill sites. Space precludes a full account of some of the better-known mills, but I have included references – and a somewhat longer section on some of the mills has been deposited at Gloucestershire Archives and the Museum in the Park, Stroud.

There are many people to thank:

The Pasold Research Fund for a grant towards research expenses which has enabled me to undertake fieldwork, visit archive collections and have maps professionally drawn.

I gratefully acknowledge my indebtedness to others who have written about the West of England woollen industry and its mills: the late Julia de Lacy Mann, the late Kenneth Ponting, Kenneth Rogers, David Jenkins, Pat Hudson, Mary Rose, Stanley Chapman, Esther Moir, Colleen Haine and Simon Mills.

All those I thanked when writing *Gloucestershire Woollen Mills*, in particular Leicester University, the (then) staff of the (then) County Record Office and Gloucester City Library; the staff of The National Archives, British Museum and Tithe Redemption Office. I am grateful to the late Mr Rex Wailes for his advice and enthusiasm (and for his distinguishing between 'dead lions' and 'live donkeys' when it came to industrial heritage); to Lionel Walrond who shared his knowledge of local mills. The late Jack Marshall was immensely kind and shared his knowledge of the woollen industry and Stanley Mill with me. I thank the anonymous person(s) who made some marginal corrections in the Gloucestershire Archives' copy of *Gloucestershire Woollen Mills*.

At Gloucestershire Archives I am greatly indebted to: Vicky Thorpe and Paul Evans; Andrew Parry for his excellent photography; and all the staff for their professional knowledge, their kindness and their smiles.

I am grateful to Nicholas Kingsley for his guidance in his various roles as archivist nationally, in Gloucester and in Birmingham.

I am grateful to my fellow Trustees of Stroudwater Textile Trust: Ian Mackintosh for his knowledge of Stroudwater mills, Anthony Burton for his knowledge of engineering history, Tony Preece for his appreciation of the mills landscape, Anna Shelley and Paul Butler for their knowledge of conservation, Tim Parry Williams and Liz Lippiatt for their contemporary textile design skills. I am grateful to members of the Heritage Committee: the above Trustees, Jane Ford, Robin Mitchell and Maureen Anderson who work to ensure that the story of the Gloucestershire woollen industry is kept alive.

I am grateful to the Museum in the Park, Stroud, for permission to reproduce illustrations and, in particular, to David Mullin, Curator, who has been unfailingly helpful and has gone out of his way to locate material and answer questions.

The staff at Birmingham Archives have been very helpful over my years of researching Boulton and Watt; I thank them.

I thank the volunteers of Nailsworth Archives for their help, and members of Nailsworth Local History Research Group.

Howard Beard has been most generous with his knowledge and permission to reproduce images from his collection. Mike Mills has been equally generous with his knowledge and permission to reproduce images from his collection. I thank them both.

I am grateful to Harry Buglass and Louise Buglass for their skill with mills maps, and especially to Harry for his patience, good humour and generosity in putting up with my changes on several occasions.

Bovis Homes kindly gave permission to view, photograph and reproduce images from the Steanbridge mural; I thank Scott Curtiss and Margaret Arnold for their help. I also thank David Calvert, whose expertise is evident in his photographs of the Steanbridge mural.

I am grateful to Revd Michael Chappell for sharing his knowledge (and putting me right) on Wotton mills; Juliet Shipman for helping me with Chalford Place and Oakridge mills; Neil Baker for sharing his enthusiasm for and knowledge of Weyhouse Mill; Paul Clutterbuck whose infectious enthusiasm for Clutterbuck history extended to him lending me his precious rare book on the subject; Audrey Penrose who has allowed me to crawl over St Mary's Mill and has shared her infectious enthusiasm for the mill estate.

I thank Renishaw plc for their courtesy in giving permission to reproduce the illustration of New Mill, Wotton, and to Chris Pockett for all his help at Inchbrook/Dyehouse mills; John Looseley for sharing his knowledge of the Stroudwater Riots; I thank Annie Blick for the sequence of photographs she took of Stanley Mill; and Roger Tann for taking many photographs in Stroudwater.

I thank Bob Jeavons of St John's Mill, Isle of Man, for permission to use the photograph of the fulling stocks there – and Danny Morris who kindly took the photograph.

I thank all owners of mills who have been welcoming, instructive and unfailingly courteous. I have tried to identify all copyright holders of images used. If, for any reason, there is an omission, I apologise.

And above all, I thank my family who have put up with my preoccupation.

PART I

CHAPTER ONE

THE MEDIEVAL WOOLLEN INDUSTRY

Woollen cloth production in medieval times was widely distributed throughout Britain. Much of it, particularly cloth produced in the countryside, was for local use. In some towns and cities, however, a thriving market-based and highly controlled production system was in place. In Gloucestershire, some of the towns noted for the wool trade, e.g. Chipping Campden, Stow on the Wold and Northleach, developed a woollen industry but, with some exceptions, notably Cirencester, the industry did not last long in those places. The main city for cloth manufacture was Bristol with its thriving guilds; to a lesser extent there was Gloucester. However, in the late twelfth century things began to change and cloth production for a wider market began to migrate to the countryside.

In the early Middle Ages woollen cloth was made by a series of hand operations. First the wool was carded or combed; next it was spun on a simple distaff; then woven on a loom worked by hand and foot, after which the cloth was fulled. This latter process, one of beating the cloth in water with fuller's earth or other cleansing agents, shrank the cloth, reducing it in both length and width giving it greater resistance to wear and weather. After fulling, the fibres were so densely packed that the pattern of weaving was usually not visible. This was an essential antecedent stage to the finishing processes – for finer cloth – of raising the nap and shearing, a process often called raising.

THE URBAN INDUSTRY

Spinners and weavers worked with English wool which was sold in market towns throughout the country or supplied by middlemen. Such was the importance of English wool that it both met the needs of domestic producers of woollen cloth, as well as being England's chief export for the cloth industries of Italy and the Low Countries – the great 'wool churches' of Northleach, Cirencester and Fairford, and others, being testament to the wealth generated by the medieval wool trade. Nonetheless there were occasions when the needs of the domestic industry prevailed over foreign sales and the export of wool was prohibited in 1244 and 1258. Some Spanish wool and yarn was imported by the mid-thirteenth century, being subject to tolls at the ports. Fuller's earth was available in England and it does not seem to have been imported. Nor would teazles have been imported – indeed, it seems that there may have been some export trade in these. In the early fourteenth century the Flemings were said to be buying up teazles, creating a shortage in England and Edward II was persuaded to prohibit their export.[1]

Some dyestuffs were grown in the Gloucestershire countryside. Weld, which produced a good yellow, was commonly grown in wheat and barley fields after harvest and cut the following year. Dyers' broom was collected from hedgerows by villagers and sold in local markets but was probably only used by smaller country weavers producing for their families and to meet local demand. Woad was the most important dyestuff, being required not only for blue cloth but also for other colours. The deepest blue dye, perse, was dyed in woad alone on white wool and mordanted with ashes. The woad plant had been cultivated in England from Saxon times and, in medieval times, was grown around Wotton-under-Edge, Dursley and Tewkesbury. Madder was the commonest English-grown red dye and it was sometimes used with woad to give different shades of blue.

Dyestuffs were, however, imported in large quantities. By the twelfth century, there were considerable imports of woad from Western Europe, particularly Picardy, much of it being sent to London.[2] It was also brought to Bristol in volume. By the fifteenth century, the Languedoc cloth industry had almost disappeared and Bristol merchants spoke of the woad trade as this 'moost chief, noblest and ponderoust merchaundys of good & able Wode'.[3]

Other red dyes – brazil, vermillion and grain (the most costly) – were wholly imported. The terms brazil, vermillion and scarlet were used in the Middle Ages both for the dyestuffs and the colour they yielded, although by the sixteenth century the terms were applied generally only to the dyestuffs. Brazil may have come from the East Indies and vermillion, a mineral derived from a crystalline substance, was said to have been found on the shores of the Red Sea. Grain, which produced a brilliant and lasting scarlet colour, was derived from the dried body of a grain-like insect found on a species of oak tree which flourished in Portugal and the Mediterranean and was used for royal robes.

A number of mordants were imported, particularly potash produced from wood ashes by leaching. Although this could be produced in England it was largely imported, particularly from the Baltic, by the mid-thirteenth century. Alum, the most commonly used mordant after potash and the sole mordant used for scarlet, was imported from Spain and southern Europe. The finest quality was said to come from the shores of the Black Sea and Genoese merchants were importing alum into England at the beginning of the fourteenth century. Verdigris and copperas were probably used as mordants; verdigris being used to fix colour, while copperas tended to deepen the colour as well as fix it. Argol, whilst not strictly a mordant, was used with other mordants for brightening colour.

Spain was the chief source of Gallipoli oil which came to England in the form of ready-made soap from Castile and was used in the manufacture of the highest quality cloths. Olive oil was imported to Bristol from Portugal which, when old, was used for dressing wool prior to spinning.[4]

Dyers were dependent upon not only their skills but also their knowledge of the best foreign supplies of the dyestuffs required for their trade. Many dyers were people of substance and influence, dealing directly with foreign merchants. Some were merchants and entrepreneurs rather than artisans, the labour of dying being undertaken by employees. It was this dependence upon imported materials, together with the need to ensure high quality, which drew dyers together in cities with others who were buying and selling in order to regulate their trade, and dyers were frequently members of the merchant guilds. There were many dyers in merchant guilds in the twelfth and thirteenth centuries and no evidence of weavers and fullers being members. The difference was between the somewhat more elevated position of town and city-based dyers who, comparatively rarely, had a designated guild (Bristol being an exception) and the weavers and fullers who were entirely outside merchant guilds and closely united in guilds of their own. Urban dyers, unlike weavers and fullers, were among the ruling class of rapidly growing towns, playing a part in civic policy.

They were sufficiently powerful to influence the Crown to prohibit the dyeing of cloth in the countryside, except that which was done for people's own use and not for sale. It appears that, in the major cloth-making towns and cities, some dyers had become entrepreneurs controlling the production of cloth and selling it at the great fairs. There is evidence of capitalism in the English cloth industry dating not from the late fourteenth and fifteenth centuries but from two centuries earlier.[5]

Bristol was one of the most significant centres of the woollen industry in the fourteenth century, it being suggested that at least 1,500 people were employed in the various craft processes. Of the nineteen city guilds, the Weavers' and Tuckers' were amongst the largest, the Dyers' and Shearmen also being large. The ordinances of the Weavers' Guild, made in 1346, demonstrate the restrictions imposed on guild members and were typical of the period. Cloth was to be woven to a specified breadth (six bonds); only the correct yarn was to be used for warp and weft; the threads were not to be too far apart or deficient in number; weavers were to receive yarn from no one but their husbands or wives. Each weaver had to make his own mark on the cloth so that, in case of faulty manufacture, the maker could be punished by the guild. There were also restrictions on how the weaver was to practise his craft; all weavers were to become burgesses of the town and be received into its liberty; tools were not to stand in upper rooms or cellars but only 'in halls or shopps next to the road in the sight of the people'. In 1389 further ordinances were made, no one being permitted to employ any servant in the craft unless it was for an entire year.[6]

The Bristol Fullers' Guild had a number of ordinances, which included a prohibition on sending cloth outside the town to be fulled. The fuller was responsible for fulling, dressing and inspecting the cloth, and was required to make amends to the owner if he damaged it in any way. In 1347 ordinances were sworn before the twelve chief fullers and 'the commons of the same mystery', a total of eighty fullers. 'Working at the stock' and sending work 'to the mill' were specifically mentioned. The Fullers' Guild's ordinances do not record any prohibitions as to the existence and setting up of fulling mills in the city. In 1381 a further ordinance required that no fuller was to receive cloth sold outside the town for 'rekkying, pleyting or amending' and all cloth washed in the town was to be washed in water and clay only. The 1346 ordinances of the Fullers' Guild distinguished between those who worked in the stocks and those who worked at the perch – the wooden bar over which cloth was hung in order for the surface to be raised with teazles.[7]

There were some general rules concerning cloth making in Bristol which were not specific to a particular guild. No wool was to be taken out of the town to be combed or spun; no oiled wool or thread could be sold on days other than Fridays and all cloth to be sold in the town was to be sold at a house in Baldwyn Street, to ensure that customs duty was paid – a similar ordinance being passed in Gloucester where 'no maner of clothe of Chapmen or merchaunt foren (was) to be sold nor were packes to be opened for sale withyn the said towne but only withyn the Bothall'.

By the beginning of the fifteenth century the Bristol guild ordinances hint at their ineffectiveness in controlling their respective craft trades. Wool textile industry regulation, albeit largely ineffectual, was perceived to be the means by which manufacture could be nurtured and standards maintained. It was also a means of controlling the number of new entrants to a craft. The remedy in Bristol (and other towns with a textile industry) was to institute more rules and enforce them more strictly. The Weavers' Guild Book records, in 1419, that trespasses against previous ordinances had led to a situation 'whereby the drapery is worse than it used to be before this time'.[8] A decline in the weaving trade was remarked upon in 1461, the blame being put on those weavers who set their wives and daughters to the loom, harming male weavers 'who go the vagaraunt and unoccupied and may not have ther labour to ther levying'.[9] During the reign of Edward IV (1461–83) weavers complained to the mayor and burgesses that only master weavers were entitled to elect the twelve special officers of the guild and that they looked after 'their owne purs'. By 1490 the master weavers had drawn up a petition liberally distributing blame for the decay of the cloth trade in Bristol, while not looking to themselves. Fullers accused merchants of sending cloth outside the city to be fulled and, in 1479, turned the accusation upon their own members:

> For as moche as the common well of the seide Worshipfull towne and contrayes adionant restith on cloth making – it be nowe ordaies the more pite is the makyng and giddying of and drapery is not alther trewist ne maner lokyst [correctly] handelid.[10]

Dyers, too, had their complaints. In particular it was asserted that men of other crafts had begun to dye cloth whereby 'the masters and apprentices of the said craft of dyeing are vagrants through want of work'.[11]

The cloth industry was well established in Winchcombe by 1327 and among the goods taxed on being brought into the town over a three-year period, in aid of paving the streets, were wool, woad, teazles and alum. Winchcombe cloth fair, to which cloth was brought from outside the town, was sufficiently important for the Abbot of Abingdon to provide a house nearby for his chamberlain, beside stabling for his horses to carry cloth bought there.[12]

A Fullers Street, recorded in Tewkesbury in 1257, indicates an established cloth industry in the town;[13] while Ralph the Weaver, Henry the Dobber (dyer) and Norman the Fellmonger in Cirencester, besides the renaming of Chepyng Street as Dyer Street, are evidence of Cirencester's growth as a centre for the industry.[14] A list of merchandise brought to Cirencester market included wool, fleeces, woad, alum, copperas and argule,[15] and a Hugh the Walker is recorded in Cheltenham in 1333.[16] Gloucester was a centre of cloth manufacture in the twelfth and thirteenth centuries; Wulward the Fuller was amongst the wealthiest burgesses in 1173 and eight different dyers were recorded c.1230. As in Bristol, Gloucester dyers were the most prosperous of the clothworkers. From 1230 onwards John the Dyer and William the Dyer both appear as witnesses to legal documents, an indication of the status of city dyers.[17]

THE MOVE TO THE COUNTRYSIDE

Fulling both felted and scoured the cloth, removing the oil with which the wool had been treated before spinning. It was a lengthy process undertaken by beating with the feet, hands, or with clubs. Fulling by foot was probably employed for heavier woollens such as those made in the West of England. The ordinances of the Fullers' Guild in medieval Bristol mention working 'in the trough'. There are some illustrations in medieval manuscripts of fullers working by foot and, in some areas such as the Hebrides, where cloth was made for local markets, this process continued until the late nineteenth century. While carding, spinning and weaving were mechanised during the Industrial Revolution of the eighteenth and nineteenth centuries, fulling was mechanised during the late eleventh century – an early example being at the Abbey of St Wandrille in Normandy between 1060 and 1080.[18] The technology is known to have reached England by the late twelfth century, although it is likely to have arrived in England earlier in the century. What has been called 'an industrial revolution of the thirteenth century'[19] initiated a transformation which was not only technological but also locational and social. The technological change translated the primitive method of fulling by foot to a mechanical action of two wooden hammers which were alternately raised and dropped on the cloth which lay in a trough, the hammers being raised by cams on a basic wheel attached to the axis of a waterwheel. This became known as a fulling mill (or tuck or walk mill) to distinguish it from a water corn or grist mill, the term mill being applied to the machinery as distinct from the building which housed it. It is possible that the first fulling mills were installed in the same building as corn milling machinery and, by the fifteenth century, there are a number of references to fulling and corn mills 'under one roof'.[20]

Fulling stocks at St John's Mill, Isle of Man; the metal box is dated 1815. (Danny Morris)

In locational terms fulling mills were sited alongside streams and rivers which provided adequate water power. Bristol was not well supplied with water power, although there is reference to a fulling mill within Bristol fullers' jurisdiction. In other towns where trades were controlled, fulling by power was explicitly prohibited. Rural fulling mills became widespread in England and the Welsh borders between the end of the twelfth and the beginning of the fourteenth centuries, and rapidly superseded primitive fulling by foot. These changes affected the distribution of the entire English woollen industry. Concentrations of fulling mills had emerged by the end of the fourteenth century in the Cotswolds, Wiltshire, Somerset, Devon and Cornwall in the South-West, the West Riding of Yorkshire, and the Lake District, in each of which areas there was adequate to good water power available. These mills were in country districts rather than towns and in the hilly areas of northern and western England rather than the flat lands of East Anglia and the South-East. This is not to say that the woollen industry disappeared from these latter areas (nor that fulling mills were not built in the flatter lands of, say, East Anglia), but that, for example, Bristol, Lincoln, Stamford, Northampton, Oxford, Winchester, Leicester and Colchester, amongst other towns, experienced a decline in textile manufacture from which, on the whole, they did not recover. Water power became the decisive factor in the location of the English woollen industry for 700 years.

The social revolution which accompanied this technological and locational one was two-fold. First, the industry, for a while, escaped the tight regulation to which it had been subjected in the towns and cities. Second, with the rise of the rural wool textile industry, poverty, unemployment and distress emerged in the older urban centres as evidenced by the complaints of various Bristol textile craftworkers in the mid-fourteenth century. Clothworkers left the towns and weavers began to relocate near to fulling mills and, with their arrival, the need for spinners increased. Rural populations grew from small hamlets to larger villages and small towns.

In Gloucestershire, as elsewhere, the religious orders appear to have been amongst the first to realise the opportunities for income generation from fulling mills.[21] They had the necessary resources to adapt their manorial corn mills and erect new buildings, as well as create the more elaborate watercourses required for larger industrial sites where fulling and corn milling were undertaken in the same building. The earliest known fulling mill in Gloucestershire was at Barton, Temple Guiting, where the Knights Templar had a mill in 1185.[22] Between 1199 and 1216 the Abbot of Winchcombe directed that the mill leat be straightened and the Winchcombe manor corn mill converted into a fulling mill.[23] One had already been built on the abbot's manor at Enstone in Oxfordshire in 1189[24] and John Blundell received 8*d* in damages when the monks diverted some of the water intended for his fulling mill to their own.[25] In 1309 Abbott Walter renewed the lease of a fulling mill in Coates, requiring that the tenant leave the pond and furnace in good condition – implying that dyeing was carried out at the same site.[26] The abbot had one other fulling mill in the vicinity of Winchcombe, at Throp, which may have been part of the manor of Sudeley.[27] In the early fourteenth century the Abbot of Winchcombe, apparently seeking to achieve a monopoly on cloth processing in the district, covenanted with the owner of a grist mill locally not to convert it into a fulling mill.[28] The abbot also had a fulling mill on his manor of Sherborne which is recorded in a pre-1224 charter.[29] Sherborne, high on the Cotswolds, was the collecting centre for the abbot's flocks at shearing time. After Easter each year the abbot and his servants travelled from Winchcombe to supervise the shearing, an account of 1468 showing that 1,900 sheep were sheared and labour hired especially for the task.[30] By 1341 several inhabitants of the manor of Sherborne were 'in mercy' for using fulling mills outside the manor.[31] One of the three corn mills in Sherborne was leased to the Lord of the Manor of Farmington, with the proviso that it should not be converted to a fulling mill without the abbot's licence.[32]

St Peter's Abbey, Gloucester, owned a fulling mill at Nether Coberley just north of Cowley and one at Hinton-on-the-Green;[33] Leonard Stanley Priory owned a mill at Millend, Eastington; Llanthony Priory owned two mills at Aylurton in the Forest of Dean, one of which was a fulling mill in the thirteenth century.[34] The Priory also owned a fulling mill at Little Barrington.[35] As at Sherborne, Barrington, high on the Cotswolds, was the collecting centre for the prior's sheep. Flaxley Abbey also owned a fulling mill in the Forest of Dean on their manor of Flaxley which was held at a higher rent than the manor corn mill.[36] Tewkesbury Abbey owned a fulling mill at Stanway, as well as three corn mills.[37]

Four abbeys located outside Gloucestershire are known to have owned fulling mills in the county. Evesham Abbey had a fulling mill at Bourton-on-the-Water in 1206, the rents of which went towards the upkeep of the monks' infirmary.[38] A fulling mill in Cowley belonged to Pershore Abbey;[39] Lacock Abbey owned a recently built fulling mill at Hatherop in 1248;[40] and Westminster Abbey owned a fulling mill at Sutton-under-Brayles before 1362.[41] Only one episcopally owned mill has been found in Gloucestershire; this was Barton Mill which by 1327 was part of the Bishop of Winchester's estates.[42]

Lay landowners were not far behind religious houses in acquiring or installing fulling mills. A tenant at Hawkesbury was fined in the reign of Edward II for having cloth fulled elsewhere than at the lord's mill;[43] there was a fulling mill at Dursley during the time of Henry de Berkeley I;[44] a fulling mill at Minchinhampton, possibly at Lodgemore, is recorded in the thirteenth century in the ownership of Thomas de Rodborough. By the early fifteenth century, John Bygge owned this mill and one on the other side of the Frome crossing at what was then called Herford, which marked the boundary between de Rodborough and Spillman lands.[45] There were fulling mills at Siddington and South Cerney on the Coln in 1285.[46] In addition, there were lay-owned mills in the Severn Valley during this period, one being recorded at Wheatenhurst on the Frome in 1336;[47] another at Bitton on the River Boyd.[48] In the latter two cases there were also corn mills, probably on the same sites, which were let at higher rents than the fulling mills. In other cases, for example at Kemerton, part of the manor of Tewkesbury, the fulling mill was worth more than twice that of the corn mill.[49] There was a fulling mill at Chedworth in 1298 and one each at Quenington in 1338 and Fairford in 1296 on the River Coln,[50] while on the Churn there was a mill at Rendcomb in the late fourteenth century.[51] West of the River Severn there was a lay-owned mill in English Bicknor, held jointly with a corn mill, which was valued at more than twice as much.[52]

Map 1 (overleaf) shows the distribution of known fulling mills in Gloucestershire between 1185 and 1399 – and there were probably a number more of which no record has been found. These early mills were widely dispersed; the rivers flowing towards the River Thames being as favoured as those which flow to the Severn. Some mills were situated on very small streams, a mile or so from the source, with scarcely sufficient water to run all year, as at Chedworth, Flaxley, Sherborne and Stanway. While the existence of monastic flocks probably influenced the acquisition or building of mills on the high Cotswolds

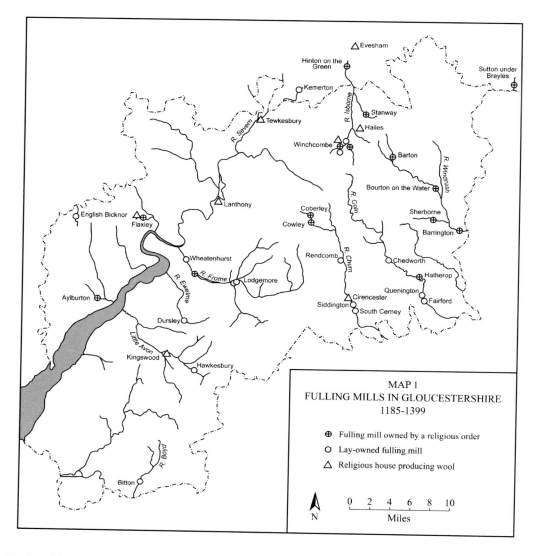

MAP 1
FULLING MILLS IN GLOUCESTERSHIRE
1185-1399

⊕ Fulling mill owned by a religious order
O Lay-owned fulling mill
△ Religious house producing wool

by the Abbot of Winchcombe and the Prior of Llanthony, the proximity of markets was also important, as at Fairford, and, above all, human initiative and the capital to accomplish building projects, with the authority to impose the requirement to use the manorial mill.

The growth of the rural cloth industry in Gloucestershire is clear from sources other than records of mill ownership or tenancies. *The Minchinhampton Custumal* of 1276 and 1306 names eight people who paid rents for digging fuller's earth, four of whom – Roger of Woodchester, Richard, Reginald and Simon – were named Fuller.[53] Court records of thefts provide some evidence of cloth manufacture, the type of cloth made and its value. In 1272 Robert the Weaver, William the Weaver of Rodborough and Roger the Fuller of Rodborough appear in the court rolls. Nicholas, son of Roger the fuller, and Robert Morice stole russet woollen cloth from Nicholas Walkare (in other words, Walker or fuller) at Frampton on Severn. Walter Taillour of Uley stole woollen cloth, priced at 13s 4d, from John Walkare of Kingswood; Richard Pyrie of Frampton on Severn broke into the house of John Whyte of Frocester and stole wool and other goods to the value of 40s; and Alice Pridie of Stroud stole three measures of white cloth worth 1s 6d from Walter Deleman of Stroud.[54]

While at the time of the Domesday survey there were five water (corn) mills on Bisley Manor and eight in Minchinhampton, by 1300 at least six mills can be identified along mid-Stroudwater,[55] although it is not clear how many were for fulling. The mill at Longfords was let by the Lord of the Manor at a rent of 15s 4½d but was being sublet for 66s 8d in the 1450s. Watercourse leases for Bisley in the mid-fifteenth century show that watercourses were being subdivided, each lease being carefully framed with reference to particular mills, bridges, meadows or houses.[56] Chalford Vale was divided between the manors of Minchinhampton and Bisley, and mills, with their related watercourses, were sometimes the focus for disputes about ownership, one dragging on for over twenty years. During one dispute a mill tenant was forcibly ejected, two men were killed and three pairs of shears stolen.[57] Some Cirencester clothiers moved into Stroudwater, leasing properties in order to have their own mill and dyehouse. John Benet of Cirencester had major business interests in Rodborough where he held a water mill. In his will he left money to the parish churches of Minchinhampton and Woodchester, as well as the chapels of Rodborough and Stroud, and to many individuals, including at least fourteen employees. Edward Haliday of Rodborough, cloth maker, whose brass is in Minchinhampton church, was the son of a fuller and also a subtenant of Rodborough where he had his mill and dyehouse.[58]

There is little evidence of the amount of cloth made in Gloucestershire at this period. By the end of the fourteenth century a far higher proportion of the domestic wool crop was being manufactured in England, and employment in the wool textile industry must have increased, despite the population as a whole having been reduced by, perhaps, one-third as a result of the plagues. An indicator of the growth of wool textile production in the fourteenth and fifteenth centuries was the recognition of its value for raising national taxes, with the imposition of a requirement for cloth to be inspected, sealed and taxed by an aulnager. It is tempting to use the aulnage records as an indicator of the size of the woollen industry[59] – Gloucestershire aulnage returns survive from 1474 to 1478 and a longer series survives for Wiltshire and Somerset. However, a detailed examination of the records by E.M. Carus-Wilson argues compellingly that they cannot be considered sufficiently reliable evidence; the aulnager 'cooked his accounts, leaving behind him works of art rather than transcriptions of fact'.[60]

MARKETS FOR ENGLISH WOOLLENS IN THE FOURTEENTH CENTURY

Gloucestershire cloth, particularly that made by the smaller clothier/weavers, would have been sold in the many local markets and fairs in the county. In all, charters for fifty-nine markets and fairs in Gloucestershire were granted in the thirteenth and fourteenth centuries, some of them being in centres of cloth manufacture such as Cirencester, Tetbury, Sodbury, Gloucester, Berkeley and King's Stanley.[61]

Most of the cloth manufactured in Gloucestershire which was destined for overseas was exported from Bristol. Despite the quantities of dyestuffs imported through Bristol, much of the cloth exported was 'white'; in other words, undyed broadcloth. It was in the fourteenth century, roughly from the outbreak of the Hundred Years War in 1337, that the export of English woollens overtook the woollen industries of Flanders and Italy, and England was transformed from being predominantly an exporter of raw materials to being an exporter of manufactured goods. English cloth was recognised as a source of income for the Crown and taxes were imposed on exports by foreigners from 1303 and on native exports from 1347, the Exchequer Customs Accounts covering all the major English ports. There was not a steady upward trend in exports but three marked periods of expansion, each of some fifteen years, separated by two sharp setbacks. The first expansion of the 1330s and 1340s – from the outbreak of the Hundred Years War to the eve of the Black Death – is harder to discern since this occurred when only alien shipments of cloth paid duty. In the 1330s and 1340s English woollens captured the home market; export volume was still comparatively small but, in the 1330s, cloth imports fell and by 1340 had virtually ceased. In addition, there were large government orders for clothing the armed forces. The five years succeeding the outbreak of the Black Death, which reached England in 1348, were years of depression for the growing cloth industry; the annual export total did not reach 2,000 cloths.[62]

The principal market for England's cloth at this time was Gascony which was part of the realm of the King of England. Three years before the Black Death arrived war struck Gascony. There were five years of respite, from 1340 to 1345, followed by renewed war, together with plague and famine. This led to a collapse in Gascony's wine production and, consequently, its purchasing power. By 1353 there was some recovery through to the end of the 1360s and England's woollen trade exports rapidly rose from less than 2,000 cloths per year to almost 16,000 in 1366–68.[63]

The English manufacturer had a considerable advantage over Flemish and Italian rivals in the relatively low cost of raw materials. The export duty on wool was high – some 33 per cent – while the export duty on cloth imposed in 1347 was only around 2 per cent. Bristol, the chief port for the Gascon trade, was exporting half or more of England's cloth and her annual exports rose from 1,300 cloths in 1353–55 to over 6,000 in 1361–63, and in 1367 they reached almost 8,000. The late 1350s and early 1360s, despite a second outbreak of plague in 1361, were a time of prosperity and not only for prosperous Bristol cloth merchants such as Robert Cheddar. But, by the end of the 1360s, there was a succession of disasters with a corresponding slump in England's foreign trade. Plague appeared again in 1368, lasting for two years, following which England suffered with severe floods and the harvest was a failure. The conflict between England and France, which had smouldered since 1363, broke out in 1369 and raged until a truce in 1375. And in these years there were further plagues, and exports of cloth from Bristol dropped severely in 1371–72.

When customs returns are again available, at the opening of Richard II's reign in 1377, they show there had already been some recovery. England was still an exporter of wool but the export of woollens had increased dramatically. Towards the end of the fourteenth century Bristol developed an active trading relationship with Portugal and the Toulouse region, whence dyers were now obtaining most of their woad, rather than from Picardy. In 1389 leave was sought for a Genoese ship to be unloaded in Bristol, the chief cargo being alum, as there was 'at present little of such merchandise in those parts and it is greatly in demand'.[64] The destinations to which Bristol ships went with cloth included Flanders, Spain, Portugal and western France.

However, in the later fifteenth century, the London-based organisation known as the Merchant Adventurers emerged and rapidly grew in power. As Bristol's cloth exporting role declined, Merchant Adventurers assumed the prime role in the export of West of England cloth, a dominance they were to maintain until the Industrial Revolution. Relations between the Merchants of the Staple (who had dominated the wool trade) and the Merchant Adventurers became more strained as the Adventurers expanded the export of English manufactured cloth, and wool exports declined. Various skirmishes ensued and wise merchants became members of both the Staplers and the Merchant Adventurers.

In 1486 an Act was passed in the City of London Common Council formally creating the Merchant Adventurers of London. In some ways this new organisation resembled a medieval guild but in other ways it was markedly different. For instance, it had no meeting place of its own but usually assembled in the hall of the Mercers. It grew in strength and actively concerned itself in the cloth trade, resembling, rather, some early form of joint stock company since each member traded individually while, as a collective, they chartered ships, fixed the freight and decided when the fleet should sail. The significance of the woollen industry to the Exchequer (and the power of its exporting merchants) is indicated by the royal protection which was both sought and granted. When there were rumours of hostile ships in the English

Channel Henry VII gave warship protection for a cloth fleet. In 1492 the Adventurers' fleet went out with the King's ships of war, and later, both Henry VII and Henry VIII, more than once, lent ships and soldiers to accompany the Merchant Adventurers fleet.[65]

A Frenchman, writing in the mid-fifteenth century, depicted the wealthier end of the woollen cloth industry:

> In England, your clothiers dwel in great femes abrode in the countrey, havying howses with commodities lyke unto gentylmen, where aswel they make cloth and kepe husbandry, and also grasse and fede shepe and cattell.[66]

By the late fifteenth century 'Stroudwaters' was a familiar word in both British and overseas markets as a collective name for a particular kind of high-quality woollen cloth. In 1487 a Hanseatic agent, discussing the purchase of cloth at Blackwell Hall, spoke of 'well dyed and costly [cloth] such as Bristols, Stroudwaters, Castlecombes and such like'.[67] Stroudwaters took their name from the River Frome which flows swiftly in a steep-sided valley between Bisley and Minchinhampton; a valley which had no adequate road along it until the early nineteenth century, communication being from upland village to upland village across the valley at the mill sites.

By the end of the Middle Ages England's woollen industry had been transformed from one based in major market towns and cities, where in some towns it was subject to guild regulations, to an industry located along river valleys, sometimes sparsely populated ones, where water power could be generated for fulling mills. The so-called Industrial Revolution of the thirteenth century had a profound effect not just in locational terms and in speeding up a tiresome process, but also for control and regulation; city guilds were powerless to intervene and, for some time, the Crown was unable to impose effective regulation. The rural industry, which had undoubtedly existed in the eleventh to thirteenth centuries meeting local needs, was changed by the adoption of the fulling mill. This created a new dynamic, providing a basis upon which the more enterprising could manufacture cloth for home and overseas markets, some becoming very wealthy in the process.

Endnotes

(All MSS references, unless otherwise attributed (e.g. Public Record Office - PRO), are to Gloucestershire Archives.)

1. E.M. Carus-Wilson, 'The English Cloth Industry in the Late Twelfth and Early Thireeenth Centuries', in *Medieval Merchant Venturers*, p.215
2. Ibid., p.217
3. E.M. Carus-Wilson, 'The Overseas Trade of Bristol in the Fifteenth Century' in Ibid., p.37
4. Ibid., pp.51–8
5. E.M. Carus-Wilson, 'The English Cloth Industry', p.235
6. F.B. Bickley (ed.), *The Little Red Book of Bristol*, II, 1900, p.12
7. E.M. Carus-Wilson, 'The Overseas Trade of Bristol', p.222
8. Bickley, II, p.119
9. Ibid., p.27
10. Ibid., p.158
11. Ibid., p.82
12. *Landboc: Sive Registrum Monasterii de Winchelcumba*, I, p.63
13. *VCH Gloucestershire*, II, 1907, p.154

14   E.A. Fuller, *Ancient Cirencester*, 1874, p.5
15   *VCH Gloucestershire*, II, p.154
16   Ibid., p.157
17   Bristol and Gloucestershire Archaeological Society, Trans. BGAS, LVII, p.107; Madoc, History of the Exchequer, in R. Perry, 'The Gloucestershire Woollen Industry 1100–1690', Trans. BGAS, LXVI, 1945, p.58
18   R.V. Lennard, 'An Early Fulling Mill', a Note, *Economoic History Review*, XVII, 2, 1947, p.150
19   E.M. Carus-Wilson, 'An Industrial Revolution in the Thirteenth Century', in *Medieval Merchant Venturers*, pp.183–210
20   Jennifer Tann, 'Multiple Mills', *Medieval Archaeology*, XI (1967), pp.253–5
21   The existence of more monastic records than manorial for this period may skew the picture
22   W. Dugdale, *Monasticon Anglicanum* (1655–73), II, p.540
23   *Landboc*, I, p.63
24   Ibid., p.195
25   E. Dent, *Annals of Winchcombe and Sudeley*, 1877, p.92
26   *Landboc*, I, p.259
27   Formerly the property of John Waupol in 1252–1314, *Landboc*, II, p.155
28   E. Dent, p.82
29   *Landboc*, II, p.274
30   R.H. Hilton, 'Winchcombe Abbey and the Manor of Sherborne', in H.P.R. Finberg (ed.), *Gloucestershire Studies*, 1957, pp.110–11
31   D678/ct roll/96
32   R.H. Hilton, p.102
33   *Historia et Cartularium Monasterii Sancti Petri Gloucestriae*, III, p.60
34   *Taxatio Ecclesiastica Nicholai*, IV, p.172
35   PRO C115/A vol 7/F 193
36   *Taxatio*, p.171
37   'tria molendia aquatica et unu(m) fullaraticu apud Stanweye', *Taxatio*, p.234
38   W. Dugdale, *Monasticon Anglicanum*, II, p.24
39   PRO E315/61/F26
40   *VCH Gloucestershire*, VII, p.92
41   D1099/T (14)
42   PRO Ministers' Accts, 845-9
43   PRO Ct Rolls 175/41
44   I.H. Jeayes, *A Descriptive Catalogue of the Charters and Muniments at Berkeley Castle*, p.102
45   C.E. Watson, *The Minchinhampton Custumal*, Trans. BGAS, LIV, p.370
46   Inquisitiones Post Mortem (IPM) IV, pp.138, 186
47   IPM, V, p.258
48   Ibid., p.184
49   Ibid., IV, p.189
50   *VCH Gloucestershire*, VII, pp.78, 125
51   Ibid., p.225
52   IPM, IV, p.231
53   C.E. Watson, op. cit. p.323
54   E.G. Kimball, *Rolls of the Gloucestershire Sessions of the Peace*, Trans. BGAS, LXII, pp.68–9, 96

55 E.M. Carus-Wilson, 'Evidences of Industrial Growth on some Fifteenth Century Manors', in *Essays in Economic History*, Vol. 2, Edward Arnold, 1962, p.154
56 PRO SC6/850/26ff
57 Ibid., p.156; M.A. Rudd, *Abstracts of deeds relating to Chalford and Colcombe*, Trans. BGAS, LI, 1929, p.218
58 E.M. Carus-Wilson, 'Evidences of Industrial Growth', p.157
59 R. Perry, 'The Woollen Industry 1100–1690', Trans. BGAS, 1945, pp.55–6; R. Perry, *The Woollen Industry in Gloucestershire to 1914*, Ivy House Books, 2003, p.6
60 E.M. Carus-Wilson, 'The Aulnage Accounts: a Criticism', in *Medieval Merchant Venturers*, p.291
61 H.P.R. Finberg, 'The Genesis of the Gloucestershire Towns', in H.P.R. Finberg (ed.), *Gloucestershire Studies*, pp.86–8
62 E.M. Carus-Wilson, 'Trends in the Export of English Woollens in the Fourteenth Century', in *Medieval Merchant Venturers*, pp.239–64
63 Ibid.
64 Ibid.
65 E.M. Carus-Wilson, 'The Origins and Early Development of the Merchant Adventurers' Organisation in London', in *Medieval Merchant Venturers*, pp.143–82
66 E.M. Carus-Wilson, 'Evidences of Industrial Growth', p.151
67 Ibid., p.152

CHAPTER TWO

THE GLOUCESTERSHIRE WOOLLEN INDUSTRY IN TUDOR TIMES

By the opening of the sixteenth century three areas stood out as being by far the most important for cloth manufacture in England: the South-West (Gloucestershire, Wiltshire, Somerset, Dorset, Devon and Cornwall), East Anglia (Norfolk, Suffolk and Essex) and West Yorkshire (including the adjacent parts of Lancashire and the North Riding). While the industry had decayed in some English towns formerly noted for its manufacture, it was still important in a number of Gloucestershire towns besides Bristol. But it was on the wane in Bristol. The medieval crafts guilds clung to their authority over members, the latter, for their part, seeking protection from competition. In 1535 there were thirty-eight members of the Tuckers' Guild.[1] Calls for protection were rife as complaints were made against shearmen who were said to be usurping the work of tuckers who had previously undertaken all the finishing, including burling, raising and shearing.[2] And, while tuckers complained about the clothiers in the city,[3] they were not all, as claimed, impoverished; James Stevens, in his will of 1553, left house property and charitable bequests.[4]

The weavers' ordinances were reframed in 1562 with the intention of protecting the trade, the number of apprentices a weaver might have being limited to two.[5] Eventually, the guild bowed to the inevitable and permitted clothiers to send chains out of Bristol to weavers in southern Gloucestershire. The Weavers' Company struggled on into the seventeenth and eighteenth centuries, passing ever more draconian measures to protect the industry which, by then, had moved firmly beyond the controlling bounds of the city.

Leland's excursions in Gloucestershire in 1538 confirm the decay of cloth making in some of the towns in the plain. Of Thornbury he wrote: 'There hath been good clothing here … but now idleness much reynethe here.'[6] Berkeley had seen better days: 'it hath very much occupied, and yet somewhat dothe clothinge.'[7] Cirencester was still a busy centre of the industry, its Weavers' Company seeking to control and exclude in much the same way as the Bristol guilds.[8] Of the freemen admitted in Gloucester between 1535 and 1545, twenty-two were weavers, two clothiers, two dyers and one was a tucker.[9] The Gloucester Weavers' Company survived into the early seventeenth century but there were few weavers in the city.[10] It would seem that there was no fulling mill within the city, for a leading burgess, John Sandford, had a fulling mill in Stonehouse and John Cooke, a wealthy mercer, built a fulling mill at Ebley.[11]

The scale of growth of the Gloucestershire woollen industry can be seen by a comparison of two lay subsidy (tax) assessments: that of 1334 and that of 1523. While neither can be taken as an indicator of absolute wealth, they are a useful comparative indicator. Gloucestershire experienced a 62 per cent increase between 1334 and 1523, compared with, for instance, Lincolnshire, a wool-growing area, where the subsidy paid was around the same at the two dates, and Norfolk, with its cloth industry, where the increase was only around 14 per cent. The subsidy rolls provide evidence of the growing concentration of the wool textile industry in the Cotswold valleys of Stroudwater, Cam and Little Avon. Winchcombe, for example, paid less in 1523 than in 1334, whereas Bisley paid thirteen times more. And Cirencester, the important wool market for the Cotswolds, paid six times as much as Bisley in 1334 but by 1523 Bisley paid more.[12]

The organisation of the rural Gloucestershire woollen industry, as in Wiltshire and Somerset, has been characterised as a domestic or putting-out system.[13] This meant the distribution of wool by a clothier to spinners and weavers who undertook the carding, spinning and weaving in their homes, the woven cloth being then brought to a mill, often owned or leased by a clothier, for fulling. Clothiers who held mills also undertook fulling on commission for those who did not. If the cloth was to be dyed it might be sent to a specialist dyer or dyed in a dyehouse adjoining the mill. This has been contrasted with the organisation of cloth production in Yorkshire where independent farmer/clothiers undertook the spinning and weaving with family members in their homes, taking the cloth to be fulled by a commission fuller and then sending the unfinished cloth to be sold at one of the cloth halls. But both of these descriptions are over-generalisations. In Gloucestershire while there were, indeed, capitalist clothiers – the forerunners of the 'Gentlemen Clothiers' of the eighteenth century – there were also small clothiers who operated from their homes, organising cloth manufacture on a small scale, probably undertaking some of the manufacturing tasks themselves. And, while there was already what might be called the beginnings of an industrial proletariat, there were also quasi-independent spinners and weavers. Spinning and weaving were carried out in homes in Stow on the Wold,[14] for example, probably for clothiers in Cirencester or at some of the small nearby country fulling mills. And, as in Bristol, dyers were an elite, distinguished by their knowledge of dyestuffs and the 'art and mystery' of their craft. The large-scale capitalist operation and the small independent craftsmen operated within the same communities in Gloucestershire.

By Tudor times, whilst clothiers probably bought locally produced wool from nearby farms, others purchased from wool farmers in the adjoining counties. A Minchinhampton clothier purchased 549 stones of wool from a Wiltshire grower early in the sixteenth century and in 1546 a Wickwar cloth maker came to a long-term agreement for a supply of wool from a Somerset grower.[15] But two sources of wool predominated: markets, which were frequently held in small towns – those at Cirencester and Tetbury being particularly important, not just for Gloucestershire clothiers but also for those from Wiltshire – and the middlemen, the wool broggers. These broggers had superseded the earlier Merchants of the Staple who had traded with the great Cotswold wool farmers, selling wool for export. The larger clothiers came to recognise that middlemen were as essential in the supply of raw materials to them as they were for smaller clothiers. Nonetheless, the brogger was frequently blamed for rises in the price of wool.[16]

Weavers worked in their own homes and the more well-to-do amongst them merged into the ranks of the smaller clothiers. Sometimes the weaver didn't own the loom which he worked, but rented it in his home or could, on occasion, go to weave in the clothier's house where there might be a workshop. But most sixteenth-century clothiers saw no advantage to collecting looms together.

The fulling mill had largely come into the occupation or ownership of the clothier by the Tudor period. Sometimes the clothier owned it outright or mortgaged but frequently he held it on lease, sometimes for long terms. The final finishing processes of fulling, scouring, tentering, raising and shearing were frequently carried out at the mill. Beside the fulling mill there was often a close of land for a rack or racks (the name Rack Close or Rack Hill sometimes being the only indicator of the former use of a known mill site). Some cloth was dyed locally; some was dyed and dressed in London, but much was shipped abroad undyed and undressed.

From the late fifteenth century the first references are found to a mechanical process for raising the nap on cloth. There are many parallels between the development of the woollen industries in Gloucestershire, Wiltshire and Somerset, but one machine seems to have been almost exclusively employed in Gloucestershire and, much later, in the late eighteenth and early nineteenth centuries, to have been a cause of considerable labour unrest in Wiltshire and Somerset. This was the gig mill. Earlier authors have dated the introduction of the gig mill in Gloucestershire to the 1770s[17] or the 1790s,[18] but one was in existence in Stonehouse by 1497[19] and, during the sixteenth century, references to them increase, one being recorded at Woodchester in 1517, one at Cirencester in 1533, one at Damery in 1544 and one at Cam in 1556.[20] The gig mill was first mentioned in a statute of the 1460s (3 & 4 Ed IV) and again in the 1550s (5 & 6 Ed VI; JF13.9) when it was noted that 'in many parts of the realm were newly and lately divised erected builded and used certain mills called gigg mills' and their use was prohibited.[21] As with other regulatory matters concerning the woollen industry, Gloucestershire clothiers seem to have ignored them when it suited, perhaps in the knowledge that prosecution was unlikely, but when times were difficult, they were more inclined to seek the protection of regulation.

Cloth, done up in packs of ten pieces, was taken to a regional market, to Bristol or to London. Sometimes the clothier took his own cloth, although he often sent it by one of the many carriers.[22] The Blackwell Hall market was opened for sale every week from noon on Thursday until Saturday morning. The sale of cloth to the merchant was frequently complex, involving intricate credit arrangements; 'the oil of credit indeed greased every cog in the machinery of the woollen industry'.[23] From 1564 onwards the commercial route for cloth shifted from Antwerp (the chief market for undyed cloth) and the line of the Rhine to that of the Elbe, to the market at Middleburg; the chief inland market for English cloth being at Frankfurt.

The wealthier of the West of England clothiers were important figures in the industrial life of Tudor England, sometimes of considerable or aspiring social standing. The clothier Thomas Sewell of Bisley was a landowner and wool grower. In his will of 1540 he left a flock of 200 sheep to his wife and some to his son. He possessed a house, fulling mill and dyehouse with 'vattes and furneys'. Thomas Gower of Wotton-under-Edge held, at his death in 1570, a leased part of Lord Berkeley's Haw Park and left nearly £900 to his children, besides legacies to the poor in seven neighbouring parishes where his spinners and weavers lived.[24] The Palling family, clothiers of Painswick, established themselves at Brownshill; Edmund Fletcher, clothier of Painswick, bought the lease of the tithes of Painswick church and also acquired the rectory tithe barn. After the Dissolution of the Monasteries the Manor of Througham, formerly in the possession of Cirencester Abbey, was sold in 1544 to William Compton, a clothier of Chalford, and sold by his son to William Stumpe of Malmesbury in 1552. The Manor of Paganhill was acquired by Richard Fowler, a Stonehouse clothier, in 1538 and Ruscombe Farm estate was leased to the Gardner family, clothiers, from 1532 and purchased by them in 1574. Edward Stephens of Eastington pulled down the old manor house and began a grand new house in about 1578,[25] in which the family lived until the early eighteenth century when they moved permanently to Chavenage.[26] Thomas Tayloe, clothier, leased the 'large and beautiful' medieval house Over Court, Bisley, in 1593 and the house remained in his family for several generations. In the Severn Vale, Millend House (later called Eastington House) was in the Clutterbuck family from 1552 and Richard Clutterbuck bought lands in King's Stanley, purchasing the borough in 1613. The power exercised by the greatest clothiers could upset markets: William Stumpe of Malmesbury was accused of seeking to usurp control of Tetbury market – said to be one of the best markets for wool and yarn in Gloucestershire – in order to destroy it.[27] Stumpe was one of the justices for both Wiltshire and Gloucestershire from 1538 and in 1545 was appointed escheator for Gloucestershire and the Marches of Wales. The accumulation of land by the wealthier clothiers was on a scale sufficient to arouse the fears of country gentry who gained an insertion into an Act of 1576 by which clothiers in Wiltshire, Somerset and Gloucestershire were not permitted to own more than 20 acres (18 Eliz c.16); however, there does not appear to have been any serious attempt at enforcement. Clothiers were sufficiently numerous and supported the livelihood of so many textile craftspeople, as well as contributing to the Royal revenues, that they could not be ignored by government. In the industrialised valleys of sixteenth-century Gloucestershire, a capitalist system had come into being, contrasting with the social and political privileges and constraints which characterised an earlier age. The government listened to the country clothiers – it could not afford not to.

MILLS OF THE RELIGIOUS ORDERS

In the early Tudor period there was some change in the distribution of fulling mills owned by religious houses (Map 2) and, in the upheavals following the Dissolution of the Monasteries between 1536 and 1540, fulling mills formerly owned by religious houses came into lay hands and were developed by the new landowners.

The monastery of St Peter's, Gloucester, no longer appears to have possessed a fulling mill at Hinton; Flaxley fulling mill is not mentioned after 1400 and neither is the Sherborne mill, and the Barton on

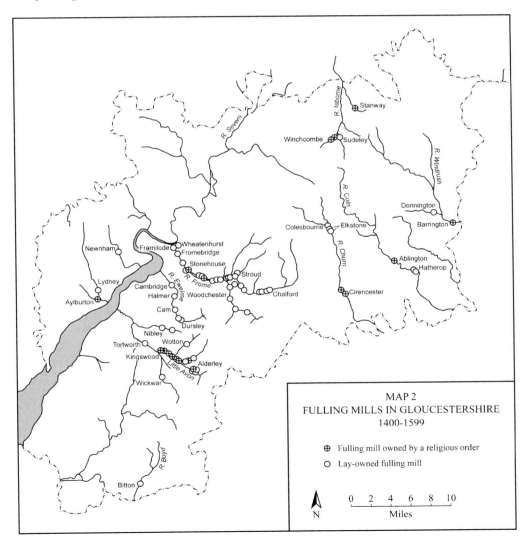

Windrush mill of the Templars was no longer used for fulling. This does not represent a decline in monastic interest in the income to be generated from cloth, but rather a redistribution of the industry, reflecting the growing dominance of the rivers flowing west towards the Severn, most of the ecclesiastical landowners mentioned above erecting new fulling mills on other manors. Llanthony Priory still retained its fulling mills at Aylurton and Little Barrington, and had, in addition, one at Colesbourne.[28] Gloucester had acquired a mill in Stonehouse during the time of the first Norman abbot (1072–1104);[29] this estate reverted to the Crown and some time later the mill was granted back to the abbey which then erected a fulling mill there.[30] Winchcombe Abbey's mill at Stanway continued as a fulling mill until the early sixteenth century and two abbey fulling mills continued to exist in Winchcombe.[31] By 1498 Fromebridge Mills, owned by the Abbey of Godstow, comprised corn and malt mills, together with fulling mills containing two stocks. In 1496 John Arundel sublet the 'cornemille a maltmylle with an orchard lyng thereto a tokemylle other wyse called a walkemylle withy to stokkys with a pasture lying thereto called Milham' to Roger Porter at a rent of £4 p.a. At the Dissolution the mills were conveyed to the Haynes family – John Haynes was digging fuller's earth at Alkerton in 1556.[32]

By the sixteenth century, Cirencester Abbey and Kingswood Abbey (Kingswood, at this time being an outlier of Wiltshire) also had fulling mills. The last abbot of Cirencester 'buildid 2 Fulling Mills … that cost a 700 Markes of Mony. They be wonderfully necessary, bycause the toun standith alle by Clothing.'[33] This 'right goodly clothing myll' was built near the town wall using stone from the ruinous Roman walls. In 1533–34 the abbot leased 'all that newe house and myll called Seynt Mary Mill with ffoure stocke and one gyggemyll theryn bylded' to Richard Fowler of Stonehouse.[34] In 1563, after the mill had passed into lay hands, the occupier was involved in a court case with the tenants of the next mill downstream. The issue, as was often the case, concerned water. Part of the stream went underground at St Mary's Mills and the exit had been blocked by a board; 'the tenants of Langley myles conferre they could gane no water … but … by giuing of money to the [four] tuckers of the newe myles'.[35] The abbot owned one other fulling mill in Cirencester known as Barton or Clark's Mill.[36] Much of the cloth fulled at these mills was probably woven in the town. Sir William Nottingham founded a hall for the 'habitations of four poor weavers' in the early fifteenth century in Cirencester, and a fraternity of weavers was founded in the town.

While Kingswood Abbey had owned a number of mills in and around the town of Kingswood from the thirteenth century onwards, there is no record of any of them being converted to fulling mills until the sixteenth century. But during this century Abbey/New Mill,[37] Bytheford (Ithell's) Mill (shear grinding and fulling),[38] Charfield/Sury/Kingswood New Mills (fulling and corn),[39] Walk Mill and Nind/Nyen Mill[40] were all involved with the woollen industry. Kingswood Abbey also owned the appropriately named Monk Mill at Wortley, Wotton-under-Edge, which by the end of the sixteenth century was also a double fulling and corn mill, and further upstream the less appropriately named Hell (Hill) Mill.[41]

LAY-OWNED MILLS

Evidence of the development and indications of concentration of the woollen industry can be seen from the increase in the number of lay-owned fulling mills (Map 2). Some mills on smaller, slower streams (for example, Chedworth and Cerney) had reverted to corn milling by the fifteenth century, although there was a cluster of two mills in Elkestone and one nearby in Colesbourne on the upper Churn.[42] There were two mills at Hatherop on the Coln and a mill at Donnington on a tributary of the Windrush.[43] Further out in the plain the woollen industry was not only vacating towns near the Severn but some of the mills in the flat lands were reverting to corn milling or becoming double mills. Wickwar Mill in the

Severn vale was a fulling mill in 1552 but, ten years later, it had reverted to corn milling. Halmer Mill was a double fulling and grist mill by the 1560s, and a corn mill at Cambridge, on the River Cam, became a double mill with fulling stocks for a short while in Elizabethan times.[44]

Already by this period the three river systems, along which the majority of fulling mills was located by the end of the eighteenth century – the Little Avon, Ewelme/Cam and Frome – had a concentration of mills along them. Leland described Alderley on the upper reaches of the Little Avon as 'a clothinge Village' and, indeed, it had four fulling mills in Tudor times. Kingswood, a mile or so downstream, had 'Some Clothyatrs in it, els a little and bare Village',[45] and, as seen above, it was densely populated with mills owned by the abbey. Wotton-under-Edge was described by Leland as 'a praty Market Towne, welle occupied with Clothiars'.[46] A survey of Wotton borough during Elizabeth I's reign records four mills, but only one, near Coombe, is described as a fulling and grist mill. Dursley, on the Cam, was 'a praty Clothinge Towne; There is ... a goodly Springe, and is as the principall Hedd of the Broke servynge the Tukking Miles about the Towne'.[47] In 1486 Henry Buscrope left to his son 'my house containing two mills, a gristmill and a fulling mill';[48] in 1513 Thomas Wolworth left a 'myll and dyeing house' to his wife and George and William Harding held a mill in Dursley in 1562–63 with a parcel of land called Rackham.[49] Downstream in the hamlet of Cam, John Fynemore received the lease of Corietts Mill, containing a grist, fulling and gig mill, from the Crown in 1522.[50] This site, later known as Cam Mill, is one of the two mills still producing woollen cloth.

Leland had nothing to say about Stroudwater but there is other evidence of the development of woollen manufacture in the valley. Court rolls become more numerous, as do property deeds, and in 1561–62 a snapshot of the Gloucestershire woollen industry is provided in an account of clothiers fined at Blackwell Hall for the manufacture of faulty cloths. While there was a cluster of fines levied on clothiers in Cirencester, including one for faulty blue cloth, and a few fines in what had, by this time, become outposts of the industry – Northleach, Tetbury, Sodbury and Berkeley – there was a significant concentration in the middle reaches of the Stroud Valley from Stroud to Chalford. And it was only in this area that clothiers were fined for faulty red cloths as well as white. In addition to clothiers in the Frome Valley, fines were also imposed on clothiers in Wotton, Alderley, Dursley and Coaley, small towns and villages which, situated in the Ewelme and Little Avon valleys, continued to grow in significance for cloth production in later centuries. In all, eighty-one cloth makers – including some bearing the names of well-known clothier families, such as Clutterbuck, Webb, Stafford, Halliday, Shewell and Ithell – were fined.[51]

From the mid-fifteenth century a number of new fulling mills, not necessarily in new buildings, but quite frequently double mills, began to appear in the Frome Valley. The former Gloucester Abbey corn mill at Framilode was a double mill by 1556.[52] Upstream of Fromebridge the River Frome flows in two channels, and in part of Stonehouse manor in three, each of which shows signs of re-routing to accommodate mills and which caused complaints in manor courts and, occasionally, cases in Chancery. In 1575 a messuage, fulling mill with two stocks and 13 acres of land, on the north branch of the river, together with the advowson of Eastington church were conveyed to four men, two of whom lived locally.[53] This was Churchend Mill which, in 1597, was conveyed by Richard Stephens, James, his son, and Katherine Sandford, in trust to Richard Browning, clothier, as a marriage settlement on James and Katherine.[54] The Stephens family became one of the major Gloucestershire clothier dynasties. John Sandford had been granted part of the manor of Leonard Stanley in 1553, including a fulling mill in the tenure of Thomas Clutterbuck.[55] Richard Clutterbuck was confirmed as tenant of the mill – Millend Mill – together with 'the liberty to dig clay for milling cloth' in 1591.[56] Here was one branch of another clothing dynasty – the Clutterbucks – which became pre-eminent at Eastington and King's Stanley. Robert Clutterbuck, in his will of 1562, bequeathed to his wife his leasehold claim in a 'tuck myll and grysh mill' in King's

Stanley.[57] By 1563, when the manor was sold to two Stonehouse clothiers, William Fowler and William Sandford, the mill at King's Stanley contained three stocks and a grist mill under one roof. Further evidence of the growing significance of the woollen industry in King's Stanley is the seven people who paid an annual fee ranging from 4*d* to 12*d* for digging fuller's earth in 1535.[58] Besides the mill in Stonehouse which belonged to St Peter's Abbey, Gloucester, a fulling mill containing two stocks, a river gate and small close in Stonehouse was leased to Richard Bence by the Lord of the Manor in 1496.[59] This is likely to have been Bond's Mill.

Upstream of Ryeford, but still within Stonehouse Manor, was a corn mill at Ebley under two tenancies, one part of which (Derehurst's Mill) had, by 1537, become a fulling mill.[60] By 1597 both mills were in the possession of Leonard Bennett and thereafter the mills were sometimes let independently and at other times jointly. A fulling mill at Dudbridge, which became incorporated later into Hawker's Dyeworks, known as Cherynges Mill, was a fulling mill by the 1480s, later becoming known as Norris's Mill – members of the Norris family holding it for about 100 years.[61] Where two manors adjoined at a good water-power site, as at Lodgemore, Stroud, mills were erected on either side of the river. This was the boundary between the lands of the Spillman and the de Rodborough families. Here there were two fulling mills respectively called Nether Latemore and Upper Latemore, and later known jointly as Lodgemore, which, by the early eighteenth century, if not before, were run as a single concern.[62] Fromehall Mill, too, dates from this time. Wallbridge Mill is known to have been a fulling mill from 1470;[63] Capel's Mill from 1513;[64] Bowbridge from 1513;[65] Arundel's from 1585;[66] Brimscombe from 1594[67] and Dark Mill from 1597.[68]

Further upstream the stretch of the River Frome which flows through Chalford drove at least five fulling mills in the seventeenth century, possibly more. St Mary's Mill, an ancient site belonging to a chantry at Minchinhampton church, was a fulling mill by 1548 and in 1594 comprised two fulling mills, a gig mill and a grist mill.[69] Woolings (Tayloe's) continued to be a fulling mill in the Hone family; and Chalford Mill, the property of Corpus Christi College, Oxford, had become a fulling mill by Tudor times; also Stoneford (Halliday's) Mill;[70] and Baker's (Twissell's) Mill may have been fulling in the sixteenth century – the Twissell family was a noted clothing dynasty which lived in some style at Frampton Place.

Three tributary valleys to the Frome already had clusters of fulling mills along them in Tudor times: the Nailsworth, Painswick, and Slad valleys (other streams were utilised later).

In the Nailsworth Valley Millbottom Mill, Horsley, a corn mill, was granted to William Webb, a clothier, in 1564; Longfords Mill, an ancient site, was leased by clothier members of the Elkington family from the mid-sixteenth century.[71] Woodchester Mill may have been an early fulling mill and certainly was by the early sixteenth century when a gig mill was recorded there.[72]

The Painswick and Slad brooks and their tributaries drove many cloth mills in the eighteenth and early nineteenth centuries, but in the sixteenth century, while there were a number of mills along these streams, no evidence has yet been found of them being fulling mills, although it is highly likely that some were in view of the evidence of material wealth from this period in Painswick. Some of the clothier dynasties had already emerged in the parish by this date. New Mills on the Slad brook is known to have been a fulling mill and was in the hands of the Watts family, who were clothiers in the mid to late sixteenth century.[73] The intensive exploitation of the water power of these streams for the woollen industry, which occurred in the heyday of the industry during the early Industrial Revolution, was aided by the water power engineering works in the form of leats, dams and tailraces, which had, in Tudor times and probably earlier, already been undertaken.

STATE REGULATION

With the growth of the woollen industry came outcries against deceitful practices by clothiers, weavers and dyers. One treatise named and shamed those areas most noted for 'false clothe', adding:

> The trewest cloth that is made in this realme is all sortes of fine clothe, especially such as be brought to the markets undressed ... and therefore for sortes like Gloyucester, Somersett and Wiltshyre whites and long Worcesters together with Kentishe cloth.[74]

The price of each pack of ten broadcloths at the London market in Blackwell Hall varied according to the quality, but also the reputation of the maker whose mark was on the cloth. A cloth mark was required by an Act of 1536 (27 Hen VIII, c.12), and the Privy Council on at least one occasion intervened to enforce this, but it is clear that cloth marks were being used a long time before they were required by statute. The cloth mark was usually a simple geometrical design which included the maker's initials. A cloth mark of repute was a prized possession and something which was not disposed of lightly. While much industrial legislation was enacted in the reign of Edward IV, an increased volume reached the statute book under the Tudors. Whether it was to control middlemen dealing in wool, or a limit set on the number of looms a single person might own, or methods of winding wool, or the use of gig mills for raising the nap, or the relationship of the clothier with his various employees – codified in the 1563 Act of Apprentices – there was hardly a process in the manufacture of cloth which escaped regulation, although the extent to which Gloucestershire clothiers adhered to the rules is debatable.

One of the main subjects of statutory regulation was that of the length, breadth and weight of broadcloth. As with other regulatory roles, aulnage was farmed out and, while it was a useful source of income for the Crown, it did little to raise or maintain standards of production. After much to-ing and fro-ing between Blackwell Hall, the Crown and Privy Council on the subject of the London aulnager, a statute was passed in 1607 which enabled clothiers to sell cloths to merchants without the intervention of the aulnager, providing that they had already been sealed at the place of manufacture. The enforcement of a seven-year apprenticeship and a limit on the number of apprentices any master weaver might keep had been imposed by law in 1552 (5 & 6 Ed VI, c.8; 3 & 4 Ed VI, c.22) and in 1563 the Statute of Artificers extended the requirements for apprenticeship to all manual occupations, although spinning and burling were not included. The separation of craftsmen and clothiers was endorsed by a decision in the Court of the King's Bench in 1580 which stated that clothiers were immune from the apprenticeship obligation, provided that they did not personally engage in manual labour.[75] On the one hand, entrepreneurial clothiers were developing businesses which, to flourish, required some autonomy in business practices, while on the other, a series of largely unenforceable restrictive acts was passed. The supervision of textile manufacture had become embroiled in a new system of government which was developing under the Tudors and which could not be effectively implemented.

TRADE FLUCTUATIONS

The reign of Henry VIII was the golden age of the English broadcloth industry. By the 1540s the annual export of undressed white cloth from London was nearly twice the amount of the beginning of the century. There had been times of depression in the past, such as 1520–22 and 1529–31, and in 1529 there was so much unsold cloth at Blackwell Hall that Cardinal Wolsey intervened and required the merchants to buy it up. In the 1550s depression set in and in 1563 the export of white undressed broadcloth from London fell to almost the lowest level it ever reached in Henry VIII's reign. It was not until 1567 that Hamburg became an alternative to the collapsed market at Antwerp. But, even after this, annual shipments of broadcloth from London only slowly rose in the late 1570s and early '80s. In 1577 Gloucestershire clothiers complained of the scarcity and high price of wool,[76] and a letter to Sir Giles Poole and others said that:

> clothiers and clothe making (in) divers townes of this realme are latelie decayed, the inhabitants and artificers living by the same greatly spoiled and impoverished.[77]

By the end of 1586 West Country clothiers were complaining to the Privy Council that the Merchant Adventurers were failing to buy their cloths, 'Whereby they were in some parte forced to decay their workmen'. The Privy Council, concerned that workmen should continue in employment (during a period of famine), 'by contyunance of draperie in the aforesaid counties' required clothiers to continue their trade and to take cloth to Blackwell Hall as formerly. Blackwell Hall merchants were ordered to purchase the unsold cloth on threat of losing their monopoly and liberty was given to any Merchant of the Staple to buy up cloth for export.[78] The crisis persisted through the spring of 1587 and there was considerable discontent in Gloucestershire. 'The poore people of Wiltshire and Gloucestershire, livinge wholly upon cloth making, in great numbers were ready to growe into a mutinie for this cause.'[79] In May 1587 the Government took steps to free the export trade in cloth from all restrictions and the market began to recover somewhat in the summer. But the crisis had ushered in a period of trade depression which did not end until the coming of the new wave of prosperity in the very late 1590s and early 1600s. In the latter years of the sixteenth century wool prices rose and clothiers sought to minimise lower profits by suppressing the wages of those who worked for them. It was said in 1613 that wages had not risen in the previous forty years.[80]

During Tudor times the woollen industry was recognised as a significant source of income for the Crown, statutes being controlling rather than enabling. Gloucestershire clothiers appear to have held an ambivalent attitude to state intervention, on the one hand petitioning for further controls during recessions, and on the other, ignoring them when times were good. And, to a large extent, they were getting away with it. The Tudor period was distinguished by the terminal decline of the woollen industry in cities, towns and villages in the Severn plain, and, while there were a number of fulling mills along the rivers flowing to the Thames, by far the greater number were in Stroudwater, the Ewelme/Cam and Little Avon flowing to the Severn. The concentration of the industry along the western flowing river valleys was confirmed in the Tudor period.

Endnotes

1. F.F. Fox & J. Taylor, *Some Account of the Guild of Weavers in Bristol*, 1889, p.67
2. Ibid., p.71
3. Ibid., p.93
4. Ibid., p.68
5. Ibid., pp.38, 52
6. Leland, *Itinerary*, V, p.100
7. Ibid., p.101
8. W.S. Harmer, *Cirencester Weavers' Company*, 1919
9. *VCH Gloucestershire, IV*, p.52
10. *VCH Gloucestershire, II*, p.154

11 Ibid., p.52. Cooke's mill may have been an investment rather than for his own use
12 E.M. Carus-Wilson, 'Evidences of Industrial growth on some Fifteenth Century Manors', in *Essays in Economic History*, II, pp.152–3
13 This characterisation is alluded to by G.D. Ramsay, *The Wiltshire Woollen Industry in the Sixteenth and Seventeenth Centuries*, p.6, and Kenneth G. Ponting, *The Woollen Industry of South-West England*, p.24
14 Joan Johnson, *Stow on the Wold*, Alan Sutton, 1980
15 Peter J. Bowden, *The Wool Trade in Tudor and Stuart England*, Macmillan, 1962, p.57
16 G.D. Ramsay, *The Wiltshire Woollen Industry in the Sixteenth and Seventeenth Centuries*, Cass, 1965, p.9
17 E.A.L. Moir, 'The Gentleman Clothiers', in H.P.R. Finberg, *Gloucestershire Studies*, University of Leicester, 1957, p.260
18 J. de L. Mann, *The Cloth Industry in the West of England from 1640 to 1880*, Oxford, Clarendon, p.138. The author uses the phrase 'for this purpose', allowing that gig mills might have been used for other purposes earlier
19 Janet Hudson, *The Early History of Two Stonehouse Mills*, Trans. BGAS, 118, 2000, p.125
20 D6746/L213 X; PRO C1/793/25
21 JF13.26 GS
22 G.D. Ramsay, p.25
23 Ibid., p.27
24 R. Perry, *The Woollen Industry in Gloucestershire to 1914*, Ivy House, 2003, p.26
25 R. Bigland, *Historical Monumental and Genealogical Collections, Relative to the County of Gloucester*, I, 1881–89, p.375
26 Nicholas Kingsley, *The Country Houses of Gloucestershire*, I, Cheltenham, Nicholas Kingsley, p.88
27 Ibid., p.33
28 *Valor Ecclesiasticus*, II, p.425; *VCH Gloucestershire*, VII, p.189
29 *Historia et Cartularium Monasterii Sancti Petri Gloucestriae*, III, p.1
30 C. Swynnerton, *The Water Mill of the Abbots of Gloucester in Stonehouse*, Trans. BGAS, XLVI, p.50
31 *Valor Ecclesiasticus*, II, p.460
32 *VCH Gloucestershire*, 10, pp.148–52
33 Leland, *Itinerary*, II, p.129
34 D674 b/T8
35 D674 b/L1; D674 b/12
36 D674 b/T9
37 *Hockaday Abstracts, Kingswood*
38 E.S. Lindley, *A Kingswood Abbey Rental*, Trans. BGAS, LXX, p.148; E.S. Lindley, *Kingswood Abbey its Lands and Mills*, Trans. BGAS, LXXIV, p.57
39 PRO SC 12/portf7/70
40 *Hockaday Abstracts, Kingswood*
41 D1086; E.S. Lindley, *Kingswood Abbey its Lands and Mills*, I, Trans. BGAS, 73, 1954, p.188
42 *VCH Gloucestershire*, VII, pp.189, 216
43 Ibid., pp.92, 153
44 D340a/T146; RF 127.23(5)gs; Berkeley MSS III, p.158
45 J. Latimer, *Leland in Gloucestershire*, Trans. BGAS, 14, 1889–90, pp.258–63
46 Ibid.
47 Ibid.
48 *Hockaday Abstracts, Dursley*
49 Ibid; RF 319.10gs

50 A.W. Hughes, *Papers on the History of Dursley*, 21149gs
51 G.D. Ramsey, 'The Distribution of the Cloth Industry 1561–62', in *English Historical Review*, Vol. 67, using PRO Exch KR Mem Roll Hil 7 Eliz 329–32
52 D149/149/172, 173; *Hockaday Abstracts, Frampton*
53 Box 29 gs
54 D1229
55 D540/T59
56 D1229
57 Trans. BGAS, LI, p.254
58 Trans. BGAS, XLV, p.243
59 Janet Hudson, pp.123–32
60 289: 2,3,9,21gs
61 D67/Z46; cf Prob.6/15
62 D149/21
63 C.E. Watson, *The Minchinhampton Custumal*, Trans. BGAS, LIV, p.370
64 *VCH Gloucestershire*, 10, Stroud
65 Ibid.
66 Ibid.
67 P.H. Fisher, *Notes and Recollections of Stroud*, p.8; E.A.L. Moir MSS, *History of Stanley & Ebley Mills*
68 *VCH Gloucestershire*, 10, Stroud
69 D67/Z/32; D1815, Clutterbuck
70 M.A. Rudd, *Historical Records of Bisley with Lypiatt*, p.309; PRO SC 6/850/31; Juliet Shipman, *Chalford Place, A History of the House and its Clothier Families*, 1979; M.A. Rudd, *Abstracts of Deeds relating to Chalford and Colcombe*, Trans. BGAS, 51, 1929, pp.211–24
71 *VCH Gloucestershire*, 10, Minchinhampton
72 D67/Z/50
73 D547a/T4/6; D547a/T65
74 Eileen Power & R.H. Tawney, *Tudor Economic Documents*, III, 1951, pp.216–17
75 G.D. Ramsay, p.60
76 Calendar State Papers Domestic (SPD) 1547–80, p.550
77 Calendar Acts of the Privy Council (APC) 1577–78, p.28
78 Cal APC 1586–87, pp.71–2, 272–5; G.D. Ramsay, p.68
79 Eileen Power & R.H. Tawney, p.284
80 G.D. Ramsay, p.69

CHAPTER THREE

ORGANISATION AND LOCATION OF THE WOOLLEN INDUSTRY FROM THE SEVENTEENTH CENTURY TO THE EVE OF THE INDUSTRIAL REVOLUTION

DISTRIBUTION AND SCALE OF THE WOOLLEN INDUSTRY

A unique picture of the distribution and scale of the early Stuart woollen industry in Gloucestershire is presented in John Smyth's muster roll of 1608.[1] Smyth, who lived in North Nibley and was steward of the Hundred of Berkeley, drew up a list of able-bodied men by hundreds, manors and tythings in the county who were deemed fit to serve in His Majesty's wars. In almost all instances the occupation of the man is given. It is reasonable to assume this to be an under-representation of able-bodied men – some may have sought exclusion by a variety of means – and there will have been some occupied in the industry who were not deemed to be able-bodied. The total of 1,786 weavers recorded is too low to have produced the quantity of cloth which was probably being made at the time and, since women and children undertook wool preparation and spinning, the total figure of people employed in the woollen industry would have been considerably higher. Nevertheless, despite these limitations, this is a unique industrial census.

The distribution of able-bodied men employed in woollen manufacture, as a proportion of the total able-bodied males on each manor or tything, is shown in Map 3 (overleaf). From this the concentration of the industry on the Cotswold scarp in the three valleys of the Frome, Ewelme, Little Avon and their tributaries is clear. Gloucester, Tewkesbury and Cirencester excepted, the total recorded population of many of these industrial villages and hamlets is higher than elsewhere. Weavers were scattered in the upland villages (for example, two at Fairford, three at Stow on the Wold and four at Northleach) and fullers were employed at the fulling mills on the Churn at Colesbourne and North Cerney, and at Ablington, Quenington and Hatherop on the Coln. But the great majority of fullers were in villages and hamlets along the western flowing Rivers Frome, Cam and Little Avon and their tributaries.

In some tythings in Stroudwater the percentage employed in branches of woollen manufacture exceeded 50 per cent of all recorded males (for example, Woodchester 64 per cent; Rodborough 52 per cent; Nailsworth 83 per cent; the Stanleys 55 per cent; Upper Lypiatt 66 per cent; and Paganhill 54 per cent), detailed in Table 1 (overleaf). In all, 746 men were recorded as clothiers, weavers, fullers or dyers in the Stroudwater area.

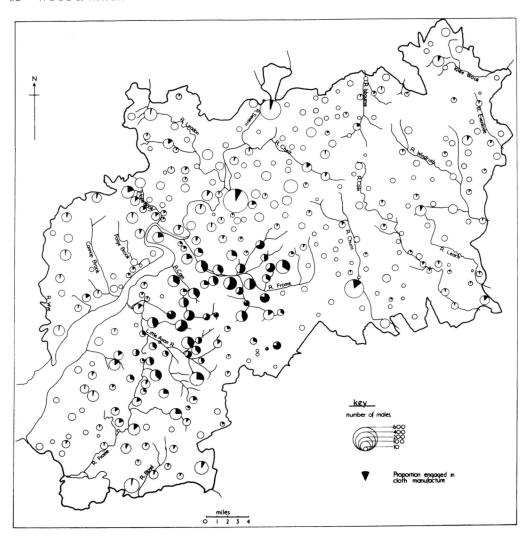

MAP 3: The percentage of males employed in the Gloucestershire woollen industry in 1608.

Table 1 – Employment in the woollen industry 1608: Stroudwater

Village	Clothiers	Weavers	Fullers	Dyers	Shearmen	Other workers	Total in cloth	Per cent of total counted
Minchinhampton	4	33	23	1	0	0	61	36
Woodchester	7	16	6	3	0	0	32	64
Rodborough	6	43	11	1	0	0	61	52
Horsley	1	37	12	1	0	2	53	43
Cherington	1	11	0	0	0	0	12	34
Avening	0	17	0	0	0	0	17	19
Nailsworth	2	10	13	0	0	0	25	83

Nympsfield	0	6	0	0	0	0	6	23
Painswick	4	33	10	0	0	0	47	29
Paganhill	5	34	4	1	0	0	44	54
Bisley	1	36	16	1	0	0	54	33
Upper Lypiatt	3	23	11	0	0	3	40	66
Nether Lypiatt	8	4	10	2	0	0	24	36
Througham	0	4	0	0	0	0	4	20
Steanbridge	0	15	7	0	0	1	23	48
Stonehouse	7	12	7	4	0	0	30	32
King's Stanley	6	62	12	0	0	0	80	56
Leonard Stanley	2	22	12	0	0	0	36	54
Frocester	0	15	0	0	0	0	15	25
Oxlynch	2	23	4	1	0	1	31	25
Alkarton	3	22	4	0	0	0	29	41
Eastington	2	14	6	0	0	0	22	41
TOTALS	64	492	168	15	0	7	746	

Source: John Smyth, Men and Armour, *1608; Perry, 1945*

Some villages were almost completely industrialised, such as Upper Lypiatt, where only thirteen of the sixty-one recorded males were not directly employed in the industry. Sixty-four clothiers were named in the Stroudwater area, only four villages being without one. Some of the later, well-known clothier families appear; five Cloterbokes are named in the Stanleys. There were 492 weavers, some described specifically as broadweavers, besides 168 fullers and tuckers. There were also fifteen dyers, two card makers, three cloth makers or clothmen and a millwright.

A high proportion of males was employed in the woollen industry in the Dursley area (e.g. Dursley 54 per cent, Woodmancote 67 per cent, Uley 57 per cent, Owlpen 82 per cent, Cam 55 per cent and Stinchcombe 79 per cent), as detailed in Table 2 (overleaf). Eleven fullers were recorded in Dursley, besides twenty-seven weavers, including Richard Typet whose descendants were making cloth in the town in the late eighteenth century. There was a high concentration of clothiers in the area – eleven in Stinchcombe, one of whom was Lord of the Manor, besides thirty-three weavers. In Cam there were seven clothiers, forty weavers and fifteen fullers. Coaley, out in the plain, had two clothiers, thirty-six weavers and five fullers, indicating a supply function for clothiers located upstream at the Cam, Dursley and Wotton mills. Altogether, 338 men were recorded as being occupied in the woollen industry in the Dursley area, comprising more than 50 per cent of the recorded able-bodied male population. This was an area even more dependent upon the woollen industry than Stroudwater.

Table 2 – Employment in the woollen industry 1608: Dursley area

Town/Village	Clothiers	Weavers	Fullers	Dyers	Shearmen	Other workers	Total in cloth	Per cent of total counted
Dursley	14	27	11	1	0	2	55	54
Woodmancote	0	24	10	4	0	0	38	67
Uley	3	29	0	0	0	0	32	57
Owlpen	0	13	1	0	0	0	14	82
Cam	7	40	15	0	0	4	66	55
Coaley	2	36	5	0	0	0	43	41
Stinchcombe	11	33	1	0	0	4	49	79
Slimbridge	0	35	6	0	0	0	41	36
TOTALS	37	237	49	5	0	10	338	

Source: John Smyth, Men and Armour, 1608; R. Perry, The Gloucestershire Woollen Industry 1100–1690, Trans. BGAS, 194.

High proportions of recorded males were employed in the woollen industry in the Little Avon area (for example North Nibley 61 per cent, Huntingford 67 per cent, Falfield 60 per cent and Cromhall 53 per cent); see Table 3. A Kingswood was an enclave of Wiltshire Smyth did not provide figures for the town in his 1608 survey, although we know that the woollen industry was well established there with seven mills, at least six of which were fulling mills by the time of Smyth's survey. There were forty-four clothiers, of whom fifteen were in Wotton, in the Little Avon Valley and its tributaries. Five clothiers were named in North Nibley and eight in Wickwar. Wotton, Coombe, and Synwell all had fifteen or more weavers; Hawkesbury had thirty-six and Wickwar thirty-four, while North Nibley had fifty-eight. Shearmen were rarely enumerated in the survey which suggests that cloth dressing was still largely undertaken by fullers. There were, however, two shearmen listed in Wotton. The highest proportions of textile workers to the total listed male populations were in Falfield (60 per cent), North Nibley (61 per cent) and Huntingford (67 per cent) Names that were well known in the trade in the eighteenth century – Adyes, Osbornes, Gayners, Purnells and Neal – were all found in the environs of Wotton. The Gayners and Ithells were also listed, descendants owning mills in the Kingswood area in later centuries.

Table 3 – Employment in the woollen industry 1608: Little Avon & district

Village	Clothiers	Weavers	Fullers	Dyers	Shearmen	Other workers	Total in cloth	Per cent of total counted
Wotton-u-Edge	15	34	6	1	2	2	60	39
Coombe	0	15	0	1	0	0	16	42
Synwell	2	26	3	1	0	0	31	44
Ozleworth	0	5	0	0	0	0	5	11
Wortley	3	11	8	0	0	0	22	48
Alderley	2	8	2	3	0	0	15	45

Boxwell	0	1	0	0	0	0	1	6
Hawkesbury	1	36	3	0	0	1	41	28
Wickwar	8	34	0	0	0	2	44	39
N. Nibley	5	58	15	0	0	0	78	61
Charfield	2	10	0	0	0	0	12	39
Tortworth	1	11	10	0	0	0	22	39
Huntingford	0	0	2	0	0	0	2	67
Falfield	0	23	1	0	0	0	24	60
Cromhall	6	25	1	0	0	0	32	53
Berkeley	0	10	1	0	0	0	11	11
Alkington	2	37	8	0	0	1	48	45
Stone	1	7	2	0	0	0	10	29
TOTALS	48	351	62	6	2	6	474	

Source: John Smyth, Men and Armour, *1608; Perry,* The Gloucestershire Woollen Industry, *Trans. BGAS, 1945*

Berkeley, out in the Severn vale, was still home to men employed in the woollen industry in 1608, Smyth listing two clothiers, forty-seven weavers and nine fullers in Berkeley and Alkington. By 1608 there was a single clothier in Thornbury, however, with twelve weavers and thirteen more in the villages of Morton and Kyneton. There was also one tucker. By 1608 cloth manufacture in Chipping Sodbury had declined relative to villages in the river valleys to the north. There were, however, six clothiers in the three Sodburys, five in nearby Yate, and thirty-nine weavers in these villages, besides two fullers in Chipping Sodbury, where there was probably a fulling mill.

Besides the concentration along the three river systems flowing west into the Severn, there were four discernable clusters of clothiers and woollen workers in the Forest of Dean. One was around Mitcheldean and Abenhall, another at Westbury on Severn, a third cluster at Purton and the fourth cluster around Newent.

In the former medieval woollen manufacturing towns five weavers, a shearman and six clothworkers were all that remained of the once-thriving industry in Winchcombe, while at Chipping Campden there were just four weavers, a fuller and a dyer. In the south, Tetbury continued to be a wool town of some importance with nine clothiers, thirty-four weavers and a tucker. Marshfield, too, had three clothiers, four weavers and three fullers. Cirencester was not only important for its wool fairs: five clothiers were recorded in the town in 1608, besides forty-one weavers, a tucker, a dyer and four card makers. The guild of weavers, known as the Weavers' Company, continued to hold control of the cloth trade, notwithstanding its somewhat dubious credentials and the claim that its regulations harked back to earlier times.[2] The woollen industry in Gloucester had declined in importance; while the Weaver's Company still existed in the city, Smyth records only four clothiers, twenty-nine weavers, a dyer and two shearmen. Presumably Gloucester cloth was sent to nearby Wheatenhurst where seven fullers are recorded. By the 1620s only two or three clothiers were left in Gloucester, whereas there had once been 'nearly 20 men of good estate who have kept great numbers of poor at work'.[3]

Stroud had become 'The Metropolitan Town of the Clothing Industry' by the early seventeenth century, with new building development, some of it speculative and much of it undertaken by clothiers or members of cloth-making dynasties such as the Webbs, Capels, Arundels, Sewells, Davis and Watts.[4]

John Webb occupied Stafford Mill in 1658/9, the Webbs had acquired Dolemanne's Ham in 1634 and, when Samuel Webb married Elizabeth Smart in 1685, his estate included a mansion house, various other houses and lands, with Ham Mill containing two stocks, a gig mill and a grist mill, besides a dyehouse, presshouse and five racks. New Mills had been in the hands of Richard Watts in the late sixteenth century, and by the seventeenth century, it was in the hands of Giles Davis who also owned Pigghouse Mill in Painswick. Davis had also purchased land to the west of Stroud in Paganhill.

A century later Sir Robert Atkyns remarked, 'it is computed that 50,000 cloths are made yearly' which, 'reckoning the fine with the coarse, may be estimated at ten pounds a cloth and with crafting in rugs, and other incidents of the clothing trade, it may very well be estimated at £500,000 a year'.[5] Defoe remarked, 'such is the greatness of this prodigious manufacture that they are said to take yearly thirty thousand packs of wool and 25,000 packs of yarn ready spun from Ireland'.[6]

INNOVATIONS – CLOTH AND MACHINERY

During the 1560s new types of cloth, known as the new draperies, had been introduced to England – permission being granted to foreigners wishing to settle for the purpose of manufacturing these cloths in 1561.[7] While some new draperies were being made in Wiltshire, most of the cloth made in Gloucestershire was broadcloth, although serge yarn (using long-combed, rather than carded, wool) was being made in Tetbury before 1635.[8] Serge makers are recorded in Tetbury in 1671 and Dursley by 1691 and 1698. There was a worsted comber in Dursley in 1709 and a woolcomber in Horsley in 1730.[9] Cloth was being 'dyed in the say' in the seventeenth century and stammel, described as a lighter cloth, was being made in Stroudwater in the mid-seventeenth century.

By 1620 a new woollen fabric, known as Spanish cloth, was being made by some Gloucestershire and Wiltshire clothiers. This was a medley cloth of two or more colours dyed in the wool and was sometimes, but not necessarily, made of Spanish wool, almost certainly mixed with some English wool. Both cloth which had been dyed in the wool and medleys had been known for a long time.[10] The novelty of Spanish cloth was that it was particularly well made, combining durability with finish and was lighter in weight than other West of England cloths. Contemporaries believed that it was first made by Benedict Webb, a well-to-do clothier of Kingswood. Webb had been apprenticed to a mercer and had studied cloth making whilst resident in France. By 1580 he was in Taunton and began to make 'a sort of medley cloth … called … Spanish clothes'.[11] He had moved to Kingswood before 1600 and by 1640 the export of Spanish cloth from London amounted to over 12,000 pieces a year. By 1630 there were complaints that some clothiers were 'falsifying' against Spanish cloth by dyeing white cloth in the say; in other words, before it was fulled; but by the 1650s say had become a recognised product which was distinguished from the improved Spanish cloth. The

Gig at Longfords, 1963. (Jennifer Tann)

manufacture of Spanish cloth was sufficiently noteworthy for John Eyles's monument in Uley church, dated 1731, to record that he died at the age of ninety-one and was 'ye first that ever made Spanish Cloath in ye P'sh'. This might date his commencement in this line to sometime around 1660. By 1691, if not before, Spanish cloth was being made in Stroudwater and was finding a good market at home.

Besides innovations in varieties of cloth, a comb for woolcombing was invented in Tetbury c.1710, a centre of spinning from the seventeenth through to the end of the eighteenth century.[12] The gig mill had become almost ubiquitous by the early eighteenth century, particularly in the Frome Valley and its tributaries. Between 1730 and 1740 a variant, the knapping mill, was invented by Sir Onesiphorous Paul in Woodchester. This raised the nap 'in little knots at regular but very small distances that gives it a singularly pleasing appearance'.[13] Paul was an innovator in other ways, too, and William Playfair commented that cloth made in the neighbourhood 'owes much of its unrivalled excellence to his ingenious and spirit improvements'.[14] By 1743 there was a knapping mill at Dudbridge (there were three by 1772), one at Inchbrook by 1758, at Brimscombe by 1760 and at Dark Mill by 1786.[15]

MILLS AND EMERGENT FACTORIES

The picture of the distribution of the woollen industry from John Smyth's *Men and Armour* of 1608 is endorsed by the evidence for the existence of fulling mills from the beginning, through to the end, of the seventeenth century. Many of them were double mills. Indeed, John Smyth used the term in describing his home village of North Nibley: 'In the easterne part of the village of Nybley arise divers springs of excellent sweete water, which united and brought into one streame make a pretty river; whereon are seated seavern tuck mills and grist mills, most of them double mills before the said streame be passed through this village; the like whereto I knowe not within this county.'[16] Evidence for the construction costs of a mill, including watercourses, is rare but it can be surmised that, particularly at long-leated sites, watercourse construction may have been as costly as the construction of a simple two-storey mill building and this probably explains, in part, the longevity of mill sites. Before the advent of additional powered textile machinery in the late eighteenth century, the power needs for a fulling mill were low and the retention or addition of a corn mill made economic sense, enabling the owner/lessee to spread his business risks. The pre-Industrial Revolution age enabled a degree of flexibility which was more difficult in the late eighteenth and early nineteenth centuries. While many fulling mills were a development of an earlier single-function corn mill, the subsequent number of stocks and milling stones in a building reflected the overall state of the trade, or the perceived conditions in a particular location. Barton Mill, Cirencester, contained two stocks and one grist mill in 1606 but by 1730 the stocks had gone and a second pair of stones had been added. Bitton Mill, in South Gloucestershire, comprised three stocks and a corn mill in 1634 but was solely a corn mill later in the century. Doynton Mill, too, was a double mill in 1647, but reverted solely to corn milling later. In the flatlands near the River Severn: Stone, Damery, Tortworth, Coaley and Halmer mills were all double fulling and grist mills in the early to mid-seventeenth century but reverted solely to corn later in the century.[17] All the mills along the Ewelme, known to have been fulling mills at some period in the seventeenth century, were double mills; some, such as Dursley and New Mill (Dursley), also containing or having adjoining dyehouses. In the Little Avon Valley from Hill Mill to Purnell's Mill – twenty-one sites – thirteen were double mills in the seventeenth century. Some of the well-known Frome Valley mills such as Stanley, Dudbridge, Fromehall and Brimscombe were double mills at some stage in the seventeenth or early to mid-eighteenth centuries. Stanley Mill, for example, comprised three fulling, one gig and one grist mill, with a warping room in the adjacent building, together with shear shops and a dyehouse.[18]

For many of these mills it is not clear whether the owner or lessee of the fulling mill also operated the corn mill or whether the corn mill was under a separate tenancy. There is evidence that the latter was the case at some sites, however. When Doynton fulling mill (with the adjoining dyehouse) was leased in 1647 the lease made clear that fulling had to give precedence to corn milling, the mill being let 'with such quantities of water as was conveniently spared from a grist mill adjacent to the said tuckmill without prejudice or hinderance to the said grist mill when water shoulde be scarce or wanting'.[19] Sometimes a fulling mill was built more or less adjacent to an existing one as at Lodgemore, Stroud, where Upper and Nether Latemore mills were located either side of the stream on different manors and in separate ownership and tenancies. Monks Mill, Alderley, 'as well the fullinge mills and gryste milles or corne milles', was in the hands of Christopher Neale in 1612; a year later Christopher Purnell, clothier of Wortley, leased 120sq. ft adjoining to 'erect there a newe ffullinge mill', the two mills, collectively known as Monk Mills, being run as separate concerns. By mid-century they were in single ownership and, thereafter, run as one business. Draycot Mill comprised 'those two fulling mills and one grist mill thereunto adjacent and belonging' with a separate adjoining fulling mill in 1696.[20] The appropriately named Grindstone Mill in Alderley comprised in 1600 'one messuage or tenement, one smiths forge one grindstone mill and one fulling mill together'. A dyehouse had been added by 1640 and by 1695 the mill, containing two stocks, was solely used for fulling, the dyehouse adjoining being leased separately.[21] In 1640 one William Stones reported on John Smyth's New Mills, Kingswood:

> I have been at Newe Mills and seene your stocks there, your stocks that Francis Parry held doe stand still and not usded by any other tenant and indeed not fit to be used beinge too out of repairors insomuch that when Parry went away he was driven to mill cloth in Simon Ladley's stocks. Alsoe your tenant Henry Nelme is now away from his parte of the mills and assigned it to his brother Xopher Nelme who now dwells there.[22]

Dyehouses were sometimes shared, too. In 1704 two clothiers, one of Wick and the other of Stinchcombe, agreed to erect a dyehouse in Wick alongside the stream running from Nibley. The agreement mentions 'the conveniency of the water for dyeing and washing their wooll' and the fact that a copper or furnace would be provided, one clothier using the dyehouse on Monday, Tuesday and Wednesday and the other the remainder of the week.[23]

In 1640 a list was made of mills above New Mills in Kingswood and adjoining parishes as far as Hill Mill, Ozleworth, by Thomas Perry, then of 'Grindston Mill' (later called Ithell's Mill).[24] What the list serves to emphasise is the contribution made by mills to the local economy, besides the importance of noting any alterations to watercourses (in case of future legal disputes), a point made by John Smyth with reference to Stone Mill:

> [the river] beinge turned out of his ancient channel and course, begot, somewhat more than a hundred years agone, the setting up of newe of a mill there; ffirst a blademill, after a corne grist mill, after a paper mill, nowe at this day both a(nn)o 1639, (such have been the alterations for profit).

It later became a fulling mill – the occupiers of the mills paid Lord Berkeley a rental for having turned the watercourse to the mill.[25]

The concentration of mills in the Frome Valley was greatest in and upstream of Stroud, where the gradient of the valley was steeper. Mills multiplied especially around Chalford which was said in 1712 to be 'a remarkable place for the great number of clothing mills and a great quantity of cloth made there and in the neighbourhood'.[26] Some mills, while appearing to have been powered by the Frome, were in fact driven by springs flowing out on either side of the valley. The density of mills in the Frome Valley demanded careful protection of water rights. An early seventeenth-century list of watercourse leases in the Frome Valley, from the lowermost mill near Stroud, lists twenty-seven mills, the majority of which can be identified.[27] Alterations to the height of weirs could have serious repercussions by flooding the mill upstream or withholding water from the mill downstream (see Chapter 6).

By the 1720s former double fulling and grist mills in the Frome Valley were increasingly becoming single-purpose fulling mills, some with associated textile processes such as gig mills, dyehouses, press shops and shear shops. Ebley Mill had a shear shop and dyehouse by 1689;[28] Dudbridge Mill had three stocks, a gig mill, grist mill and dyehouse;[29] Lodgemore Mill was a fulling mill with dyehouse and press-house in 1749;[30] Arundel Mill had three stocks, a gig mill and dyehouse in the same year;[31] the grist mill had gone from Brimscombe by 1703; St Mary's had four stocks and (probably) a gig mill by 1741;[32] but Badbrook Mill at the Stroud end of the Slad Valley had a grist mill added to the stocks and gig between 1728 and 1733.[33] Almost every mill between Stanley and Seville's Mill, Chalford, had a gig mill by 1720.

During the early to mid-seventeenth century the Nailsworth Valley was increasingly developed, including some mill sites which, in the late eighteenth and early nineteenth centuries, were to become among the largest woollen factories in Gloucestershire. Woodchester Mill comprised a fulling mill with two stocks (the grist mill had gone), a dyehouse, presshouse, madder house with lofts above and a 'little stove ... used for drying of cloth and other things' by 1744. By 1748 a gig mill had been added.[34] Longfords Mill was in the Pinfold family during the early eighteenth century; by 1783 it was tenanted by Thomas Playne, whose widow Martha ran the business from 1789 until their son Peter was old enough to run it. Martha Playne was exceptional in her day in being prepared to run the mill, and was clearly capable of the task.

The Painswick Valley had numerous mills along it in the seventeenth century and with ten fullers in 1608, some mills were highly likely to have been fulling mills. Painswick town itself has evidence of considerable wealth accumulation – fine buildings, some with known associations with clothing families, and elegant tombs of members of these families in the churchyard. By the eighteenth century the documentary evidence presents a picture of a well-established industrial valley, at least eleven sites known to have been fulling mills in the early to mid-eighteenth century.

There is evidence of six fulling mills in the Slad Valley in the eighteenth century (see Part II), including the ancient Vatch Mill site, owned by members of the Clissold family for many years, which was to be highly developed in the nineteenth century. And, in the Toadsmoor Valley, the site collectively known as Toadsmoor Mills was a fulling and gig mill in 1608 and continued as such in the eighteenth century.

Some vestiges of cloth manufacture survived in the north of the county at Winchcombe and Stanway on the River Isbourne; in the east fulling mills continued to exist at Barrington and Windrush on the River Windrush; Coln St Aldwyn, Quenington and Arlington survived on the River Coln, and Coberley and Cirencester on the River Churn.

Seventeenth-century Eyles Mill, Uley, 1963. (Jennifer Tann)

Dursley Mill, a small pre-Industrial Revolution fulling mill, 1963. (Jennifer Tann)

Few fulling mill buildings survive from the pre- or proto-Industrial Revolution period; those that do are usually ones which either reverted to corn milling, or were subsequently extended in clearly defined phases of building, the older part being retained; or else they were converted to a dwelling. Eyles Mill, Uley, a small two-storey building located on a spring, was later incorporated into a farm and is one of the oldest surviving woollen mills in Gloucestershire. Dursley Mill, a small two-storey square building behind the fine double-gabled, seventeenth-century clothier's house, was a double fulling and grist mill with dyehouse in 1699. It was still a double mill with dyehouse in 1732, but by the late eighteenth century it had reverted to corn milling. Quenington Mill, a single-function fulling mill in 1726, was a fulling and paper mill by 1748 and has been converted to a house. Inchbrook Mill, a small two-storey late seventeenth-century or early eighteenth-century building, is conserved at Renishaw's Woodchester site; Egypt Mill, Nailsworth, is an example of a mill extended laterally, the earlier eighteenth-century western section having a later addition of a matching eastern range in the nineteenth century. Cricketty Mill, Bisley, a two-/three-storey continuous range of small mill and house, by a small brook flowing from Bisley parish to the Toadsmoor Valley, has survived as a house, being beyond the range of post-textile industrial development which later ensured continued use of bigger mills in the main valleys.

WEALTH ACQUISITION

Daniel Defoe described the typical West of England organisation in the early eighteenth century:

> the spinning work of all this manufacture is performed by the poor people, the master clothiers who generally live in the greater towns,[35] sending out the wool weekly to their houses by their servants and horses, and at the same time bringing back the yarn that they had spun and finish which is then fixed to the loom.

He remarked that:

> it was no extraordinary thing to have clothiers in that county worth from ten thousand to forty thousand pounds a man, and many of the great families, who now pass for gentry in these counties have been originally raised from and built up by this truly noble manufacture.[36]

Benedict Biddle, who could not sign his will of 1692, left his grist mill and two fulling mills in Cam (Corrietts Mills) and various parcels of land of more than 3 acres to a kinsman in Wotton, and his goods to his wife, together with various modest cash sums to be paid within one and two years of his

Unnamed mill in Chalford – a mid-nineteenth-century picture of a sixteenth-century mill; the building survives somewhat altered on Chalford Industrial Estate. (Museum in the Park, Stroud)

Steanbridge House, late seventeenth- or early eighteenth-century home of the Townsend family; the centre of a mill estate; note the mill on the right of the picture. (Bovis Homes; David Calvert)

Steanbridge Lower Mill. (Bovis Homes; David Calvert)

death.[37] The larger Gloucestershire clothier was very much an industrial capitalist as the 1701 inventory of Jonathan Witchell shows. With an estate of movable items valued at £3,066 11s 10d, he carried out some parts of the manufacture under his own roof, and reference is made to goods in the mill, although it is by no means clear whether he held a mill or whether he sent his goods to be fulled on commission. As his inventory makes clear, he lived in some comfort. The house comprised a hall, parlour, kitchen and cellar with four 'chambers' above; the hall containing six 'Turkey chairs', a sideboard and 'one joynt chair'; with 'brass Andirons … six cain chairs [and] two cases of knives' in the parlour; while in the kitchen there were brass candlesticks, two spice mortars and a warming pan and six chairs. Reference is made to the warping loft and a wool loft which contained wool, a warping bar, cards, and remnants of warp and weft; there were items for chains and spinning as well as wool at the dyehouse valued at £20, as well as tools, including a cloth press, teazles and handles for raising the nap of the cloth.[38] Clothiers quite frequently had numbers of pairs of shears in a shear shop, sometimes located at a mill, for finishing the cloth. And it was not necessarily the eldest surviving son who inherited his father's clothing business. Jasper, a clothier, was the eighth son of Richard Clutterbuck of King's Stanley who, when he died in 1627, possessed 'partly by inheritance, partly by purchase … a very considerable amount of property'.[39] Thomas Clutterbuck, clothier, of Peck Street, King's Stanley, was the third son of Richard, his elder brother Richard becoming an attorney.[40]

One way in which wealth accrued in the woollen industry was retained and enhanced was by the inter-marriage of clothier families. Jasper Clutterbuck of King's Stanley, who died in 1782, had married Martha Roberts, daughter of a Rodborough clothier, and his father had married a daughter of Giles Nash, a clothier of Stonehouse.[41] Richard Osborne, who had been mayor of Wotton-under-Edge in 1697, moved out of the town, bought and rebuilt Wortley House and then became owner of Monks Mill, Alderley. His son, another clothier, married a daughter of the Blagdens, Kingswood clothiers, who had moved up into the ranks of the gentry. The Osbornes had already acquired about 1,000 acres of land and four more mills. The grandson John Osborne greatly expanded the business at Monks Mill and, after his death in 1770, his daughter and heiress married Samuel Yeats who, after a bad start in business on his own account, moved to Monks Mill and managed very successfully, repaying a loan within a few years. The Adey family were clothiers for several generations, their prosperity beginning with Daniel (d. 1752), a clothier of Coombe. He married, in succession, a Moore, a Purnell and a Blagden, each of whom probably brought additional wealth into the Adey family.[42] A number of the Clutterbucks had two marriages, perhaps indicating a wife's death in childbirth. But the most upwardly mobile marriage was between Nathaniel Clutterbuck and Mary Clifford, co-heiress of Frampton Court.[43]

There was also the smaller clothier who did not own or rent a mill but who owned tools and carried out some of the processes himself. He, too, owned the materials at all stages of manufacture but sent his cloth to be fulled and dyed on commission. Edward Wilson, whose possessions were sold in 1755, had household goods amounting to £40 and over £100 invested in implements and cloth.[44] Nathaniel Hodges 'usually carried about a broadcloth to sell in a retail way'.[45] Some smaller clothiers owned land sufficient to support their families. Nicholas Dangerfield was a clothier of Randwick, whose probate inventory of 1705 shows him to have been more like the archetypical farmer-clothier of Yorkshire. He owned goods to the value of £858 16s and was owed £326 6s for cloth sold in London. He had shears and other working tools, cloth and yarn worth £485, and, in addition, ten sheep, six lambs, two pigs, three cows, one calf, and one mare.[46] A marriage settlement of 1734 illustrates the tools of a North Nibley clothier who did not own or rent a mill and who, presumably, carried on the trade with a small capital. The unusual feature is that the clothier in question was a woman who, on her marriage, sold 'all and singular her woollen clothe, wool yarn and goods now being in the shear shop wool loft, the roomes – and all her woollen clothe in London or elsewhere'. She owned a shear shop containing fourteen pairs of shears, two shearboards, two dubbing boards, one stage of handles and two cloth racks. In the press shop there was one cloth press and twenty-five dozen press papers.[47]

The weaver often had few possessions. Living in his own or a rented cottage he worked long hours and often walked to the clothier's mill or workshop to collect a heavy chain of yarn. John Gainy of Stroud, a broadweaver, left £11 to one daughter, £5 to another daughter and his house and loom to his son, with another £7 10s to be divided amongst other beneficiaries.[48] He was comparatively well off compared with the Painswick weaver who had to mortgage his two looms, three beds and other goods to repay a debt of £20.[49] James Barton of Stroud, a scribbler, left his house with the appurtenances to his wife and 1s to each of his children.[50] These workpeople probably had small gardens attached to their cottages which would have been inadequate for feeding a family.

Much of the broadcloth sold was in its white or undyed state, although Stroudwater reds were famous from the fifteenth century. John Smyth records only thirty-eight dyers for the whole county, more being in the valleys of the Frome, Little Avon and Ewelme. There were no dyers along the Coln Valley, for instance, although there were several fulling mills. While some cloth was dyed by clothiers or their employees at the mill, much was dyed by specialist commission dyers who were accountable to the clothier if they spoiled a cloth. In 1717, for example, a case involved Nathanial Beard, clothier, and Thomas Allen and Anne Hawker. Beard had had his cloth dyed by Robert Hawker, Anne's late husband, and had apparently demanded repayment for some spoiled cloth. Henry Dudbridge of Woodchester and Samuel Butt of Stroud, both dyers, were asked what the custom was regarding damaged cloth. Dudbridge said that he had been on occasion obliged 'to take nothing and sometimes but half price for dyeing and sometimes to take the cloth so spoiled and pay the clothier for it'. Samuel Butt added that it was usual for dyers to keep an account of the work done and the prices charged. Clothiers could then peruse them and compare the dyer's books with their own.[51]

MARKETS AND MARKETING

The Turkey Company was an important buyer of Gloucestershire cloth; so, too, was the East India Company which purchased much cloth made in the Chalford area, an inn at Chalford being known as the Company's Arms. Cloth for the home market was sold either in London, Bristol, or locally within the county. The two cloth fairs at Cirencester were important in Defoe's day and doubtless much of the cloth from lesser clothiers was sold there. 'Superfines, seconds, forests, drabs, naps, duffels, and all that variety to be found in a well-stored draper's shop' were sold in local markets.[52] But this trade required clothiers to pay close heed to changes in fashion and to respond rapidly. Cloth destined for overseas markets went by wagonload to London, while some of the wealthier clothiers employed agents who travelled with samples round England and the Continent taking orders. In the 1770s Rudder estimated that more than half of the Gloucestershire cloth was sold in foreign markets and by this time the Blackwell Hall factors had a virtual monopoly of the London market.[53] They played a significant role as middlemen between clothiers and foreign merchants and were depended upon to give credit until such time as cloth was sold. Daniel Packer of Painswick sent much of his cloth to London to be dyed and finished in the 1760s. When a fault was discovered, after finishing, it was a debatable issue as to where the responsibility lay. In 1762 Packer sent a protesting letter:

> If there had been any fault in the cloth it would have shewe'd itself when it was boil'd and should not have been gran'd. If every Dyer and Cloth Drawer is to do as he please wth Clothiers Property I think it be high time to have done with them.[54]

And, six years later, Packer reported that he could get cloth dyed more cheaply in the country than in London; he used the Partridges at Bowbridge so it is likely that the quality was good. The year 1768 was particularly bad for trade: 'we shall have fewer clothiers another year'; one was sent to Gloucester gaol and another shot himself in the head – 'he was deeply in debt for wooll'.[55] Packer depended on his London factor Thomas Misenor to intercede for him with mercantile firms: 'I have left the price of all my cloths to you who are on ye spott, and must know how best ye price of Cloths is this year.'[56] He sent samples of cloth to his factor, seeking advice as to what was most likely to sell:

> If the Turkey gentlemen would give a reasonable price I should be very glad to have my clos' sold to them, but not on such shameful terms. I had rather they lay and toock the chance of the Company's next buying.[57]

In the 1780s the Dursley clothier John Wallington was encouraged by a London factor to:

> have on Eye to Turn som of your Most ingenious hands on the Fancy Trade which … has answered verry well in Point of Profit to all the Manufacturers.[58]

THE PHELPS OF DURSLEY

The organisation of the Gloucestershire woollen industry on the eve of the Industrial Revolution is well illustrated by the Phelps of Dursley.[59] John Phelps owned Townsend Mill in the 1740s and by the 1750s it was in the hands of William and John Dela Field Phelps. William sold cloth both on his own account and as a partner in Phelps & Co. Fulling, dyeing, gigging and burling were undertaken at the mill. Most of the other operations were put out to workers in their own homes. Wool was mainly bought from Bristol merchants, although some long staple wool was purchased from Edward Tugwell, a wool stapler of Tetbury. Alum and dyestuffs were purchased from Bristol merchants and Phelps acted as wholesaler to some of the smaller clothiers of the Dursley district.

Some varieties of cloth were dyed in the wool and this was contracted out to other commission clothiers and dyers, together with any cloth that had to be woaded. Richard Hawker of Dudbridge, Camm Gyde, possibly of the dyehouse near Skinner's Mill, Painswick,[60] the Adeys of Dursley and the Gainey family of Uley all executed orders for woading and blacking wool. The Gaineys also dyed list wool. After being oiled the wool was mainly scribbled and carded at the mill, although some was contracted out to employees of Edward Tugwell of Tetbury.[61] There are no references to spinning (although in 1761 Richard Hawker undertook some twisting of list yarn) and no records of weaving either, these activities probably being covered in the 'Cloth Book', which is missing from the records.

Some fulling and cloth dressing was undertaken on commission for other clothiers – teazles were purchased from William Plaister and from Thomas and Edward Bennett of Somerset, as well as from Isaac Palmer of Eastington. In June 1772, 73,251 cloths were rowed and 221 mosed. William Phelps required some cloths to be knapped and these were sent either to Thomas Blackwell of Nailsworth or to Richard Hawker of Dudbridge. Cloth was almost certainly sheared on the mill premises by the 1770s for the flocks were sold on to Joseph Lenton of Dursley.[62]

After fulling the cloth was dried on racks or in a stove; Phelps & Co. rented five racks at £2 10s p.a. and a stove for £3 p.a. from Mrs Esther Phelps. Some cloth was drawn on the premises and some contracted out to John Ferebee of Uley, but much cloth was drawn by their factors, 2s per cloth being the usual rate charged. Some cloth was pressed by Thomas Tippetts in Dursley and at other times it was sub-contracted in London by the Blackwell Hall factors, the sum of 30s per month usually being deducted for this by the factors.

Much of the piece dyeing was done at Townsend Mill, while scarlets were sometimes sent to be dyed by John Partridge of Bowbridge or Isaac Watts, commission dyers who specialised in this highly skilled process. The Phelps also did commission dyeing for other clothiers, including Samuel Ricards of Wortley, John Dauncey of Wotton and John Holbrow and Co. of Uley.[63] The Dyebook gives an indication of the range of cloth being made in mid-Gloucestershire in the 1770s. Whilst commissions for dyeing superfine broadcloth occur most frequently, there were also beavers, narrows, shalloons, medleys, say-dyed drabs and thin medinas.

In common with other medium-sized cloth businesses, most of the Phelps' cloth was sold through Blackwell Hall factors. When the cloth arrived in London it was stored at Blackwell Hall for which a halleage charge of 2s 6d per cloth was made. Cloth was minutely examined by the factors, who made deductions for faults. The clothier was not usually paid a lump sum after the sale of a batch of cloth, but rather drew on the factor at intervals during the year. This was to the advantage of both parties. For the clothier it provided funds as needed and the factor was not required to pay out a larger lump sum. Moreover, the factor provided credit[64] for which interest was charged. Phelps was charged between 2.5 per cent and 5 per cent and he frequently owed £20 or more interest on a year's credit. In addition, the clothier paid commission to the factor for cloth sold: 'Commission on cloth sold ... £103 0s 8d.'[65] Clothiers felt at the mercy of the factors. Daniel Packer of Painswick told his factor that if cloths were not sold on better terms:

> more makers must be utterly ruined ... if the Gentlemen factors had ask'd a better price ... last year the company would not have bought one cloth the less, and been content the Makers shou'd make a living profit.[66]

The clothier might set a minimum price on his cloth and Hanson & Mills told Phelps that they would 'endeavour to get you some good orders upon your terms'. But, when Phelps threatened to take his cloth elsewhere, believing he was in a position to demand a higher price, he received a reprimand: 'We are concerned to observe a mode of addressing us we do not think we have given occasion for.'[67] Another clothier, William Hood, was told 'if the end will not fetch the price you have fix'd shall take the liberty of sending back to you as you do not approve of it being sold lower'. J. Wight of Painswick bowed to the factor's verdict: 'you must use your own judgment as usual.'[68]

John Phelps and Phelps & Co. dealt with three firms of Blackwell Hall factors, only one firm being common to both. (Smaller clothiers usually only dealt with one.) By the 1770s Phelps & Co. and William Phelps had become accustomed to sending patterns of cloth, or responding to a specific order. Blackwell Hall factors expected a rapid response to an order which, considering the many production processes involved in, for example, a high-quality scarlet or black, it may not have been possible to fulfil, unless such cloths were already at an advanced stage of production. On one occasion Hanson & Mills requested:

> Will you in a post or two send us a Pattern letter containing abt. 6 or 8 Greens, as many Blues … a few blacks, Whites and Scarlets in good press and I will endeavour to get you some good orders upon your terms.[69]

Sometimes the factor might ask for a cloth sample to be matched – a task requiring great skill on the part of the dyer:

> Annexed are some patterns of Cloths wanted for a foreign order to which, or such part as you can do, your best attention and answer is requested. The Greens are requested pretty exact.

Seven months later it was clear that Phelps had failed to impress: 'the Greens will not do … we are extremely disappointed in finding you so wide of ye colour ordered.'[70] On another occasion a batch of scarlets was returned, being 'such bad Colours'. When trade was bad the clothiers resorted to sending cloth to London as a speculation but this was not always successful if the market was flooded and they were out of touch with fashion changes, as in 1795:

> The rage for French Cloths is so far abated that scarcely anyone will pay the smaller advance for them … when the Town fills at winter they may sell better.

Samuel Rudder was clear who gained most in the relations between clothier and factor:

> Instead of finding [the factor] a mere agent to the clothier which he originally was, and still ought to be, it will appear that he is now, by insensible degrees, become a very different kind of person … By this expedient [the factor] hath virtually, in many instances, the power of the principal, yet still preserves the appearance, the indemnity, and every other advantage of the agent.[71]

The expenses involved in carrying on a large trade were considerable and the profits erratic. John Phelps drew up an account of his expenses for 1741 which totalled £1,071 19s 8d, the largest items being his payments to Excise. His receipts that year were £2,624 15s 3d and he had 157 cloths remaining unsold in London. In 1742 his expenses were £1,910 3s 10d and his receipts £4,000 7s 1d, with 89 cloths unsold. The following year he only made a net profit of £508 17s 7d. By the 1770s William Phelps and Phelps & Co. were making very little profit. In 1771 William Phelps' expenses with one factor were £2,027

9s 9d, while the receipts were £2,194 18s 4d; in the following year his profit from this firm of factors was £12 8s 6d on 233 cloths sold and 333 resting unsold. Phelps & Co. was overdrawn on the factor William Partridge by £3,449 11s 6d in 1776. This debt was settled in cash save for £1,395 11s 6d which was debited from his account. William Phelps was overdrawn on Fryer, Webb & Fryer by £2,923 2s 8d in 1779. The account was settled by his bond for that amount and no further entries were made in the book. He continued to make and sell cloth into the 1790s, dealing with Hanson & Mills, but little profit was made and cloth was sometimes sold at a loss. In common with some other cloth manufacturers of Dursley, William Phelps retired from the trade before it was too late. As a Parliamentary report of the nineteenth century commented: 'The trade did not decline in Dursley for want of capital, inasmuch as the manufacturers were all opulent men, and are now living, retired from trade, in good estate.'[72] William and John Dela Field Phelps endorsed their status as gentlemen and retired from the woollen industry.

CLOTHIERS' HOUSES

Many seventeenth- and early eighteenth-century clothiers lived in houses near to their mills. Some seventeenth-century houses are elegantly proportioned gabled buildings, such as Salmons Mill House (Stroud), rebuilt in 1593 by George Fletcher who added a porch in 1607 with his initials and cloth mark; and Egypt Mill House (Nailsworth), with its two gables on each of the four faces and the date stone RW 1698 (Richard Webb); Upper Pitchcombe Mill House was built by Thomas Palling, clothier of Upper Pitchcombe Mill, in 1743 as a three-storey house of classical design with a hipped stone roof behind a parapet, the forecourt having four piers with stone vases atop. The early eighteenth-century Belvedere Mill House in Chalford was elegantly remodelled in 1789, while the clothiers at Seville's Mill, who had lived, at first, in the attractively proportioned mid-seventeenth-century Seville's House, moved to the larger seventeenth-century Green Court (a new classical front of c.1810 was added by the Innell clothier family). Kingshill House, Dursley, was erected in 1706 for Thomas Purnell, a clothier, and was re-fronted in the late eighteenth century. (It was much altered between 1864 and 1885 and is now the Dursley RDC offices.)[73] Some middle-range clothiers' houses were grander, such as the late sixteenth-century St Mary's Mill House, Chalford, re-fronted c.1720 for the clothier Samuel Peach (low wings being added c.1820 for Samuel Clutterbuck). When Wortley House, formerly the home of the Osborne family who owned Monks Mill, Alderley, was advertised for sale in 1776 it was said to be 'fit for a gentleman or a clothier'.[74] Clothiers became collectors of works of art, good furniture and silver; the sale of the effects of Thomas Tippetts of Dursley in 1789 took thirteen days.[75]

In contrast to 'a respectable man' who might provide work for thirty to forty looms, a 'gentleman clothier' might have up to 100 or more weavers in his employ. Thomas Atkyns wrote, in 1712, of the good estates they held and, early in the nineteenth century, Timothy Exell described them as 'rich and opulent men; they were not only worth their thousands but their tens of thousands and their scores of tens of thousands'.[76] The King's Stanley, Eastington and Frampton branch of the Clutterbuck family illustrates the acquisition of wealth in the woollen industry and the purchase of estates through the generations. When Thomas Clutterbuck of King's Stanley died in 1614 he held the manors of Eastleach Turville and Siddington in Gloucestershire, and East Greenwich in Kent, together with property in eight Gloucestershire parishes.[77] Jasper Clutterbuck, clothier (d.1627), possessed 'a very considerable amount of property in King Stanley'.[78] Another Jasper, who lived in King's Stanley House, had a son who became a Blackwell Hall factor.[79] A number of clothier members of the family married clothiers' daughters;[80] others became 'citizen clothworkers' in London, yeomen or attorneys.[81] Richard Clutterbuck (son of Nathaniel) rebuilt Frampton Court in 1731.[82] Sir Onesiphorous Paul of Southfield Mill, Woodchester,

Valley Mill House, Chalford, a typical late seventeenth-century gabled clothier's house of the sort frequently found adjacent to a mill. (Roger Tann)

Green Court, Chalford, the clothier's house for Seville's Mill. (Roger Tann)

Stonehouse Court. Built c.1601 by Daniel Fowler, it passed through the hands of many well-known clothier families. (Howard Beard)

entertained the Prince of Wales, became High Sheriff of the county and, after presenting a loyal address to George III on his accession to the throne, was rewarded with a baronetcy. He and others who aspired to live the lives of gentlemen were, on occasion, the butt of social mirth:

> Sir Onesiphorous Paul and his Lady are the finest couple that has ever been seen here since Bath was built. They have bespoke two whole-length pictures, which some time or another will divert us. His dress and manner are beyond my painting; however, they may come within Mr Gainsborough's.[83]

He built Hill House, Rodborough, a 'beautiful villa situated on an eminence, with a pleasant prospect of the river'[84] described by Fosbrooke as a 'superb residence' – later known as Rodborough Manor.[85] He invested in his son's education at Oxford and the Grand Tour (where he seems to have succumbed to gambling). Stonehouse Court, having been purchased by two local clothiers in 1558, was rebuilt by Daniel Fowler between 1599 and 1601. It remained in the hands of the Fowler family for about a century. Stanley House, the 'mansion house' which belonged to Jasper Clutterbuck in the mid-seventeenth century, passed successively into the hands of the Holbrow, Peach, Cooper, Davies and Harris families, all clothiers. The Paul family, clothiers of Woodchester, King's Stanley and Tetbury in the seventeenth and eighteenth centuries, made some advantageous marriages and John Paul Paul became High Sheriff of Wiltshire.[86] Highgrove House, Tetbury, 'a handsome if plain rectangular box',[87] was built for John Paul Paul between 1796 and 1798. Rookwoods, on the edge of Bisley parish, was owned by a succession of families who had made their money from cloth manufacture. John Butler (d.1716) was probably the wealthiest of the clothier Butlers, but thereafter, the family was unable to maintain their lifestyle by the cloth trade. In 1793 another clothier, William Tyler, bought Rookwoods at the age of 27 and built a small fulling mill in the valley below the house.[88] Stouts Hill, Uley, was acquired in 1697 by Timothy Gyde, a clothier. In 1716 the property was settled on his son Thomas and father and son continued as partners in the woollen industry, the profits of which went to support the development of a landed estate. When Thomas died in 1743 he owned a significant proportion of the parish of Uley. The new house, commissioned by Timothy Gyde II, was in existence by 1755 and is a 'delightfully light-hearted rococo gothic extravaganza' with 'an air of open, uncorrupted innocence and pure joy in life's pleasures that is extremely endearing'.[89] The architect was almost certainly William Halfpenny of Bristol who published a pattern book in 1752.

One of the most interesting clothier's houses is New Mills, Stroud. It is unusual in that house and mill were a continuous range and deliberately designed such that the boundaries between the one and the other were not obvious.[90] New Mills gives the appearance of an elegant country seat, an impression endorsed (in a contemporary print) by the image of two young women, who could be daughters of the house, being serenaded in the grounds. It is a grand example of the factory as country house. New Mills was built by Thomas Bayliss in 1766 and passed to his son Daniel in 1799.[91]

As, later, in the Industrial Revolution, when the sons of manufacturers made the transition from wealthy man of industry to man of leisure (sometimes engaging in public roles such as Justice or High Sheriff; sometimes as a man of country pursuits), so, too, did some of the sons of seventeenth- and early eighteenth-century Gloucestershire clothiers. Sir George Onesiphorous Paul did not run the cloth business at Southfield Mill on his father's death, but let it to a cousin, Obadiah Paul, and devoted himself to the life of a landowner (and, later, to prison reform).[92] He built up an art collection and became a prominent

New Mills, Stroud, built in 1766 for Thomas Bayliss; the factory and country house are all in one, as depicted in Rudder's *A New History of Gloucestershire*. (Gloucestershire Archives)

member of the Bath & West Agricultural Society. Members of the Paul, Wathen, Clutterbuck, Peach, Playne and other families moved away from the industrial valleys to landed estates in the countryside.[93]

The seventeenth and early eighteenth centuries mark the emergence of the gentlemen clothiers, but there were also many middle-range and smaller clothiers. The best independent dyers were highly skilled in seeking good quality dyestuffs and in colour matching. Weaving was a skilled occupation of poor males; while carding and spinning were undertaken by women and children. By the beginning of the seventeenth century, Gloucestershire was noted for a range of woollen cloth; while broadcloth predominated, especially in Stroudwater, some lighter fabrics, notably medleys, and Spanish cloth were made. The woollen industry was even more firmly located by the three main western running river systems, new mills confirming an already established pattern of distribution. With the increased density of mills in the valleys, disputes over water rights grew in frequency. More of the mills were double mills and many had gig mills, thereby consolidating finishing processes at the mill site. Some larger mills had shear shops and a dyehouse at the site and were poised for further development when the increased mechanisation of manufacturing processes began in the Industrial Revolution.

Endnotes

1. John Smyth, *Men and Armour for Gloucestershire*, 1608
2. That the Company had had a medieval charter was fiction and Perry alleges that the Philip & Mary Charter confirmed by Elizabeth – quoted at length by Rudder – was a forgery: Perry, Trans. BGAS, VOL 1945, p.94)
3. *VCH Gloucestershire, IV*, p.75
4. Ian Mackintosh, 'Laying the Foundation; Stroud in the Sixteenth Century', *GSIA Journal*, 1985, pp.29–38; Ian Mackintosh, 'The Metropolitan Town of the Clothing Trade', *GSIA Journal*, 1986, pp.29–39
5. R. Atkyns, *Ancient and Present State of Gloucestershire*, 1712, p.42
6. D. Defoe, *A Tour Through the Whole Island of Great Britain*, 1724–27
7. Cal SPD 1547–80, p.129
8. *Hist. MSS Comm.*, H. of Lords MSS, 1692–3,42; *Glos N & Q*, III, 406
9. DC/E46/2; boxes 23 (bdl 8)gs, 24 (bdl 11), 34,67gs
10. J. de L. Mann, *The Cloth Industry in the West of England from 1640–1880*, Oxford, Clarendon Press, p.xv
11. E.A.L. Moir, 'Benedict Webb, Clothier', *Economic History Review*, 10, 1957, p.256
12. P328/1 IN 1/1
13. S. Rudder, *A New History of Gloucestershire*, 1779, p.629
14. Playfair, *British Family Antiquity*, VII, 1811, Appendix p.ix
15. See Part II
16. Berkeley MSS, III, p.269
17. Coaley Mill later becoming an ironworks
18. D873b; D1347 acc. 134
19. Box 65gs
20. AT 19919/37gs
21. D1086, box 18
22. 16533 fol. 107
23. Box 7gs
24. Smyth MSS, 16533 fol. 107
25. Berkeley MSS, III, p.362

26 Rudder, p.629
27 PRO E315/394/1
28 289(64)gs
29 D1159
30 D67/2/46
31 D1347 acc 1603
32 D1815
33 D1512
34 D44/A(b); D149/375/954
35 This is not the most appropriate word for the 'clothing villages' which characterised Gloucestershire at this time
36 Defoe, I, p.280
37 Stroudwater Textile Trust (STT) MSS, copy of will of Benedict Biddle
38 D333/F26
39 M.E.N. Witchell & C.R. Hudlestone (eds), *Account of the Principal Branches of the Family of Clutterbuck from the Sixteenth Century to the Present Time*, Gloucester, privately printed, 1924, p.75
40 Ibid., p.77
41 Ibid.
42 R. Perry, *Wotton-under-Edge: Times Past – Time Present*, 1986, pp.76–7
43 M.E.N. Witchell & C.R. Hudlestone, pp.75–7
44 Gloucester Journal (GJ), 10 November 1756
45 Ibid., 26 March 1734
46 D149/263/466
47 D654/11/25/F1
48 D1159
49 DC/E3
50 Ibid.
51 *41st Rept, Deputy Keeper of the Public Records*, 515, 1880
52 S. Rudder, p.61
53 Ibid., pp.61–2
54 D149: F11.4
55 Ibid., 18 August 1768
56 D149: F11.4, 18 May 1768
57 Ibid., 28 January 1769
58 D149:997, M. Cowley to John Wallington, 17 April 1788
59 Jennifer Tann, 'Some Account Books of the Phelps Family of Dursley', Trans. BGAS, 86, 1967, pp.107–17
60 Jennifer Tann, *Gloucestershire Woollen Mills*, 1967, p.209
61 Eight Dursley card makers were named in Bailey's British Directory of 1784
62 Flock engines were installed in some mills as a sideline
63 Hawker & Richards of Bowbridge dyed cloth for a number of clothiers, not only those of the Stroud and Nailsworth valleys, but also for Timothy Larton of Alderley New Mills, William Phelps of Dursley and the Wallingtons of Dursley, besides the Phelps'
64 Contrary to E.A.L. Moir's assertion in 'The Gentleman Clothiers', in H.P.R. Finberg (ed.), *Gloucestershire Studies*, University of Leicester, 1957, p.236
65 19965gs

66 D149: 980, D. Packer to Thomas Misenor, 22 April 1769
67 PRO C113 16–18
68 Stroud Museum MSS
69 PRO C113 16–18
70 Ibid., 3 October 1795; Moir, pp.234–5
71 S. Rudder, p.61
72 PP 1840, XXIV, p.454
73 N. Kingsley, *The Country Houses of Gloucestershire*, Vol. 2, Phillimore, 1992, pp.5, 50
74 GJ, 26 February 1776
75 R115.118gs
76 *A Brief History of the Weavers of Gloucestershire*, J13.8gs
77 M.E.N. Witchell & C.R. Hudlestone, p.45
78 Ibid., p.75
79 Ibid., p.77
80 Jasper, who died in 1782, married Martha Roberts, daughter of Thomas Roberts, clothier of Rodborough; his father Jasper married Catherine, daughter of Giles Nash, clothier of Stonehouse; his son Jasper became a Smyrna merchant with his cousins, ibid., p.88
81 Ibid., p.76
82 Ibid., p.90
83 Moir, p.196
84 Rudder, p.629
85 T.D. Fosbroke, *Abstracts of Records and Manuscripts respecting the County of Gloucester*, 2 vols, Gloucester, 1807, p.365
86 Kingsley, p.159
87 Ibid.
88 By 1813 it was a grist mill and by 1841 had become a cottage; Juliet Shipman, *Rookwoods*, pp.55–62
89 Ibid., p.240
90 In a similar way that, on a landed estate, the ha-ha was designed so as to provide a sense of 'uninterrupted property' when, in fact, it may have been the boundary to the estate
91 He was bankrupt in 1812
92 Moir, op. cit., pp.195–224
93 A curiosity is the fact that, after clothiers had left, the names of their former houses were sometimes changed, indicating a post hoc perception of the lifestyle of the gentleman clothier. Longfords House, built in 1801 for the Playne family, was subsequently renamed Stroud Court; Avening Lodge was redesigned for Edmund Clutterbuck *c*.1825 and in the 1920s renamed Avening Park; Hill House, Rodborough, was renamed Rodborough Manor in the late nineteenth century

CHAPTER FOUR

WOOLLEN TRADE FLUCTUATIONS AND STATE INTERVENTION IN THE SEVENTEENTH AND EIGHTEENTH CENTURIES

State intervention in the seventeenth and early to mid-eighteenth centuries was perceived by King and Parliament to be the best means by which the woollen industry was fostered – and income generated through taxes. The form it took was protection and regulation, but in a manner which harked back to medieval and Tudor times, and failed to recognise the needs of entrepreneurial clothiers – although they were not averse to seeking protection when trade was depressed. On one occasion the state intervened at the behest of a London merchant, and this was to have a disastrous effect on the woollen industry from which it struggled to recover for much of the seventeenth century.

In 1614 the export of wool, wool fells and fuller's earth was banned as being injurious to cloth manufacture, a remedy which favoured woollen manufacturers but was to the detriment of Merchants of the Staple.[1] This was followed, in the same year, by the Cockayne project which brought the English broadcloth industry to its knees. In order to foster the dyeing and dressing of broadcloth in England rather than in Holland and Germany, the privileges of the Merchant Adventurers were suspended in 1614 and their place taken by a new company, the 'King's Merchant Adventurers', whose leading member was Alderman (Sir William) Cockayne. The export of unfinished cloth was prohibited and cloth was required to be dyed and dressed in London. The two unproven assumptions behind this plan were that first, London dyers were sufficiently skilled and could cope with the increased volume of white cloth for dyeing, and second, that the overseas markets for white cloth would be prepared to purchase English-dyed broadcloth instead. The project met with hostility in overseas markets where the dyeing and dressing of British cloth was a significant occupation. Gloucestershire clothiers who produced large quantities of white, undyed broadcloth were hit badly and complained vociferously. In 1616 Toby Chapman and William Warner complained to the Council of State on their own behalf and on behalf of Gloucestershire clothiers that merchants who had previously purchased their cloth weekly in London had ceased to do so and 'the loss of their trade is the ruin of themselves' – and those who worked for them.[2] Another complaint 'touching the decay of their clothing trade' was met by the King's Merchant Adventurers' retort that the deceitful making of Gloucestershire cloth was much complained about and this prevented their purchasing more.[3] The Council of State, not heeding the King's Merchant Adventurers' excuses, ordered them to buy up the Gloucestershire cloth and, in August 1616, the King, after thanking the Council of State for attending to the clothiers' petition, summoned Sir William Cockayne, remonstrated with him on his failure to ensure the purchase of surplus cloth and ordered him to report daily on his progress.

Cockayne and his fellow merchants, having 'reduced the clothing trade from a flourishing state to decay', were ordered immediately to buy up all Gloucestershire cloth 'and such as shall weekly be brought in'.[4] The standstill was further exacerbated by Holland, which had been a large market for unfinished cloth, refusing to buy cloth finished in England and, by September 1616, there were large quantities of cloth remaining in store in Gloucestershire, Wiltshire and Worcester.[5]

The project collapsed in 1617 when the old Merchant Adventurers had their former privileges reinstated. But the whole cloth trade had been badly knocked and to make matters worse, the Thirty Years War broke out in 1618 and the German market shrank disastrously. Cries of distress from the West Country reached the Privy Council and it was noted that in Gloucestershire, Wiltshire and Worcestershire between a half and a third of the looms were out of use. In August 1618 a petition, signed by several JPs, was sent to the Privy Council to secure special protection for Benedict Webb, a significant clothier of Kingswood; he then engaged in a lawsuit, in order that his numerous employees would not be thrown out of work, the JPs being nervous of the social fallout were Webb to be declared bankrupt.[6] There was further distress in the cloth-making areas in 1620 (exacerbated in 1622 and 1623 by the scarcity of corn). The situation was sufficiently serious for the Privy Council to consult with clothiers, justices and Merchant Adventurers about the cause of the crisis and its possible remedies. Blame was liberally distributed – the Merchant Adventurers claimed that local cloth searchers failed to perform their work effectively, with the result that the standards of cloth making had fallen, while clothiers chiefly blamed wool dealers.[7] It is possible that imports of dyestuffs had been disrupted during the Cockayne project, for clothiers complained in 1621 about the high price of logwood 'considering the greate stand of cloth'.[8] By the following year money was said to have been so scarce that the 'poor [had] assembled in troops of forty or fifty and gone to the houses of the rich and demanded meat and money which [had] been given through fear'.[9] The Council of State wrote to the Sheriff of Gloucestershire about reported assemblies and riots 'partly occasioned through want of employment for the poorer sorte by the decay of cloathing'. Each parish overseer was requested to ensure either that the poor had some form of work, or parish relief, even if this meant a higher levy on the better off.[10] The Gloucestershire justices reported in March 1622 that they had called the clothiers together and had ordered them to employ their workpeople one month longer in order to prevent mutinies. The clothiers responded that they were 'so reduced by the decay of trade as to be quite unable to employ their workpeople'. They alleged that there were 1,500 looms in Gloucestershire, each of which, directly or indirectly, provided work for sixteen people who received little more than 1s per week. Again, the same multiple reasons for the decay of the woollen industry were rehearsed: the Dutch taking up the former British market for undyed cloth in Holland, 'false' making of cloth, British wool exports, various impositions levied on British cloth, the scarcity of coin, the increased wearing of silks and stuffs instead of woollen cloth[11] and the fact that war made traffic unsafe for merchants.[12]

Gloucestershire clothiers were blamed for the decay of the Coventry woollen industry in the early 1620s, being accused of sending shorter undyed cloth to Coventry for dyeing and dressing, which was then passed off as Coventry cloth. The number which could be sent per year was fixed at 300 and, not surprisingly, complaints were made that this was greatly exceeded.[13] Difficulties were further exacerbated in 1625 by London being considered unsafe for clothiers, 'by reason of the exceeding great infeccion at that time there'.[14] Gloucestershire clothiers, together with their counterparts in Wiltshire, Somerset and Berkshire, asked the Merchant Adventurers to move the staple out of London, either to Reading or Southampton while the plague raged.

Gloucestershire clothiers, as with their counterparts in Wiltshire, Somerset and other clothing regions, were having to adapt to new circumstances. However, while small clothiers failed and profits were low, it was still possible for the more enterprising of the larger clothiers to make considerable profits, some moving into the lower ranks of the gentry. Those clothiers who survived the depression which commenced with the Cockayne crisis and was exacerbated by the Thirty Years War, recognised that new products and new markets were the way to survival and profits.

In 1631 a commission was set up for the reformation of the trade in white cloth made in Oxfordshire, Gloucestershire and Wiltshire and the eastern limits of Somerset. The purpose was to investigate whether the regulatory statutes for the cloth industry – of which there were many – were recognised and correctly implemented. Anthony Wither was sent to enquire, having given the Council of State eight reasons why gig mills should be suppressed: 'no nation useth any gig mills but England and in England no county but Gloucestershire and in that shire not about eight or 10 clothiers that stand up for them.' A petition by weavers, spinners and fullers of Leonard Stanley and King's Stanley stated that 800 people had been employed:

> but now they are never likely to be employed again – not only these parishes but the whole country is maintained by the trade of clothing which, standing still, must bring impoverishment upon the whole whereby many thousands of poor people will be utterly undone.

The weavers claimed that, since the King's proclamation, their masters had not been able to dress their cloth as formerly 'and are constrained to put away two parts of their workfolks'.[15] Clothiers making red cloth explained that without the use of their mosing mills, they could not dress half the amount of cloth that they had formerly done. And, if they were required to do the work by hand, king teazles could not be provided in less than two years.[16] Clothiers making white broadcloth claimed that there were so many penalties to which they could be subject, that they threatened to give up the trade altogether.

The King commanded that clothiers use only one mark on their cloths, but clothiers responded that there were different marks for different kinds of cloth and these marks had become well known abroad. To alter them would be of great disadvantage.[17] Makers of red cloth wrote twice more to the Lords pointing out that they had received no reply to their petition concerning mosing mills, and in December 1633 the Council of State ruled that no more gig mills should be set up and that a trial should be made of the cloth dressed in these mills. No record of the results of the trial survives but the proclamation concerning gig mills was not retracted;[18] nor, as on other occasions of state intervention in their industry, did clothiers operate in accordance with state regulations: they continued to use gig mills.

Complaints and restrictions rumbled on. In June 1634 four clothiers complained to the JPs that their cloth taken to Blackwell Hall had been seized, although it had been 'good and lawful'. They threatened to put off workpeople and JPs wrote to the Council of State in some anxiety, explaining that 'the welfare and in a sort the peace of their country relies much upon the employment of great numbers of poor people and they upon the passage of the clothiers trade'[19] – such was the anger occasioned by depressed trade, together with regulation at a level which could not be effectively enforced, and, if it had, would have served to suffocate the industry.[20]

In September 1638 a Commission of Inquiry into the entire woollen industry of England was established with full powers of investigation vested in the commissioners, one of whom was the unpopular Anthony Wither.[21] The commissioners made their report in 1640.[22] The stated causes for the decline of the industry much resemble those contained in the report of 1622, an additional reason being the frequent use of gig mills which had 'bin suppresst in all places thereof except in Gloucestershire about Stroudwater, where they dayly more and more increase, to the great disgrace, preiudice and danger of the ruining of the manufacturer there if not in time prevented'.[23] The Commission recommended that twelve months' notice be given to owners of gig mills for them to be destroyed. In order to supervise these new restrictions, it was suggested that the chief clothing towns be incorporated and that Gloucester and Stroudwater would benefit from becoming corporations. These recommendations went unheeded;

the desire to re-establish a central administrative system for woollen cloth was doomed to failure and the 1640 Commission Report marked the end of government attempts to set the whole broadcloth industry in order. Meanwhile, the Civil War was imminent and King and Parliament were concerned with other matters.

Clothiers were suspected of favouring the Parliamentary side, probably with some justification, having suffered at the hands of the King's regulatory and fiscal policies. With Gloucester and Bristol (until 1643) being held for Parliament, while the King's headquarters were at Oxford, there was much military activity in the county. Clothiers suffered from marauding Royalist troops, and in February 1642 the King authorised Prince Rupert to commandeer all cloth in the chief cloth-making areas of Gloucestershire and have it sent to Cirencester:

> Whereas wee are credibly informed that at Cirencester, Stroud, Minchinhampton, Tetbury, Dursley, Wotton under Edge, and Chipping Sudbury great quantities of cloth canvas and locharame are to be had for supplying ye gt necessities our souldiers have of suits, wee … doe hereby pay you to send a competent party of Horse under ye command of some able person to visit these severall places wch lye not farr asunder, and to bring from thence all such cloth canvasse and locherame as they shall find there to Cirencester.[24]

Whether the clothiers were ever paid seems doubtful since they were required to go to Oxford for reimbursement. In February 1642 Samuel Webb of Ham Mill received a letter of protection from Prince Maurice following his 'ample satisfaction' of allegiance to the King, so that no one shall 'presume to plunder any [of] your houses or take or carry away any [of] your goods … upon any pretence whatsoever'. And in August 1643 a similar letter of safeguard was signed by Prince Rupert.[25] Prince Maurice's protection note was signed eight days after Cirencester had been assaulted and taken by the King's forces, the Parliamentary side reporting that the Royalists 'took away cloth, wool and yarn, besides other goods from the clothiers about Stroudwater to the utter undoing of them and theirs'.[26] Clothiers begged for protection from the King:

> some of us have lost our whole estates, and many of us have the residue that is left now in the hands of merchants within the city of London without which we cannot keepe our work-folkes in work as is commanded and as we desire, our credit being lost with our goods and/or cloth being made, we have no way of safety to keepe it, noe meanes to vent it whereby we are in a most distressed condition.

In response they were told that the King had 'a deep sense of the importance of the trade of those parts', and that arrangements would be made for clothiers to send their cloth out through Bristol and other western ports in the King's hands, and clothiers could collect their money from London.[27] In the event, clothiers could only collect their money from London once Parliamentary forces had captured the West, and manufacture would continue to be interrupted until the end of the war.

Gloucestershire benefited from a large order of cloth for the army in 1649, the Council of State ordering 16,000 coats and breeches for the soldiers in Ireland, 'the coats to be made of Coventry or Gloucester Cloth of Venice colour red, shrunk in cold water'.[28] Export difficulties were experienced in the 1650s, Merchant Adventurers asking for a convoy to escort them to Hamburg.[29] In 1664 Gloucestershire clothiers, together with those from other parts of England, petitioned the House of Commons concerning 'the present decay of the trade', blame being laid on the Act regulating Blackwell Hall and Leaden Hall which required clothiers to spend time in London, plus increased halleage fees;[30] the House of

Commons' response was to order a committee to be set up to consider the possibility of suspending the Merchant Adventurers' charter.[31]

The situation did not improve in the latter seventeenth century. In 1677 John Smyth the younger of Nibley reported on the situation in the Kingswood district:

> it formerly depended much upon clothinge, but is now decayed with the trade. And in the opinion of my Lord Hale, to me, about a month before his death, who was a neighbour to the place, the estates of the parish would not mayntayne the poor theare in twelve yeares.[32]

And in 1691 Gloucestershire justices wrote to the Privy Council drawing attention to the starving condition of unemployed workmen 'as their faces (to our griefs) do manifest'.[33]

Trade revived at the end of the seventeenth century and continued to improve until 1710 when there was financial panic and great speculation in the industry. While some clothiers were said to be making 1,000 pieces a year, many made less. Between 1718 and 1722 William Palling of Painswick sent about 230 cloths a year to London for export to India and the Levant.[34] Estimates of Gloucestershire's total cloth production range from 23,102 pieces p.a. between 1713 and 1719[35] to 50,000 in 1712;[36] even if this is an exaggeration in comparison with Yorkshire's estimated annual production of, at most, 2,660 pieces p.a., it is an indication of relative scale.

Depression set in from 1726–27, and cloth exports fell. Two Acts of 1726 and 1727 to regulate the woollen industry forbade combinations of workmen, prohibited payment in truck and legislated for the fixing of wage rates by magistrates. Gloucestershire workers petitioned the Privy Council for regulation of their wages but clothiers did not comply. The rates were confirmed in 1729 and 1732 but the clothiers 'treated the said order with the greatest contempt, not anyways complying therewith'.[37] A further appeal to Parliament was made in 1756 when, two years into the Seven Years War, trade depression caused major difficulties in the woollen industry. In the ten years between 1745 and 1756 there were eighteen bankruptcies (compared with seven in the previous eleven years).[38] And, between 1751 and 1756, it was said that £50,000 had been lost by 'bankruptcies, fall of goods, and want of sales'.[39] The weavers pressed for confirmation of the wage rates which had been set in 1728 and claimed they 'labour under great hardships by being paid in goods, or by way of Truck, contrary to the said Act'.[40] On investigation it was found that the clothiers had ignored the laws when it suited them and a new Act was passed in 1756 (29 Geo II c.33) empowering the justices to establish piece rates for weavers and others employed in the woollen industry.[41] The clothiers responded by trying to persuade the workers to sign an agreement to say they were satisfied with their wages. The weavers threatened to throw any weaver who signed such an agreement into his master's mill pond and, when violence began to occur, the justices granted the wage list. Clothiers objected that such a system was not only unworkable, but 'repugnant to the liberties of a free people and the interests of the trade'; moreover, 'the weavers by this Act will be rendered more our masters than we are now theirs'.[42] They emphasised the need for less regulation, trade being 'a Tender plant that can only be nursed up in liberty'.[43]

The weavers struck work for six weeks, the resulting loss being estimated by William Dallaway, a Stroud clothier, to have been from £15–20,000.[44] At a meeting held on 11 October 1756, between 2,000 and 4,000 weavers gathered in Stroud and presented a proposed set of wages to the clothiers. Inclined as before to resist, the clothiers, recognising the potential danger of the situation and being in a room more or less surrounded by the mob, agreed. But not before the rioters surged in and several clothiers jumped from windows. Major General James Wolfe was sent in command of six companies of infantry to restore order, and related:

those who are most oppressed have seized the tools, and broke the looms of others that would work if they could. I am afraid they will proceed to some extravagancies, and force the magistrates to use our weapons against them.[45]

The wage rate was formally accepted in the next Quarter Sessions and the soldiers were withdrawn. But the clothiers petitioned Parliament and in February 1757 a House of Commons Committee heard evidence,[46] as a result of which an Act (30 Geo II c.12) was passed repealing the Act of the previous year which legislated for wage rates to be fixed. Subsequent petitions from 'distressed Broad Cloth Weavers' and, separately, from 'Gentlemen and Landholders' of clothing parishes which warned of 'bad consequences' were not heeded. The petition from the latter group suggests that, notwithstanding the wealth and social aspirations of larger clothiers, they were not recognised as gentry. Unlike Wiltshire, where clothiers had become magistrates, Gloucestershire magistrates were landed gentry and had few, if any, clothiers in their number and workers met no great hostility from the bench. However, the clothiers were a greater political force in Gloucestershire than they were elsewhere. Many were freeholders and had a vote in county elections and there were enough of them to make their influence felt. By 1750 they were described as 'an opulent and significant body of men'.[47]

Three factors affected the adequacy of weavers' earnings: first, whether they had regular employment; second, whether they could employ family members or unpaid apprentices; and third, the price of food, particularly bread. John Osborne, clothier of Monks Mill, Alderley, noted in his diary in October 1766:

the Greatest Riot … as ever was known in this county when they assembled in and about Wotton to the number of 800 or 1,000 men, women etc – pulled down and destroyed all the Boulting mills in this neighbourhood, plundered Farmer Collect of Alderley … under pretence of lowering the price of provisions.

Three men were executed, many were sent to Gloucester gaol, 'the rest reprieved for transportation'.[48]

Yorkshire competition became a more serious threat from the 1760s. Bath coatings, which had been made for some years by a Gloucestershire clothier, were dismissed in favour of Yorkshire ones at a considerably lower price and Daniel Packer wrote to his factor in 1768 that he had met so many losses that he thought of relinquishing the clothing trade altogether.[49] Gloucestershire seems to have had its fair share of demand for uniforms after the outbreak of the American War in 1775, but there were many setbacks. After 1773 the East India Company's exports fell to under 10,000 pieces a year and it is this which must account for the decline in trade in Chalford and possibly Horsley.[50] A severe winter in 1783/84 and another slump in East India Company exports produced the most severe depression Gloucestershire had experienced in years. Nearly 15,000 destitute people were estimated to be in the fifteen Stroudwater parishes[51] and 'the complete disappearance of our trade' was referred to as relief committees were set up in Stroud and elsewhere.

Samuel Rudder observed that 'the cloathing manufacture' was the pre-eminent industry of the county, but that it advanced by unequal steps, being 'affected by the situation of our public affairs'. Rudder was clear that the woollen industry benefited in wartime by the demand for army clothing. And, if war was with the French, this was an added advantage for, being rivals in woollen manufacture, the superior British fleet was better able to escort trading vessels to overseas markets formerly supplied by the French. Having consulted 'the most intelligent clothiers', Rudder divided the industry, by market, into four categories:

1. The inland trade (not the home market), governed by 'fashion and fancy', had only been introduced into Gloucestershire within the previous fifty or sixty years. Clothiers and their employees travelled

through Britain and Ireland and also sold to merchants who exported to the colonies and other overseas markets and 'some large fortunes have been made', although in 1779 it was in 'a languid condition'. The returns of this sector of the trade were estimated to be £250,000 p.a.

2. The army trade and trade with the London drapers which was estimated to be worth £100,000 p.a.

3. The Turkey trade which had, in large measure, been lost to the French, although Gloucestershire superfines still found a market. Overall it was estimated to be worth £50,000 p.a.

4. The East India Company trade was 'the most considerable branch of our foreign trade' yet, 'tis far from being advantageous to the clothier'. This was the 'coarse trade' and very few had succeeded in it, although many had ruined their fortunes 'without making a bad debt'. The blame was laid at the feet of the Blackwell Hall factors.[52] From the late 1760s Gloucestershire clothiers had 'been the losers'. Bankruptcies had become more frequent and the factor claimed all the bankrupt's goods in his possession, to the great detriment of the many small traders who were also creditors. Therefore, 'many clothiers of property have declined the [East India] company trade and struck into other employments'. This trade was estimated to be worth £200,000 p.a.[53]

Relative calm was established until the depression of 1792, when eighty weavers gathered near the house of a Horsley weaver, charging him with taking work below the usual rates, forcing him to return the chain to Daniel Lloyd, the clothier for whom he worked. When the weaver escaped to his brother-in-law the mob began to un-roof and demolish the house. The weavers were ordered to make a public apology, but in 1793 the trouble became more serious in Uley and the military arrived.[54] Threats to pull down clothiers' houses were not idle and precautions were made to withstand attacks. But by this time the situation had changed. Clothiers were adopting new machinery and were building larger mills and factories into which more of the manufacturing operations were brought.

There is no disputing the severity of trade depressions in the woollen industry during the seventeenth and eighteenth centuries. However, although clothiers complained vociferously, and while there were some bankruptcies, they did not go out of business in such large numbers as they were to do, and with such finality, in the nineteenth century. The mention of their 'estates' in one of the petitions is an indicator of the degree of wealth which the larger clothiers had accumulated and, with this capital behind them, they could survive even severe setbacks. The families of Clutterbuck, Peach, Wathen, Paul, Halliday and others were not put out of business, indeed they prospered. John Osborne of Monks Mill wrote more in his diary about lifestyle than business. He noted in 1759, 'I erected the gig mill this year' at a cost of between £90 and £100 and in 1766 noted that the mill was 'put into good repair', but he was more animated in recording in 1759 that he had bought 'a pair of Chaise Horses £40' and a new post chaise 'large and roomy' for £74. In 1764 he raised the south front of Monk Mill House, removed the entire roof and raised a parapet wall. Two years later he erected a portico to the house and 'a new road for carriages thro Wortley field to ye House'.[55] The people who suffered most were, inevitably, the smaller clothiers and textile workmen who, if put out of work, had no capital to fall back on. While some had smallholdings from which they could support their families in times of hardship, small garden plots were insufficient and these were all that the smaller clothiers, weavers, spinners, scribblers and dyers had to survive on.

It was not until the early years of the nineteenth century that either masters or men recognised their regional role as players within a larger negotiating system and, in taking action, acknowledged that issues which concerned them affected others in the same sector elsewhere.

Endnotes

1. Calendar State Papers Domestic (SPD) 1614, p.253
2. SPD 1611–18, p.389
3. SPD 1616, p.389
4. SPD 1616, p.390
5. SPD 1616, pp.390, 394
6. SPD Jas I, XCVIII, p.81
7. APC 1619–21, p.197
8. APC 1621–2, p.162
9. SPD 1622, p.346
10. Cal Acts of the Privy Council (APC) 1622, pp.224–5
11. This led to a proclamation for the general wearing of woollen cloth and for burials in woollen cloth – APC 1621–22, p.153
12. SPD 1622, p.389; APC 1622, p.190
13. APC 1621–23, p.265; ibid., 1627–28, pp.152–3; ibid., 1628–29, p.80
14. APC 1625–26, pp.135, 161. Many more than ten clothiers had gig mills
15. SPD 1633, I, IV, p.1
16. SPD 1633, p.36
17. SPD 1633, pp.108, 165–6
18. SPD 1633–34, p.316; ibid., 1633, p.249
19. SPD 1633–34, pp.4, 57
20. SPD 1625–49, p.488
21. He had been removed from office in 1634, charged with attempted dishonesty in Wiltshire
22. *Hist. MSS Comm.*, Duke of Portland's MSS, VIII, 2–3, cited G.D. Ramsay, *Wiltshire Woollen Industry in the Sixteenth and Seventeenth Centuries*, Cass, 1965, p.98
23. *VCH Gloucestershire*, II, p.160
24. *Trans Newcomen Society*, V, p.89
25. These letters are printed in P.H. Fisher, *Notes and Recollections of Stroud*, 1891, repr. 1975, p.212
26. Ibid., p.213
27. 10952 (27)gs
28. SPD 1649–50, p.343
29. SPD 1653, p.230; 1653–54, p.31
30. SPD 1663–64, p.535
31. D225/212
32. *Hist. MSS Comm., Various Colls*, I, 154
33. F.A. Hyett, *Glimpses of the History of Painswick*, Gloucester, 1928, p.69
34. Palling Papers, quoted in J. de L. Mann, *The Cloth Industry in the West of England from 1640 to 1880*, p.34
35. Mann, p.32
36. R. Atkyns, *Ancient & Present State of Gloucestershire*, 1712, p.78
37. 2885gs, A State of the Case ..., 1757, p.9; W.E. Minchinton, 'The Beginning of Trade Unionism in the Gloucestershire Woollen Industry', Trans. BGAS, LXX, p.134
38. Mann, p.42
39. A State of the Case ...
40. JHC XXVII, p.468, cited in Moir, p.255
41. W.E. Minchinton, p.134
42. State of the Case ..., p.14

43 A State of the Case …
44 JHC XXVII, p.730
45 Quoted in Minchinton, p.131
46 From a clothier and four workmen, probably 'carefully selected'
47 Mann, p.118
48 D2930/1
49 Mann, p.49; D149 F114, Packer to Marsh & Hudson, 31 May 1768
50 Mann, p.56
51 GJ, 12, 26 January 1784
52 S. Rudder, *A New History of Gloucestershire*, 1779, p.61
53 Ibid., pp.61–3
54 E.A.L. Moir, 'The Gentlemen Clothiers', in H.P.R. Finberg (ed.), *Gloucestershire Studies*, p.257
55 D2930/1

CHAPTER FIVE

INDUSTRIAL REVOLUTION 1790–1835: THE ADVENT OF MACHINERY

The period 1790–1835 was one of contrasts: innovation and risk-taking, business optimism and expansion – and also failure. In these years the Gloucestershire woollen industry was transformed from being one in which the barriers to entry were low, when reasonably prosperous broadweavers could join the ranks of the smaller clothiers, to one in which the capital required for setting up in business, the organisation skills for acquiring machinery and managing its operation, and the need for a knowledge of markets was such that, without these, failure was likely. It was possible, even given these attributes. The years 1790–1835 marked the emergence of the factory system.

MANUFACTURING PROCESSES

Until the 1770s and through to the 1790s the typical Gloucestershire woollen mill contained fulling stocks and a gig mill (perhaps with an adjoining grist mill). There was sometimes a dyehouse at the premises and there were also some independent specialist dyers who had made their name in scarlet, black or blue dyeing. Wool had been washed and warps sized on mill premises for many years and larger clothiers had begun to bring the finishing process on site and to have shear shops and numbers of pairs of shears, as well as a burling shop, often on the top floor of the fulling mill. Production expansion took place by the multiplication of these units at a single site. These were emergent factories which, in some cases, required new or extended mill buildings. Even after larger clothiers were introducing new machinery for wool preparation, and later for spinning, such was the optimism of new entrants to the industry that they adapted older corn mills or built new small, water-powered mills in the remoter upper reaches of valleys, sometimes on extraordinarily small tributaries and springs, for fulling and gigging. Hound Mill at Barrington, a fulling mill in the monastic period, had been converted to corn milling by the early seventeenth century, but by *c.*1700 a pair of fulling stocks had been added and by the late eighteenth-century corn milling had ceased.[1] On the whole, such mills never did become fully fledged factories and, when hard times came, they were amongst the first to fail. Few clothiers gathered handloom weaving into their mills – the perceived advantages of supervision and reduction of raw material theft were outweighed by the high capital cost of the space required to accommodate handlooms. When John Anstie wrote his *Observations on the Importance and Necessity of Introducing Improved Machinery into the Woollen Manufactory* in 1803, he suggested that there had been little opposition to 'improvements in manufacture' until shearing frames were introduced, and he drew attention to the cotton industry in which much machinery originated 'and its very existence … depends not merely on the *continuation* of the present improvements but also on *progressive improvement* [his italics], for enabling the Manufacturers to contend against the advantage from the lower price of labour possessed by other nations'.[2]

The first authoritative and detailed account of the processes employed in the Gloucestershire wool textile industry is that of William Partridge – indeed, no other area has an equivalent from as early in the nineteenth century. Partridge was a member of the family based at Bowbridge Mills/Dyeworks. He immigrated to America and published his *Treatise* in 1823, although the period he was describing was probably *c*.1810. Textile inventions occurred in the different regions of England but Partridge commented that manufacturers rarely employed operatives from other regions on account of the different cloths made, the machinery used and methods of working (besides regional terms for both processes and machinery), added to which 'the workmen have strong prejudices against each other'.[3] He highlighted the skills of Yorkshire workmen in carding, slubbing and spinning, besides which they were 'more conversant in the general concerns of a factory', but West of England workmen were said to be more skilled in the making and finishing branches of the trade in fine goods.[4] There is little evidence to suggest that Gloucestershire clothiers were reluctant to adopt new machinery in the early Industrial Revolution, nor were they ignorant of new technologies and machinery suppliers. Processes and inventions for cloth manufacture can be considered under five headings:

1. Wool preparation: sorting, scouring, drying, picking, burring (dyeing), blending
2. Yarn manufacture: willying, oiling, scribbling, carding, slubbing, spinning
3. Weaving: warping, sizing, weaving
4. Finishing: scouring, fulling, gigging, mosing, shearing
5. Dyeing

1. Wool preparation

By the Industrial Revolution wool beating was usually undertaken by the wool supplier. Scouring, however, took place on the clothier's premises. A wool scourer was, according to Partridge, given better wages than those in any other branch of factory employment. Scouring with seg (urine) took place in a shallow, conical iron or copper furnace set in brick; the liquor (two parts water to one part urine) was heated to between 125 and 130 degrees Fahrenheit. Urine 'from persons living on a plain diet' rather than 'from luxurious livers' was matured in a large closed vat. (Cider and gin drinkers provided a lower quality of seg than beer drinkers.) The wool was stirred back and forth before being washed and then spread out to dry.

Wool was dried by three different methods: by open exposure to the sun and air, by a fire stove or by an air stove. Partridge comments that it might appear that wool drying was so simple an operation as not to merit description. Yet wool exposed to strong sun after scouring (or to too great a heat in a fire stove) was likely to be spoiled. It was common practice in summer for wool to be left outside until partly dry and then brought into an air stove until completely dry; this was said to spin better and to make a softer cloth than wool left outside until totally dry. The air stove was usually a long, narrow building with vertical narrow apertures around 2–3ft long at regular intervals in each of the long walls to ensure a cross draught. Such buildings were used for both wool and cloth drying and can often be identified at former mill or dyehouse sites.

Partridge described a fire stove as a small, circular stone or brick building around 16ft in diameter and two or three storeys high. The building was heated by a cast-iron closed fire from the top of which a cast-iron pipe passed through the building, terminating around 3–4ft above the roof. Wool was hung on wooden arms protruding from a framework fixed to the walls. A wool stove, properly managed, could dry 240lb of wool in 12 hours. There are several surviving round towers in Gloucestershire – the best examples being at Frogmarsh and New Mills, Wotton-under-Edge. Playne[5] referred to the 'strange round towers still to be seen in some places' in which wool was dried. They present a puzzle and have been a focus of some controversy for a number of years.⁶ They did not by any means exist at all mills and were frequently sited away from the rest of the buildings at a mill or dyehouse site. There are examples of semicircular buildings let into the hillside. Why would wool drying have presented such a risk that the building be located at a distance? 'Stoving' was the term for bleaching in which sulphur was burned, producing asphyxiating fumes, and this may have required bleaching stoves to be located away from workpeople. There is some oral evidence for the use of sulphur at the site of one semicircular building located in the rear grounds of a dyehouse. It is entirely possible that they could have been used for both drying and bleaching purposes. Surviving stoves almost all antedate the mill to which they belong.

2. Yarn manufacture

After scouring or dyeing, the wool had to be opened and cleaned. This was formerly done by women beating the wool on hurdles with rods, subsequently picking out all foreign matter by hand, but by the early nineteenth century this operation was more generally done by the willy, twilly or devil. This was the first Industrial Revolution machine to be adopted in the Gloucestershire woollen industry and was, at first, a simple hand-operated drum set with iron hooks which tore the wool apart and, if it was dyed in the wool, loosened 'the filth of the dye'.⁷ The willy was in use by the mid-eighteenth century, although not on a widespread scale. Scribbling was being done by hand on what was known as a 'horse' – the scribbler sat astride a bench and worked the wool with a hand card against a fixed, slightly curved, vertical base. It is not clear when the scribbling horse came into being – its use may have been required when medley cloth began to be made and wools of different colours were mixed. Scribbling was often done in the clothier's mill or workshop.⁸ By 1775 scribbling was being carried out on the premises at Stanley Mill, although it is not clear whether this was being done by machine.⁹ The carding machine had been adapted for scribbling by the 1780s in Yorkshire but, following riots in Wiltshire, its introduction to Gloucestershire may have been delayed. One riot took place at Woodchester in 1792 but the rioters came from elsewhere, for the workpeople supported the clothier being 'universally convinced of the advantages [of] the machines … and how [the industry] owed its flourishing state to their introduction'.¹⁰ A petition was sent to Parliament from Gloucestershire in 1794 to limit scribbling engines 'so as to have enough work for others', for prior to the introduction of the scribbling engine, scribbling had provided employment for men 'advanced in Age or by Accident or Misfortune disabled from working in other occupations which require bodily Exertions'.¹¹

Wool was spread with Gallipoli oil – 2–3lb of oil being required for 20lb of wool – and it was carded. Under the domestic system carding was often done by female spinners at home. A carding machine, consisting of rollers covered with card wire, was patented in 1748 (Daniel Bourn no.628), but it was not until the 1790s that improved versions were widely adopted. By 1792 several Dursley clothiers had carding machines in their mills.

The introduction of machinery to the Gloucestershire woollen industry met with sporadic outbursts of protest but not on the scale experienced elsewhere. In 1776 an experimental spinning machine was set up by Somerset clothiers in Shepton Mallet workhouse. A mob, mainly from Frome and Warminster, destroyed it and, in some alarm, clothiers from Somerset, Wiltshire and Gloucestershire met in Bristol and issued two statements. The first recommended that experimental machines be set up in areas where cloth manufacture took place, proposing that meetings be held in various places and reports made on local reactions at a future meeting to be held in Bath. The second statement affirmed the Shepton Mallet clothiers for their initiative. Of the signatories which can be identified, fifty-three were from Wiltshire and Somerset and only eight from Gloucestershire.¹² After the next clothiers' meeting in Bath the clothiers issued a statement in which they expressed determination to encourage the use of machinery and took issue with those who alleged that workers would be deprived of the means of earning a livelihood.¹³

This has prompted the suggestion that Gloucestershire clothiers may not have been as enthusiastic about machinery innovations as those in the neighbouring clothing counties.[14] An alternative explanation is that Gloucestershire clothiers anticipated less adverse reaction to machinery on account of their long practice of using gig mills. In the March Assizes in 1776, Thomas Eddles and around thirty others were indicted for a riot and breaking a spinning machine belonging to Samuel Heaven. While this would seem to refer to the Shepton Mallet riot, there is a note in a Gloucestershire scrapbook which says, 'and which spinning machines were at that time beginning to be used by the clothiers in this county'.[15]

Whereas the smaller Yorkshire clothier could be supplied with sufficient yarn by his family, Gloucestershire clothiers had to put wool out to communities many miles from their mills and factories. Some established spinning stations to organise and co-ordinate the work of remote spinners. The spinning jenny was one of the first textile industry innovations and, because Hargreaves did not patent it, there were no legal hindrances to its adoption. When the scribbling machine and jenny were adopted, they displaced a significant number of workers – estimates suggesting nine in ten weft spinners and six in seven warp spinners.[16] The early jenny was a hand-operated machine intended for domestic use, but it was not long before clothiers began installing them at the mill or in adjoining workshops.

Most of the spinning jennies in Gloucestershire during Partridge's time had eighty spindles, some seventy, Partridge commenting that 'A jenny ought never to contain less than seventy spindles, and is still better with eighty'.[17] Four women operating eighty-spindle jennies could turn out as much yarn per day, of any given fineness, as five could on sixty-spindle jennies, and there were some 100-spindle jennies worked by water.[18]

It has been suggested[19] that, initially, the jenny could not be used successfully in the fine cloth trade and this may have inhibited its adoption in Gloucestershire; it was fairly rapidly adopted in the coarser woollen trade of Yorkshire and in parts of south Wiltshire and Somerset. In 1788 Thomas Turner of Ebley was bankrupt and, amongst his effects, were several spinning machines. Thomas Tippetts of New Mills, Dursley, bankrupt in 1789, had 'one spinning machine complete' and both had slubbing which suggests they may also have had billies.[20] By the 1790s the problems with the use of the jenny in finer spinning appear to have been overcome and more frequent references to its use in Gloucestershire appear. A writer in the mid-nineteenth century, recalling the woollen industry of his youth, remarked that jennies were 'hailed with delight by masters and weavers' who had previously experienced delays in yarn supplies from remoter villages.[21] The jenny could be installed in workshops above mills or in adjoining premises and its relatively low cost, coupled with its greatly increased output compared with the large Saxony spinning wheel, provided an opportunity for people to set up in business on their own account. Some small-scale spinners, such as S&S Sparrow of Stonehouse, farmer-spinners, only had jennies in their workshop in 1806 and so, presumably, purchased slubbings.[22]

By 1792, when a valuation survey of property in Dursley was undertaken, the adoption of machinery had gathered pace. Howard's Mill contained seven billies, two scribbling machines, six carding machines and fifteen jennies, besides two stocks and a gig mill; Townsend Mill contained four billies, one scribbling machine, four carding machines and six jennies in the 'New Workshops' which adjoined the mill and which also contained a shear shop. New Mills was the highest rated mill in Dursley and contained one scribbling machine, four billies, four carding machines and five jennies, besides a 'double gigg mill on a new construction', shear shops, burling rooms, cloth drawing rooms, and dyehouse, as well as fulling stocks.[23] Power was applied to the jenny in the second decade of the nineteenth century and jennies with more spindles became common in the second and third decades, although it was said that yarn which was twisted by hand was preferred, particularly for warp.[24] Crompton's mule, combining spindle and roller drafting, was not suitable for woollen spinning, but features of it could be combined with features of the spinning jenny to make a mule suitable for wool. This hybrid machine, with the appearance of a mule, the carriage of which was pushed in by hand, was available from the 1790s, although not widely adopted. A machine which seems to have been unique to Gloucestershire, termed a 'jack', was more like a billy with a mule-like moving carriage. The jack was described as a machine 'which twists and lengthens the rollers into a long weak thread', thereby making the slubbing easier to weave by giving it more twist and by winding the sliver directly onto a bobbin for the weaver.[25] The jack is mentioned in advertisements from 1808 until the 1840s; an 1811 inventory of Lightpill Mill listed a sixty-spindle and an eighty-spindle jack.[26]

3. Weaving

A key technology lagged behind: weaving. And even by the 1840s, power looms were not numerous in Gloucestershire cloth mills. While other processes were being mechanised and brought into the mills, handloom weaving remained a cottage industry – even if the weaver had several looms. However, the fly shuttle,[27] which could be propelled from side to side by one person, both speeded up work on a narrow loom and, importantly, reduced the need for two weavers on a broadloom. The fly shuttle was slowly adopted in Yorkshire from c.1763 and, while there was some antagonism at first, it was 'completely accepted' by the 1780s.[28] The fly shuttle does not appear to have reached Gloucestershire until the early 1790s when a Stonehouse clothier, Watts, sought to adopt it and there was widespread alarm amongst weavers. Other clothiers, fearing riots, persuaded him to withdraw it, although some were sold to weavers.[29] One weaver said, in 1803, that he had used it for nine or ten years. The upturn in the woollen trade, and weavers enlisting in the army c.1798, prompted the majority of weavers to adopt it.[30] By 1803 it was almost universal amongst broadweavers in Gloucestershire, although not in the more complex cassimere weaving which required more than two treadles. Its attempted introduction in Wiltshire and Somerset was accompanied by greater unrest and the fly shuttle was by no means common in those counties by 1803; in Wiltshire it was said to be 'not allowed'.[31] With the fly shuttle a broadweaver could save between one-quarter and one-third of the time taken to weave a cloth.[32]

Partridge noted that 'good workmanship in weaving was of more importance than in any other branch of the manufactory'.[33] The warp had to be kept tight on the loom and the weft woven with an even beat, resulting in a consistent number of threads per inch of cloth. Weft yarn was frequently wetted before being wound onto bobbins to enable a sufficient quantity of weft to be beaten into a given measurement of cloth. Partridge enumerated any number of hazards to be avoided in weaving, some of which resulted in defects such as the cloth being 'rowy' or 'pin-rowy' (when one thread was larger than the other).

By the mid-1790s some clothiers had installed handlooms in their premises. When Nathaniel and Daniel Lloyd of Uley insured their factory buildings for £700 in 1795, the premises comprised a dwelling house with wool lofts and press shops above, 'a building consisting of a stove on the lower floor and a shear shop, weaving shop and picking loft above'.[34] There were, apparently, no factories housing looms in and around Stroud by 1802, although Peghouse Mill in Painswick had four broadlooms, Samuel

Britannia Mill, Wotton-under-Edge, in 1963; it is now demolished. (Jennifer Tann)

Wathen had twelve narrow looms in his factory and John Wallington of Dursley had some kerseymere looms.[35] Wallington told the Parliamentary Inquiry of 1803 that he would build no more loom factories on account of the expense – a room 45ft by 20ft being needed for six broadlooms, whereas thirty-two shearmen could work in the equivalent space. He added: 'We have lately been under the necessity of building an entire new place but we did not make provision for a single loom.'[36] Edward Sheppard of Uley agreed: 'A factory for weaving would be the furthest from my inclinations and intentions.'[37] It was not very long, however, before the Great Factory had been built by him especially for weavers – and one weaver giving evidence in 1802 had already heard a rumour of Sheppard's intention. By 1806 at least nine loom 'factories' existed in Gloucestershire, the largest two housing twenty and twenty-three looms respectively, two more having ten and eleven, and the other five having eight or fewer looms. The majority were situated in and around Dursley and Wotton, but some were in Painswick, Woodchester and Stonehouse. There were none in Stroud, although there was anxiety that further loom shops would be installed.[38] By 1809, when Parliament finally repealed the early restrictive statutes which were, by then, largely ignored in Gloucestershire, the factory system was well on the way to becoming established. Contemporaries were not to know that it would take until the 1840s for weaving to become a fully fledged factory process with the adoption of the power loom.

4. Finishing

Significant developments in cloth finishing were made in the early to mid-nineteenth century. After the cloth had been removed from the loom it was scoured (brayed) to remove the oil and size. This was where seg appeared again, but this time mixed 3:1 parts with hog's dung. The stale mixture was poured over the folded cloth in the fulling stocks and the stocks were operated slowly for around 5 minutes. The cloth was then removed and thrown into a heap until the urine had combined with the oil. It was then returned to the stocks, more liquor added and the stocks operated slowly (so as not to shrink the cloth) for an hour; then water was poured in a continuous stream on to the cloth, the plug at the base of the stocks removed and the operation continued until the water ran perfectly clear. If some oil remained, fuller's earth was added. Partridge was at pains to emphasise that braying was a chemical procedure in which the liquor fermented in a process accelerated by the glue used in sizing; the ammonias of the urine combining with the oil formed a soap which was readily washed out by water.

    The fulling process remained virtually unchanged until the 1840s, despite the fact that, in 1743, a Wiltshire clothier, Richard Brooks, published a broadsheet, *Observations on Milling Broad and Narrow Cloth*, in which he advocated the use of soap rather than fuller's earth. He particularly directed his attention to uneven shrinking between the lists and the middle of a cloth, and promoted his regulator by which the progressive felting of the cloth could be achieved and 'Millwrinkles and baggy Places' prevented.[39] There was certainly awareness of the invention in Gloucestershire but no record of its adoption has been found.[40] Partridge asserts that fulling, despite the term, was mainly undertaken with soap or, occasionally, urine, rather than fuller's earth; however, Playnes of Longfords were using it into mid-century – they ordered 10 tons in 1842.[41] A 36-ell (54-yard) cloth weighing 71lb required 5lb of soap for white cloth, 6lb for blue, and 7lb or 8lb for other colours in which alum mordant or tin liquor had been employed. Cloth needed to be kept moist during fulling and had to be removed several times to prevent it becoming either too hot or needing to be realigned. After 2 hours the cloth was removed, the lists pulled square and the width measured to ensure consistency in fulling. The process of fulling, removing, stretching and measuring was repeated at 3- to 4-hour intervals over a total of 16–18 hours. If fulling was undertaken with urine the process took 24 hours. 'The proof of good fulling is to have the fabric when finished stout and firm in the ground, like leather and of equal breadth in all parts.' A 54-yard piece should be reduced to 42 yards, 'but this will depend altogether upon the perfection of the workmanship'.[42]

Fulling stocks were usually made of wood in the early Industrial Revolution – and wooden stocks were still in occasional use in the early twentieth century. Iron 'legs' to stocks had been introduced by the early nineteenth century (illus) and complete iron stocks existed by the 1830s (for example at Toghill's Mill, Chalford), but it is not clear where they were made or how common they were. In the 1840s Ferrabee was making them at Thrupp.

William Lewis patented (4013) a fulling machine in 1816 in which the cloth ran between rollers 'which are made to press it with any desirable force ... [and in which] the work is also performed more equally, with much less mechanical power, and more speedily'.[43] This sounds like an antecedent to the successful Ferrabee fuller but no record of its adoption has been found.

Partridge emphasised that care had to be taken in drying so as not to expose part of a cloth to the sun while the other part was in the shade. 'This extraordinary fact is well known by every manufacturer in the West of England but I believe has never been noticed by any writer on the subject.' Manufacturers and dyers who used racks needed to make sure that the cloth was hung with the 'wrong' side facing the sun and some never allowed their cloth to be dried facing the sun at all.[44]

The process of raising became an art, particularly, it might be assumed, in Gloucestershire, for Wiltshire and Somerset clothiers had largely conformed to the statute of Edward VI which prohibited the use of gig mills. As Thomas Plomer reported in 1804, 'the whole trade of Gloucestershire is now carried on by means of it ... all the clothiers make use of it, I believe, without exception'.[45] Wiltshire clothiers encountered considerable physical opposition to the use of the gig mill (culminating in riots in 1802, and in 1812 in Yorkshire) and had more or less given up its use; indeed, upon recognising that Gloucestershire cloth was better finished than that from Wiltshire, clothiers in the latter county began to send cloth to Gloucestershire to be gigged. Plomer commented:

> all the cloths manufactured in Gloucestershire ... examined as they carefully are by the manufacturer and by the buyer ... are found uniformly to exceed those dressed by hand in strength, beauty and perfection.[46]

Raising required great attention to detail. Partridge describes work being done in 'runnings up', six of which were done in one direction and six starting from the opposite end. The handles of teazles used for the first course had their points 'quite tender' with previous work and every successive course was undertaken with better and better teazles, the handles of the last course being set one-third with new teazles. Raising, in Partridge's estimation, took 8 hours.[47] John Lewis, clothier, William Lewis, dyer, and William Davis, engineer, patented 'certain improvements in the application of pointed wires and other pointed substances ... for the purpose of raising the pile or face of woollen or other cloths' (no.4379). This sounds like the wire gig mill which Gloucestershire clothiers protested that they did not use, and no reference has been found to its adoption.

After the surface of the cloth had been raised it was cut by hand with shears. Shearing required great skill and shearmen were the highest paid of all textile craftspeople in the Industrial Revolution. While there were some independent shearmen who worked on commission, the majority of shear shops were associated with mills, the shearmen being accommodated in an upper room of the mill or in adjoining workshops.

'The art of shearing' involved moving the shears 'with great regularity and parallelism'.[48] Shearmen worked in pairs from list to list, one starting at the middle and working outwards, the other working towards the middle. Shears needed to be regularly sharpened and the major cloth-making valleys had one or more shear grinding mills, usually on a small tributary stream such as Weybridge Mill on the Miry Brook, which enters the Frome at Bowbridge.

The first patents taken out for shearing improvements were by John Harmer, a Sheffield nonconformist minister who held two patents, one in 1787 (no.1595) and the second in 1794 (no.1982). These were for machines which worked hand shears. Harmer's second patented machine was more effective and was the only one of the two suitable for fine cloth. Harmer shearing frames were sufficiently notable to be recorded in sales of equipment in the late eighteenth and early nineteenth centuries. They were said to do the work more evenly than all but the best shearmen, but not much more quickly. Paul Wathen had one at Woodchester Mill in 1801 and its adoption caused a threatening letter in 1802, but no violence ensued. A local man, I. Sanford, patented a shearing machine in 1801 (no.2258) but production ceased in 1809, probably with the maker's death.[49] In 1810 J.C. Daniell patented a shearing machine (no.3348) and this was advertised in the *Gloucester Journal* two years later (GJ, 14 September 1812). While this machine was satisfactory for narrow cloth, it was less so for broadcloth. It was another local man, John Lewis, who produced the first workable rotary shearing machine in 1815 (no. 3945), which sheared along the length of the cloth 'without any intermission of that operation being necessary before the whole piece is shorn'.[50] Lewis's patent was followed within a month by one taken out by a Stroud engineer, Stephen Price (no.3951). In 1818 Lewis and his brother William, together with their engineer William Davis, took out a patent for cutting from list to list (no.4196), the best way of shearing fine cloth – and the machine could be adapted to shear lengthways.[51] In the following year the Lewis brothers bought out Davis, who had taken out two more shearing patents (nos 4487 and 4820) and left Gloucestershire for Leeds. Davis then bought out Price's patent. These machines were probably superseded by the Miles cutter, an American invention, which was patented in England by Thomas Miles of Dudbridge in 1823 and fairly widely adopted. The Lewis cutter made most headway in Yorkshire, while Gloucestershire continued to be a focus for inventions in cutting, and in 1824 Gardner & Herbert of Leonard Stanley took out a patent for improvements (no.5059). Although it has become clear that Lewis, like Arkwright, deserves more credit for making things work than for machine originality, Lewis won two cases for infringement by William Davis of Nailsworth and N.S. Marling, both of whom were using the machine.[52] Gloucestershire woollen manufacturers seem to have had a variety of cutters on their premises: there were cutters by Harmer and Lewis, besides hand shears at Upper Cam Mill in 1826, and in 1844 Lower Mill, Dursley, contained cutting machines by Davies, Lewis and Gardner & Herbert.[53] George Oldland of Hillsley also patented a cutter in 1830 and 1832 (nos 5960 and 6236), which was praised in the *Encylopaedia Britannica*, but by this date there were fewer potential customers.

One of the final tasks in the finishing process was hot pressing to give a gloss to the cloth and to remove creases. Pressing was a heavy and hot process involving cloth being folded back and forth by the yard, glazed paper being placed between each fold, and heated metal plates at every 20 yards on the board of a screw press. Thin, unheated iron plates were placed above and below the hot ones to moderate the heat. The screw was operated by up to five men in order to exert sufficient pressure to turn it. The cloth was then allowed to cool in the press. Iron press plates and papers occur quite frequently in inventories from the early eighteenth century and by the early nineteenth century cloth was said not to be saleable without having been hot pressed.[54] John and William Lewis and William Davis patented a process for 'laying, smoothing and polishing the pill or face of woollen or other cloth'.[55] Roll boiling and, particularly, potting, in which steam was blown through the cloth, were slowly adopted mid-century.

5. Dyeing (and bleaching)

An aspect of the wool textile industry in the Industrial Revolution which has received less attention is bleaching and dyeing, both of which were usually carried out at the same place – either the larger manufacturer's mill or an independent dyeworks. Coal was required in both processes (as it was, increasingly, for wool and cloth drying). Claude Berthollet, the French chemist, remarked that 'Fuel is one of the chief articles of expense in dyeing' and it was a clothier who, in petitioning for the Stroudwater Navigation, gave as his first reason the fact that decreasing availability of wood made the supply of coal more important.[56] With the adoption of steam power clothiers had a further demand for coal and also the potential for employing surplus steam from the boiler(s) to heat liquids – and factory buildings. By the 1790s brewers, sugar refiners and salt boilers, besides Lancashire and Yorkshire dyers, bleachers and printers, had begun to enquire about the possibilities of steam heating. The manufacturer was faced with a dilemma, however, for to inject steam into a liquid would dilute it, yet, if the liquid were to be heated by a metal pipe of coils containing steam, would it become sufficiently hot? In 1800 Hawker & Richards, scarlet dyers of Dudbridge, one of the very first dyers in Britain to use tin chloride in scarlet dyeing (in 1795),[57] ordered a steam system from Boulton & Watt to heat eight 6ft 5in vessels, each of which held 500 gallons.[58] Hawker & Richards estimated their previous coal consumption, when heating vats by fires, to be 20 tons per week. And the fuel consumption using steam was 4 tons 1 cwt, a saving of over 15 tons per week. No other Gloucestershire dyers adopted the system, although three Wiltshire manufacturers enquired and one certainly installed a system. The financial case was stronger for a large dyer or for manufacturers who had a steam engine with a somewhat larger boiler. For the smaller operation the annual fuel savings did not equal, let alone exceed, the interest on the capital expended in adopting the system.[59]

In the seventeenth and early eighteenth centuries much of the cloth sent to London for sale was 'white' – in other words, undyed and unbleached – but there was also demand for whitened, therefore bleached cloth by the mid-seventeenth century. The traditional way of bleaching was by burning sulphur (known as stoving), a process which continued at Bowbridge until the end of the nineteenth century. Sulphur was lit in dishes set on the ground of the stove with the door sealed, wet wool being hung around the walls on iron hooks. While it seems clear that the round wool stoves (Frogmarsh being the best-known example) were latterly used for wool drying, it is possible that they were formerly used for bleaching.[60] Some thirty to forty round stoves have been identified, nine in Painswick, all but one being associated with mills.[61] James Winchcombe had

Frogmarsh wool stove, 1963, prior to its conversion into a house. (Jennifer Tann)

Bowbridge Dyeworks. (Museum in the Park, Stroud)

a stock of brimstone at his workshops in 1780 and Hawker & Richards had some sulphur when their stock was valued in 1805.[62]

By the late seventeenth century, much Gloucestershire cloth was dyed before being sent to London.[63] Dyeing was undertaken by some of the larger clothiers and by independent dyers such as the Bowbridge Dyeworks, Peter Watts of Wallbridge, and Hawkers at Blue Row, Dudbridge,[64] who dyed on commission, particularly scarlet. The skills were usually passed from father to son. Larger specialist dyers would be expected to match a pattern sent with the cloth to be dyed such that, when dyed, the manufacturer 'cannot perceive the least shade of difference between it and the patterns sent'.[65] A large dyeing establishment might receive up to 100 patterns in a day, all different, and colour the cloth exactly to each. There were locality specialities, too – blues in Uley and scarlets in Stroud. A painting in the Museum in the Park, Stroud, depicts scarlets stretched out on racks behind Watts' dyehouse at Wallbridge.

Clothiers in the Ewelme, Cam and Little Avon valleys tended to manufacture cloth dyed in the wool, whereas in the Stroudwater area cloth was more generally dyed in the piece, the only way in which bright and light colours could be obtained. Dyeing in the piece required particular skills in order to ensure that the colour penetrated the fabric – for broadcloth was thick. Bancroft commented that wool, when dyed in the fleece, took up more colour than when cloth was dyed in the piece, 'the very finest cloth is never thoroughly dyed scarlet, it always being found white within'. But he commended Nash & Dymock of Gloucestershire when using yarn that was little twisted in the spinning so that the dye penetrated further into the cloth which contributed to 'the remarkable beauty of their scarlets'.[66] Many dyehouses used spring water for dyeing – it would not have been polluted like river water; moreover, one of the essential elements of high-quality dyeing was consistency in the chemical components of the water, and spring water was far more consistent in this respect. Dyeing was not only a highly developed skill but a site-specific one; it being necessary, in Partridge's view, for the dyer to continue to practise at the same location 'with one kind of water'.[67]

John Hawker, son of Richard Hawker of Dudbridge, had three separate businesses at the beginning of the nineteenth century: dyeing, knapping and banking, each with a different partner. In his will Richard had left the dyeing and knapping business to his son, with a mill containing two stocks, a gig mill and three knapping mills.[68] John Hawker's cashbook summarised his business turnover in 1805 at £38,639, with a profit of £2,939 and capital of around £26,664. Profits fell in 1806 and 1807 to £1,207 and £1,293 respectively. It is not clear which business this refers to or whether it was the sum of all his business profits. If it was the dyeing business only, the large turnover might be explained by the business of dyeing on commission.

HARD OR SOFT WATER?

Besides the need for plentiful supplies of water for power, water was required for washing, braying, dyeing, fulling and also for damping cloth at various stages, for example in raising. Several authors have written of the requirement for soft water for woollen manufacture and some have attributed this quality to the rivers Frome, Ewelme, Little Avon and their tributaries;[69] Perry, for example, wrote of the Cotteswold Sands which act as a filter for the hard water issuing from the oolitic limestone strata above.[70] Both Dugdale and Rudder wrote of the need for plentiful water supplies, particularly for wool washing, but neither mentions the need for soft water. Rudder, in his description of Chalford, wrote of the 'petrifying effects of water [which are] particularly observable on the axis and other parts of a mill wheel'; Fisher noted, 'I have seen the water mill and wheel over which the brooklet falls encrusted with a thick white stony deposit from the water'.[71] At several springs in Minchinhampton, Chalford and Dursley water is so hard as to petrify leaves and ferns over which it runs. Partridge alluded to the idea of earlier writers which had been 'reiterated by every one to the present day', namely that 'none but soft water is fit to be used in dyeing'. Yet, as he points out, this was denied by the practice of eminent dyers who had 'all the time been making use of spring water that was very hard'.[72] The key, Partridge contended, was not the water's hardness but its clarity and consistency. 'Water that is variable in its property can never be used with any prospect of success; it is on this account that springs are better calculated for the purpose than mill-streams.'[73] Partridge recommended soft water for wool scouring but hard water for later washing processes, a point contested by Ponting who pointed out that while hard water appears to dissolve soap, it is actually held in suspension.[74] With respect to dyeing, different dyestuffs react differently with hard and soft water. Partridge recommended soft water for blues, since it encouraged the necessary fermentation to take place. For blacks, on the other hand, a colour for which Gloucestershire dyers were noted, the lime in solution was said to enhance the colour from logwood. Partridge drew attention to the fact that the water of Gloucestershire was 'more celebrated for dying black than in any other part of England' and contained 'carbonate of lime in solution'.[75] And Ponting noted that greater solubility with many dyes was obtained with hard water. Partridge, once again, emphasised the importance of tacit knowledge: 'dyers to become eminent … must continue to practice in one situation, with one kind of water … by which means alone they can be able to obtain perfection in the art.'[76]

MANAGEMENT

Effective management was essential in cloth production, it being:

> the most difficult (of any manufacture) to establish, and the most tardy in being brought to perfection ... The manager of a woollen factory should always be in the business ... A factory that is under the direction of a man well versed in the business will be constantly improving.[77]

Partridge was ambivalent about the 'small locality' from which employees were drawn, 'fathers and grand-fathers [having] followed the same business', for 'by this means local habits and prejudices are acquired which operate more powerfully to confine them to their district and calling'. When trade was flourishing this pool of skilled labour was of great benefit to the industry. However, when innovation might have benefited the industry both clothiers and employees tended to stick to what they knew best, whether or not there was a market for it. Partridge emphasised the importance of hands-on supervision by the owner/master manufacturer: 'so long as managers are content to leave [direct supervision of weavers] and other important branches to the direction of under bosses, so long will their concerns go on imperfectly'. The factory master 'gives directions as to what has to be done, and he personally inspects the work, so we see that it is executed agreeably to his instructions'.[78]

VOLUME OF PRODUCTION

Despite the introduction of machinery, the volume of cloth produced did not increase markedly between the 1820s and 1830s. While there was some variation between categories of cloth produced, and a peak in overall production in 1834, the total produced in 1838 was over 177,000 yards less than in 1823. There were a number of reasons for this, not least some spectacular bankruptcies, but what is clear is that demand for the high-cost cloth produced in the west was limited. It was the cheaper cloths of Yorkshire which found readier sales.

FACTORY BUILDERS

Innovators in the Gloucestershire woollen industry were often not the most wealthy 'gentlemen clothiers', but were those who had risen from the ranks of middling clothiers who did not necessarily aspire to emulate the lifestyle of the merchant clothiers. They were open to new ideas, brought different attitudes to the industry and had different social motivations. 'These nascent manufacturers were to force the old controlling class either to join them in building the factory system or to abandon the trade.'[79] It was at this stage that some of the older clothing families retired to take up the role of full-time landowner, magistrate or lawyer. Some, however, managed both simultaneously; Henry Hicks of Eastington bought Churchend Mill in 1799 and rebuilt it, focusing much of his attention on new developments in woollen manufacturing technology here. Hicks employed Thomas Hewes as millwright and in this role he would have been concerned to maximise the water-power potential of the site, besides ensuring optimal power transmission within the mill complex. A few years later Hicks rebuilt Millend Mill and built Meadow Mill in 1811. In addition to his business activities he became Lord of the Manor of Eastington in 1806.[80] His son, John Phillimore Hicks, was less attentive to the business, his diaries showing 'a distinct reluctance to concentrate on business matters and social affairs appear to dominate his life'.[81] There were arguments

Churchend Mill, Eastington. (Howard Beard)

Bayliss's Upper/Lodge Mill, Painswick, in the 1820s. (Museum in the Park, Stroud)

with his father over the adoption of new machinery and, perhaps to inculcate a different attitude, his father dispatched him to Yorkshire to report on the introduction of new machinery in the mills there.

The beginning of the Industrial Revolution was marked at Longfords Mill by the highly unusual situation of the business being led by a woman, Martha Playne. Mother of eleven children she was widowed in 1788 and her late husband's affairs seem to have been in 'a state of confusion', such that the estate could not support all his bequests. The business, renamed Martha Playne & Son, flourished and the liabilities were discharged; capital employed in 1791 being £4,861, increasing to over £12,000 by 1801. When Martha retired in 1797, the business was on a sound footing and was renamed William Playne & Co., becoming William & Peter Playne in 1801.

It has been suggested that the mills housing new processes were small affairs, 'often converted tenements'.[82] However, while first-wave and later factory building in Gloucestershire was largely focused on the sites of earlier fulling mills, these were often rebuilt or extended vertically and laterally. Humphrey Austin of Wotton-under-Edge had rented Ithell's Mill, Kingswood, in 1795, built Steep Mill in Wotton in 1800, bought Abbey and Langford Mills in Kingswood soon afterwards and then transformed Sury Mill, Wotton, into New Mills in 1802, and bought Alderley New Mills in 1806. Austin also had the dyehouse at Dyer's Brook, Wotton, a share in Strange's Mill, Wotton, and workshops at Church House.[83] None of these was small; New Mills, Wotton, being a considerable factory. Dunkirk Mill, Nailsworth, was extended laterally, parallel to the mill leat in a series of five-storey contiguous ranges and associated buildings, in 1798, 1818, 1819, 1820, 1821 and 1827; the original Nailsworth Mill was built across the valley prior to 1730, later additions being made in 1806 and 1814.[84] Longfords Mill, Avening, was built in stages across the valley becoming a large mill estate, a new 150-yard-long dam being built across the valley in 1806 and Lake Mill constructed.

In some of the tributary valleys, such as the Painswick and Slad valleys, there was a noticeable mix of modest eighteenth- and early nineteenth-century mills, which always remained small (and were, on the whole, the first ones to fail), as well as some larger late eighteenth- or early nineteenth-century developments. Upper Mill, Painswick, was rebuilt in c.1807 and Palling's Mill, Painswick, in 1818, both on a modest scale; but Vatch Mill, which burned down in 1827, was rebuilt on a large scale, an adjoining mill having been rebuilt in 1823 with a third in 1830. New Mills, Stroud, built on the Slad brook in 1776, was large for its time, while further up the valley on the Steanbridge estate, the Lower Mill, a modest sixteenth- to seventeenth-century building, was erected near the clothier's house and does not seem to have been extended. The Upper Mill, known as the Jenny Mill, was a three-storey building further up the brook, probably dating from the late eighteenth century. While Gigg Mill, Nailsworth, remained a small mill, Lock Mill, Nailsworth, was an impressive, four-storey factory development. It was still possible for new entrants to the industry to start from small beginnings and numerous small mills were built on springs and tributaries of the three main river systems between the 1790s and 1815.[85] And there was what might be considered to be the occasional aberration. In 1799 John Radway, a wool stapler, leased North Cerney Mill (which had been a grist mill for many years following the decline of the woollen industry on the rivers flowing to the Thames) and converted it to a cloth mill. In c.1824 Giles Radway built a new cloth mill south of the older mill and was still working it in 1837.[86]

Notwithstanding the trade fluctuations during the first quarter of the nineteenth century, there is clear evidence of new capital investment in the Gloucestershire woollen industry during these years. At a clothiers' meeting at The Bear of Rodborough, held in 1812, the gauntlet was thrown down:

> Have not the great number of manufacturers expended very large sums of money in erecting mills and buildings for the purpose of their trade? … the numbers of clothiers has very much increased and consequently the spirit of competition obliges every manufacturer to pay redoubled attention to equal and excel his rival in the credit of his trade.[87]

A spate of new building took place in and around Kingswood and Wotton: Edward Jackson transferred to a mill in Uley described as 'newly erected' in 1828; Waterloo Mill was built at Wotton-under-Edge in 1815; Llewellin Ford's mill in Wotton was built in 1817; Charfield Mills comprised three distinct buildings erected between 1815 and 1829; and Corrietts (Cam) Mill was rebuilt in 1818. The greatest investment was made by Edward Sheppard at Uley who was reputed to have expended £50,000 on his Great Factory and associated buildings by 1833.[88]

Jenny Mill, Steanbridge. (Bovis Homes; David Calvert)

Jenny Mill, Steanbridge, 2012. (David Calvert)

Halliday's Mill, Chalford. An early two-storey mill, probably late seventeenth or early eighteenth century on the right with a nineteenth-century extension. (Mike Mills)

In the Frome Valley the Hooper family rebuilt part of their mills in Eastington in c.1798 and 1808. There was a major rebuild at Stonehouse Mills between 1790 and 1800. Stanley Mill was demolished and rebuilt on a unique fireproof principle in 1813. The new mill at Ebley, fed by a large reservoir, was built in 1818, older mill buildings having been erected around 1800. Lodgemore Mill was rebuilt between 1814 and 1815, with further mill building in 1828. Bowbridge Mills comprised four distinct factories built in 1780, 1795, 1802 and a 'very old' mill altered in 1824. Hope Mill was rebuilt between 1828 and 1829. Stafford Mill was partly erected in c.1790, with later building in 1811, 1818, 1823, 1826, 1829 and 1831. Ham Mill was advertised as 'a capital mill and factory of great power and extent' containing several stocks, a gig, a 'commodious' dyehouse, and a large wool store in 1813, later building taking place in 1814, 1825 and 1832. St Mary's Mill, Chalford, was demolished and rebuilt in 1820.[89]

FACTORY BUILDINGS

Four categories of questions which could be asked about the design of mills, each with a number of subsets, have been defined by Keith Falconer. They all deserve more attention than can be given here and Falconer's preparedness to speculate, for instance, on who the mill designers were, is to be welcomed.[90] The first question concerns the design of mills and their evolution; the second is scale and the extent to which there was progression from small to large; the third is motive power and the fourth structure (see Table 4).

Table 4 – Questions concerning textile factories

Category	Questions
Mill Design	How were issues of plan, form and function resolved?
	How did the mill evolve from a small-scale vernacular form?
	How did mills built as integrated factories reflect form & function?
	How did mill complexes evolve, e.g. with weaving sheds, power systems?
	How do mills show the transition from domestic to factory production?
	Are different forms of ownership reflected in mill design?

	To what extent were architects, engineers, contractors involved?
	Were copybooks, treatises etc. influential?
	Which parts of mills were embellished? Relationship to size, date, ownership?
Scale	Was there a progression from early small to late large mills?
	Did small mills survive in the textile industry to a later date and did their function change?
Power	How was the pressure on the original power system resolved at the mill?
	How did water-power systems (wheels, watercourse design) contribute?
	How long did water power persist?
	Was the addition of steam related to the expansion of the mill's functions?
	Was the adoption of steam triggered by the need for dependability?
	How closely was power related to mill size?
Structure	How were traditional materials used to cope with the new scale of building?
	What use was made of cast and wrought iron for columns, beams, roof trusses?
	When/where/why/how was fireproof construction used? If not, why not?

Source: Based on Keith A. Falconer, Industrial Archaeology Review, *1993*

With regard to plan, form and function, water-powered fulling mills evolved in design over hundreds of years, sometimes as double mills. The millwork for a fulling mill was simple, being a direct drive from a waterwheel via a tappitt wheel; the stocks were, therefore, generally located adjoining the waterwheel. Fulling stocks were heavy pieces of equipment and fulling would have created a great vibration, necessitating the process to be sited on the ground floor. Fulling mills could be small, somewhat insignificant buildings and, when obsolete in the remoter valleys where the sites were not further developed in the late eighteenth or early nineteenth centuries, quickly fell into disrepair and/or were demolished. Dursley Mill is a small, squarish, two-storey stone building located to the rear of a magnificent gabled clothier's house (an indicator that a clothier's wealth was usually attained by being in control of all stages of production). Eyles Mill, dating from the mid to late seventeenth century, is a low, two-storey stone building with round and oval windows, formerly driven by a spring. Damsell's Mill, Painswick, is a small seventeenth-century, two-storey building with horizontal-headed windows divided by two mullions. Fromehall Mill, now four storeys but probably originally three, is mid-eighteenth century in date with horizontal-headed windows, some divided by two and some by three mullions.

While not problematic in the proto-industrial era, the clothier's dependence on water power and the consequent location of the industry largely, but not entirely, in the western valleys of the Cotswolds by the early eighteenth century, was a factor in determining the alignment of buildings and the development of water power. The earlier Dunkirk Mill, then called New Mill, was built across the valley, but when the site was expanded in a series of stages in the late eighteenth and early nineteenth centuries, these later buildings were set alongside the mill leat in view of the narrowness of the valley. The old Longfords Mill was across the valley, as was a later building of the 1850s, but when the lake was constructed in 1806 there was some anxiety about building a new mill up against the dam and Lake Mill was erected in 1809 parallel to the valley, fed by a leat. Bliss Mill, Chalford, was erected at a point where the Frome Valley opened out and there was plentiful land on which later mill buildings could be erected. In the plain, there may have been other deterrents such as landownership or river meanderings, but valley width was not an

Frogmarsh Mill, Woodchester; note the circular wool stove some distance from the mill buildings. (Howard Beard)

inhibitor to mill layout. While tributary valleys such as the Painswick, Slad and Toadsmoor valleys tended to have smaller mills along them, even here there were exceptions where the land allowed; Vatch Mill, on a tributary of the Slad Valley, was a major steam-powered industrial site in the mid-nineteenth century, while Longfords and Dunkirk mills, both in the Nailsworth/Avening valleys, were amongst the largest in Gloucestershire.

Since the Gloucestershire putting-out clothier frequently owned a fulling mill from before the Industrial Revolution, the transition from domestic to factory production was, in industrial building terms, dominated by adherence to water power and continuity of occupation at older mill sites. The evolution of the small water-powered mill to an early Industrial Revolution factory generally, therefore, took place in a piecemeal fashion. The eighteenth-century, two-storey Egypt Mill was extended by a more or less matching two-storey, nineteenth-century addition. The two-storey Halliday's Mill, Chalford, was extended by a four-storey, nineteenth-century block, continuing on the same building line. Where land and capital allowed, a series of mill additions was made in a continuous line as at Dunkirk, enabling greater use to be made of the mill leat. Nind Mill, Kingswood, became an extremely large industrial development through piecemeal additions, as did Monks Mill, Alderley. But, from the early nineteenth century onwards, some cloth manufacturers elected to demolish and begin again at the same site, as at New Mills, Wotton, in 1802. St Mary's Mill, Chalford, was demolished and rebuilt at the same location but not precisely on the same site in 1820; Stanley Mill was demolished and rebuilt in 1813 and the same was the case at Stonehouse Upper and Lower mills and a number of others. The five-storey Ebley New Mill was built in 1819/20 as a water-powered factory across the canal from the original site. These newly built nineteenth-century mills were generally four storeys high, with a continuous dormer or top-lit attic, and almost all were built as integrated factories to house all processes, except weaving, and to be driven by water power. Despite their preference for water power, Gloucestershire mill owners were rarely able to adopt the most efficient waterwheels (overshot and, later, suspension wheels) due to the inadequacies

Cloth air stove at Holcombe Mill, 1963. (Jennifer Tann)

of the falls available. Power was optimised instead by the multiplication of wheels as, for example, at Ebley and Stanley mills. Several steam mills were built in Wotton and these were considerably smaller than the largest water-powered mills in the Severn Plain – Old Town Factory, for example, being of five bays and three storeys high.

In the major Gloucestershire mill (re)building phase, neither mule spinning nor power looms had been adopted and by no means had all cloth manufacturers brought handlooms on site. Thus the new factories mainly accommodated preparation and finishing processes. Attention to the need for good light for burling and mending can be seen in the relatively large windows on the uppermost floor of some mills. When weaving was brought on site, it was sometimes accommodated in a separate, dedicated, two- or three-storey building at the mill site, the typical single-storey loom shed for power looms being a feature of the late nineteenth or early twentieth centuries. Some handloom shops were also built away from mill sites, at least four being known on Chalford Hill, and one at Chestnut Hill, Nailsworth, for example. Two other functions are associated with distinctive buildings: dyeing and cloth drying. A dyehouse was, typically, a long single-storey building – one could be seen at Longfords until the 1990s; and David Farrar's Dyehouse, located behind Tayloe's Mill, Chalford, was a fine example. A map of Bowbridge Dyeworks shows a similar pattern, and the single-storey range alongside the millpond between Upper and Lower mills at Brimscombe may have been a dyehouse. A large dyeworks such as Bowbridge became a cramped site over time as the nineteenth-century illustration shows (see p.72).

Cloth drying had taken place for hundreds of years on the hillsides and Rack Closes adjoining the mills. But strong sunlight and wet weather were, in different ways, problematic. A solution was the air stove – a long, often single-storey building with vertical openings in the walls at regular intervals to facilitate air circulation (see illustration above). A feature of mills more often seen in Wiltshire than Gloucestershire was the handle house with alternate stones or bricks removed to allow air circulation. There is, however, one at Brimscombe and there were others. The adoption of steam power was evidenced by a boiler chimney (a particularly fine example being at Holcombe Mill), but the choice by a number of clothiers of Boulton & Watt's boat land engine, which was much less cumbersome than a beam crank engine, meant that steam power had a less intrusive appearance at mill sites. Those manufacturers who adopted steam power in the mid-nineteenth century more often erected a distinct engine and boiler house, sometimes, as at Stanley Mill, in the courtyard. Churchend Mill, Eastington, was heated by steam and, in due course, this innovation was extended to heating dye vats.[91]

The typical Gloucestershire wool textile factory of the late eighteenth and early nineteenth centuries was a well-proportioned stone building. Early eighteenth-century mills were generally two storeys high; early nineteenth-century ones sometimes two but often four or five. Late eighteenth-century mills had a characteristic fenestration of horizontal stone window heads, the windows divided by central stone mullions (for example, the 'old mill' Longfords; Fromehall 'old mill'). The 1818 Ebley Mill has pairs of

St Mary's Mill estate, Chalford – a clothier's elegant house and mill. The core of the house antedates the mill on the left which replaces an earlier mill. (Gloucestershire Archives)

St Mary's Mill, 2012. (Roger Tann)

flat-arched windows with stone mullions, echoing the windows of the 'old' c.1800 mill, now demolished. These were, perhaps, the inspiration for the windows at St Mary's Mill, Chalford, which are of similar design. Where there was decoration it was often a pediment and central clock as at Lock Mill, Nailsworth; sometimes a cupola as at Woodchester Mill or, originally, St Mary's. These mills were all early nineteenth century in date. It would be tempting to suggest that the more consciously decorated mills were those that formed part of a clothier's estate, with a nearby well-designed house. This is certainly the case at St Mary's Mill, Chalford, which has an elegant house, extended by Samuel Clutterbuck who rebuilt the mill. But Southfield Mill, near the elegant mill house in Woodchester, was a plain affair. There are only two brick mills of any note dating from the early nineteenth century, both built in the Severn Plain, and both handsome in different ways: New Mills, Wotton, and Stanley Mill. New Mills was built in 1802 but some older stone buildings, including a wool stove, were retained. It is five storeys high with a later central tower and well-proportioned iron-framed windows; Stanley Mill was designed with its stone-quoined brick-front elevation facing the mill owner's house, the dividing members of its Venetian windows being cast-iron Doric columns. No evidence has been found of architects being employed to design Gloucestershire mills until G.F. Bodley's work at Ebley Mill in 1865 for S.S. Marling, who had commissioned him to undertake the design of Selsley church in 1858 (built 1860–62); even then it was extremely rare. Pattern books were, however, widely available and a builder could be instructed to incorporate features such as cupolas, pediments and Venetian windows from them.

The chief glory of Stanley Mill is its unique arcaded cast-iron-framed structure, visible from the interior. This is, indeed, a cathedral of industry. Stanley Mill is beautiful, remarkable and intriguing; it features in most books on industrial architecture and was the subject of a detailed study by Stratton & Trinder.[92] Not only is Stanley Mill's structure unique in the world, it is the only iron-framed woollen mill in Gloucestershire. It is a tantalising puzzle. There are no records to throw light on why the owners decided on a fireproof mill or this design, or who they consulted for it. Moreover, as Keith Falconer writes, 'Stanley Mill does not fit comfortably into any of the strands of development identified in the literature on fireproof mills'.[93] By the acquisition of land near the intended site of the new mill and the creation of a 5-acre reservoir, the owners were able to realise the full water-power potential of the site with five waterwheels and a 16ft fall, capable of generating 130hp.[94] The new mill 'demonstrably of a single build' comprised a Z shape, two ranges being fireproof and the third not. This makes sense, for fireproof construction was some 25 per cent more costly than conventional building.[95] The fireproof ranges were built to accommodate the waterwheels, provision being made for power transmission to four floors. The non-fireproof range probably accommodated hand processes. Unique features include the longitudinal cast-iron arches on the ground floors, the central arched arcade and the provision for power transmission. Falconer suggests that the designer assembled these elements as he thought fit and with little regard to precedents. He draws a parallel with I.K. Brunel's approach to railway construction some time later. The designer of Stanley Mill 'seems to have wanted to apply the most modern techniques of construction to a mill' in an area with different organisation of production and product from that in the North of England and distant from the network of contacts presumed to have existed between designers in the north. One of the new owners, Donald Maclean, was from London; might he have sought a London-based designer? Falconer suggests that the idiosyncratic fanlights over the doorways in the fireproof ranges, with their centripetal rather than centrifugal tracery, have no English medieval parallels. Falconer proposes that Benjamin Dean Wyatt or Charles Augustus Busby might be candidates, each of whom had experience which would make them a possibility. Another candidate is John Rennie who, on his move to London, regretted his lack of mill design work. He was closely associated with Samuel Wyatt's Albion Corn Mill at Blackfriars and employed cast iron in the repairs at the West India Dock warehouses. The 1815 stove building at Stanley resembles naval buildings at dockyards where Rennie was working.[96] For the time being, at least, the designer of Stanley Mill remains elusive. Where cast-iron columns were installed in Gloucestershire textile mills, they were later additions to provide added strength.

In some instances mills were divided between family members, as at Longfords in 1813, when the brothers William and Peter Playne divided the property, William taking the original mills and workshops and Peter having Lake Mill, the dyehouses and stove.[97] Bowbridge Mills was already an extensive mill and dyeworks site by 1806 when the mills were duplicated so that Nathaniel and John Partridge could occupy one part and John Junior and Joseph Partridge the other. The part of the site occupied by the latter, together with part of the site occupied by Nathaniel and John, was leased to Samuel Clutterbuck in 1810, leaving the north and south-east ends of the ground floor for the lessor.[98] John Potter, who had extended Cloud Mill at Wotton-under-Edge, leased these additions, with half the mill wheel and gearing, to William Potter in 1812;[99] in the same year, the Strange family leased a loft and a room which had been added to their mill to Austin & Co., allowing them use of the steam engine.[100] In 1818 John Blagden Hale leased part of Nind Mill, comprising six lofts, one stock and a gig mill (the old part of the mills), to John Cooper and Abraham Owen.[101]

Stanley Mill, 1813; a Grade I listed unique fireproof mill. (Annie Blick)

Interior of Stanley Mill showing a graceful arcaded cast-iron frame. (Annie Blick)

INDUSTRIAL REVOLUTION 1790–1835: THE ADVENT OF MACHINERY

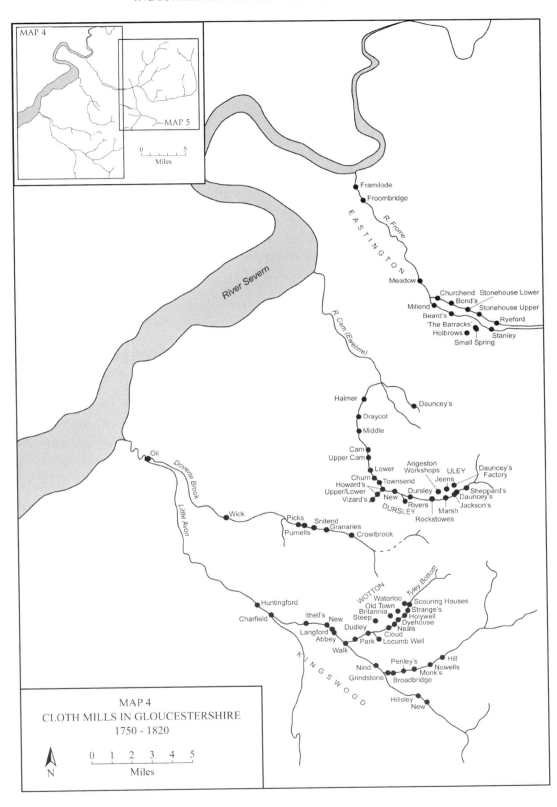

MAP 4
CLOTH MILLS IN GLOUCESTERSHIRE
1750 - 1820

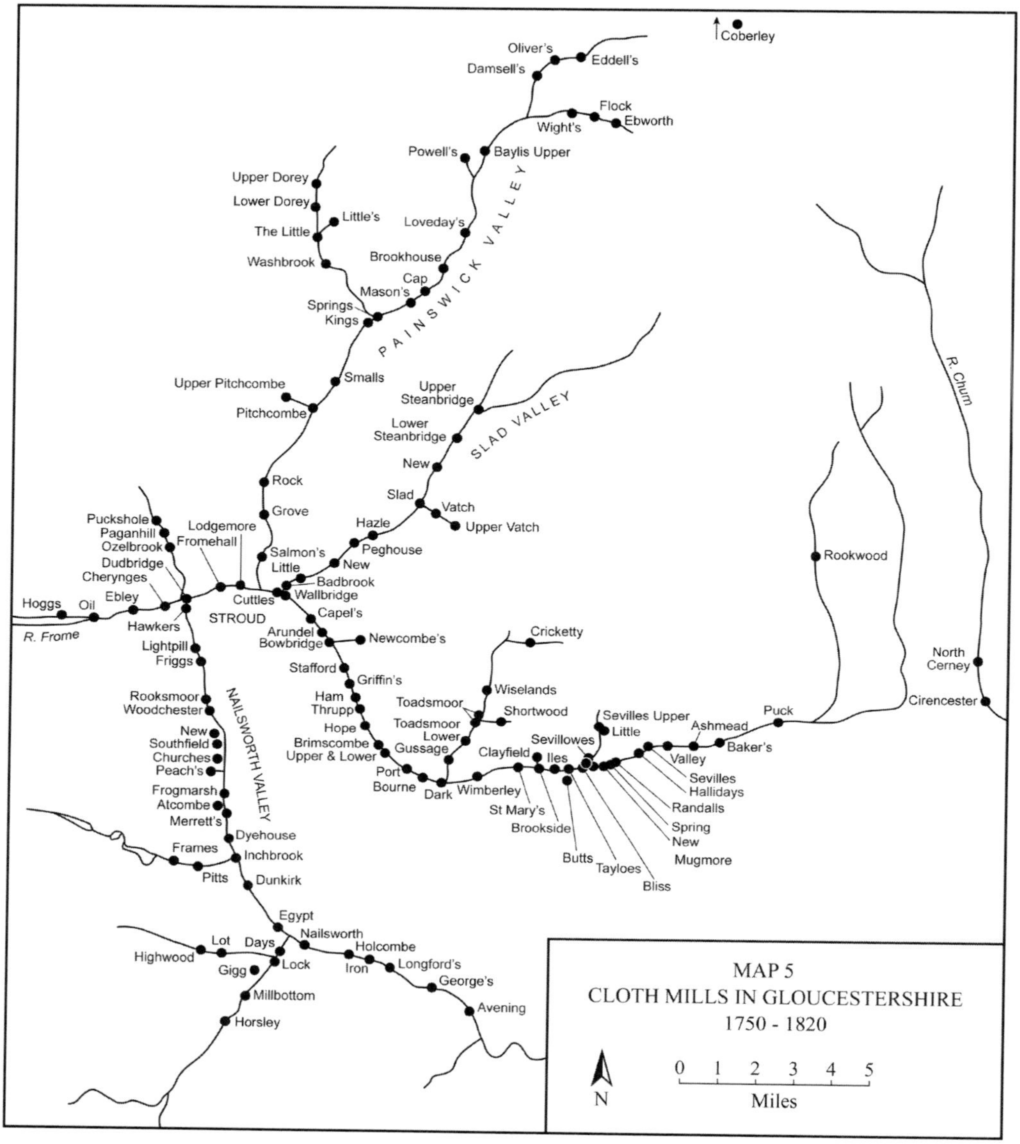

MAP 5
CLOTH MILLS IN GLOUCESTERSHIRE
1750 - 1820

The period 1770–1820 saw the greatest number of woollen mills in operation, although not necessarily all of them over the whole period. Maps 4 and 5 show 197 mills and there may have been others. The tight clustering in Stroudwater and on the River Cam from Uley to Cam Mill and on the Little Avon, particularly at Wotton-under-Edge in its tributary valley Tyley Bottom, suggests great pressure upon water-power facilities, mills lower down a stream having to wait until midday for water which had turned many waterwheels upstream first. The downstream advantage was a greater volume of water, but the upstream benefit, despite the stream being smaller, was access to power earlier in the day. It is little wonder that water-power engineering had become something of an art by the Industrial Revolution, both in waterwheel design and millwrighting, and in the engineering works, to optimise the fall and volume at each site.

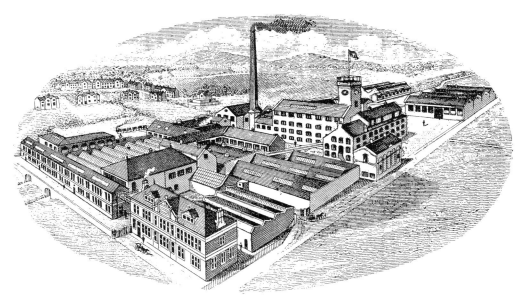

CAM MILLS, DURSLEY. (Hunt & Winterbotham, Limited).

Cam Mill, nineteenth-century buildings on an ancient mill site. (*Industrial Gloucestershire*, 1904)

Endnotes

1. *VCH Gloucestershire*, 6, 1965, p.23
2. John Anstie, *Observations on the Importance and Necessity of Introducing Improved Machinery into the Woollen Manufactory*, 1803, p.60
3. William Partridge, *A Practical Treatise on Dyeing*, New York, 1823, repr. Pasold, 1973, p.19
4. Ibid.
5. A.T. Playne, *History of Minchinhampton and Avening*, p.26
6. G.N. Crawford, 'The Woodchester Roundhouse', *GSIA Journal*, 1982; K.G. Ponting, 'Some Questions about Round Towers', *Ind. Arch.*, Vol. 4, I, p.62; Ray Wilson, *GSIA Journal*, 1989; Colleen Haine, 'Wool Drying Stoves along the Painswick Stream', *GSIA Journal*, 1991, pp.30–1; M. Palmer & P. Neaverson, *The Textile Industry of South-West England*, 2005, p.36; J. Tann, 'The Bleaching of Woollen and Worsted Goods', *Textile History*, I, no.2, 1970
7. K.H. Rogers, *Woollen Industry Processes: II, The Factory Industry*, Trowbridge, 2008, p.8
8. Ibid.
9. Stroud Museum, Clutterbuck Diary
10. *Stroud Journal* (SJ), 4 July 1868; I am indebted to J. de L. Mann for this reference
11. Journals of the House of Commons (JHC) XLIV, pp.599–600; JHC XLIX, p.599
12. *Gloucester Journal* (GJ), 16 September 1776; I owe this reference to J. de L. Mann
13. GJ, 14 November 1776
14. J. de L. Mann, *The Cloth Industry in the West of England from 1640 to 1880*, 1971, p.124
15. D4693/14
16. Adrian Randall, *Before the Luddites*, Cambridge University Press, 1991, p.230
17. Partridge, p.43
18. Ibid., p.22

19 Mann, p.126; Rogers, p.16
20 GJ, 8 September 1788; R115.118 gs
21 SJ, 4 July 1868
22 RR289.1gs
23 VE 1/1gs
24 Partridge, p.22
25 Mann, p.291
26 GJ, 24 October 1808; 20 March 1841; K. Hudson, *The Industrial Archaeology of Southern England*, Appendix
27 Or spring loom as it was called in the West
28 Mann, p.140
29 Evidence on Woollen Trade Bill, 1803, p.15
30 Ibid., pp.299–300; Mann, p.140
31 Partridge, p.58
32 Mann, p.229
33 Partridge, p.55
34 Box 52 bdl 6gs
35 PP 1802–3, VII, p.16
36 Ibid., p.299
37 Ibid., p.249
38 PP 1806, III, p.334
39 Richard Brooks, *Observations on Milling Broad and Narrow Cloth*, 1743, copy in Trowbridge Museum
40 GJ, 4 October 1743; 19 March 1743/4
41 D4644 4/1
42 Partridge p.69
43 Stroud Museum, Lewis MSS, 4 October 1816
44 Partridge, p.64
45 Speech of Thomas Plomer Esq., addressed to the Committee of the H. of Commons on the Woollen Trade, 1804
46 Ibid.
47 Partridge, pp.78–80
48 K.H. Rogers, *Woollen Industry Processes: I, The Domestic Industry*, Trowbridge, 2006, pp.29–32
49 GJ, 2 January 1809
50 He was not the first to do so but he made a machine which worked; Stroud Museum MSS, Lewis Patent 27 July 1815
51 See Mann, p.304 for a description of other shearing inventions
52 Mann, p.305
53 R65.7gs; Box 28 bdl 19gs
54 Evidence on the Woollen Trade Bill, 1802–03, p.381
55 Stroud Museum, Lewis MSS, 1819
56 C.L. & A.B. Berthollet, *Elements of the Art of Dyeing*, 2nd edn, Transl. A. Ure, London, 1824, p.265; John Dallaway, *A Scheme to Make the River Stroudwater Navigable ...*, 1755
57 See Part II, Dudbridge, p.233; Edward Bancroft, *Experimental Researches Concerning the Philosophy of Permanent Colours ...*, London, 1813, p.515
58 Birmingham Central Library, Boulton & Watt Papers, portf. 1330
59 Jennifer Tann, 'Fuel Saving in the Process Industries during the Industrial Revolution: A Study in Technological Diffusion', *Business History*, 15, 1973

60 Jennifer Tann, 'The Bleaching of Woollen and Worsted Goods, 1740–1860', *Textile History*, I, 1970
61 Tithe map evidence; C. Haine, 'Wool Drying Stoves along the Painswick Stream', *GSIA Journal*, 1981
62 D149/F20; D1181
63 Mann, p.9
64 Some of Hawkers' early nineteenth-century account books survive, GRO D1181
65 Partridge, p.93
66 Edward Bancroft, *Experimental Researches Concerning the Philosophy of Permanent Colours*, Vol. 1, 1794; Vol. 2, 1814, pp.67, 68n. The copy in Stroud Museum was presented to F.C. Bassett (of Bowbridge) by Dr Thomas and Mrs Partridge in 1871
67 Partridge, pp.87–91
68 D873 T75
69 E.A.L. Moir, 'The Gentlemen Clothiers', in H.P.R. Finberg (ed.), *Gloucestershire Studies*, 1957, pp.226–7
70 R. Perry, *The Woollen Industry in Gloucestershire to 1914*, Shrewsbury, 2003, p.14
71 Rudder, *A New History of Gloucestershire*, 1779, p.468. The author took numerous water samples from different places in the three river systems in the 1960s. The variation in hardness was relatively little and it was uniformly hard
72 Partridge, pp.87–8
73 Ibid., p.88
74 Ponting's technical notes to Ibid., p.237
75 Partridge, p.91
76 Ibid., pp.237, 92
77 Partridge, p.217
78 Ibid., p.55
79 Adrian Randall, *Before the Luddites*, p.42
80 Stephen Mills & Pierce Riemer, *The Mills of Gloucestershire*, Barracuda Books, 1989, p.38
81 Stephen Mills & Henry Hicks, 'A Man of Wide Horizons', *GSIA Journal*, 2008, p.8
82 Mann, p.131; Randall, p.43
83 R. Perry, *Wotton under Edge: Times Past – Time Present*, 1986, p.103
84 PP 1834, XX, pp.259–61; Ian Mackintosh, 'Dunkirk Mills: a Brief History', STT, 2002
85 Stroudwater Textile Trust, Nailsworth Mills Walk 1
86 D182/111/203; D2525; GDR T1/45
87 JF 13.27gs
88 See references in Part II
89 PP 1834, XX, pp.254–90
90 Keith A. Falconer, 'Textile Mills and the RCHME', *Industrial Archaeology Review*, XVI, 1, 1993, pp.5–10
91 Stephen Mills & Henry Hicks, 'Man of Wide Horizons', pp.14–15
92 M. Stratton & B. Trinder, 'Stanley Mill, Gloucestershire', *Post Medieval Archaeology*, 22, 1988
93 Keith A. Falconer, 'Mills of the Stroud Valley', *Industrial Archaeology Review*, XVI, pp.62–81
94 PP 1834, XXIV; elsewhere this is claimed to be 200hp
95 Jennifer Tann, *The Development of the Factory*, 1970, p.147
96 Keith A. Falconer, 'Mills of the Stroud Valley', pp.79–80
97 D1229 acc 9196, box 317; Ian Mackintosh, 'Power and the Playnes', STT typescript
98 Box 69gs
99 Box 60 bdle 2
100 D1759
101 D654/111/73

CHAPTER SIX

THE AGE OF MACHINERY: SOURCES OF POWER

WATER

The changing location of the Gloucestershire woollen industry from one scattered over the county to one largely concentrated along three west-flowing river systems (Stroudwater, Ewelme and Little Avon) demonstrates the significance of power as a dominant (but not sole) locational factor. By 1790 these valleys were densely packed with mills – almost 200 (see Maps 4 & 5). The builders of the earlier mills had more choice of specific location and were astute in, for example, identifying a point at which a river gradient suddenly steepened, as at Halliday's Mill, Chalford, or a site where the water supply was enhanced by springs (which also provided purer water for dyeing), as in Dursley where:

> on the south east side of the church yard, some springs rise out of the ground, like boiling water, in so copious a manner as to drive a fulling mill at about a hundred yards below and are never known to diminish in quantity.[1]

Two small springs running steeply by the side of Dark Lane and Millswood in Chalford drove four or five small mills in the early nineteenth century, while the Black Gutter, emerging from the opposite side of the valley, drove at least three mills which were described as being 'exceedingly well supplied with water'.[2] Griffin Mill, Stroud, was said to possess 'water power unequalled by any mill on the stream'; N.S. Marling told the factory inspectors that the supply at Vatch Mill on a tributary of the Slad brook was regular; Crowlbrook Mill at North Nibley had 'a constant supply of water', while Walk Mills on the Little Avon was supplied with 'a powerful stream of water which has never been known to fail'.[3] An anonymous writer discussing the proposed Stroudwater Navigation Act in 1775 stated:

> The time of short water in the year is very inconsiderable, the present act supposes it to be two months but 'tis in … few summers that it is felt for so long … and all the pools from Wallbridge to Severn might in a general way be filled in 6 hours.[4]

The fact that some small fulling mills were in seemingly bizarre locations on small springs, whereas corn mills were more often located in main valleys, suggests that fulling mills did not require large amounts of power: 2–3hp was considered adequate per pair of fulling stocks, compared with 8hp per pair of 4ft corn grinding stones.[5] Part of the reason for this is that fulling stocks were directly driven by a tappitt wheel from the waterwheel, whereas corn milling required gearing, through which power was lost by friction. Few fulling mills had only one pair of stocks, however. At King's Mill, Painswick, for example, the power

Nind Mill, Wotton-under-Edge; one of the largest mill complexes in the southern area, partly water-driven into the early twentieth century. (*Industrial Gloucestershire*, 1904)

was 'irregular during the summer season, there being several mills above; the one immediately above is a grist mill which occasions a greater irregularity in the power than would occur if it was employed as a woollen manufactory'.[6] Fulling was a continuous process over many hours and cloth could be spoiled if there was an interruption caused by a shortage of water power. Consistency of water supply was more important than volume until the Age of Machinery.

Water-driven fulling mills were held by some clothier dynasties over several generations – the Wathens, Clutterbucks, Playnes and Marlings, for example. It was these mills which became amongst the first Gloucestershire proto-factories and later became integrated factory complexes. The transfer to the factory system took place at water-powered sites, some of which had been the sites of medieval or later fulling mills (only a few workshops and factories were built away from rivers in Uley, Wotton-under-Edge and Stroudwater). The water-powered sites with most potential for development, for instance those on a main stream which was supplemented by water from tributary streams or springs, had usually been developed earlier, leaving sites with less power potential for later entrants to the trade. Nind Mill on the Little Avon had one of the largest ponds in Gloucestershire, receiving the combined waters of the Kilcot and Ozleworth valleys, as well as the Ham's Gully brook, which together drove five waterwheels. The present Stanley Mill, built from 1813, was erected on a much earlier mill site and was entirely driven by the five waterwheels until 1822.[7] The Playnes of Longfords Mill, Avening, instead of investigating the possibilities of steam power in 1806, built a 30ft-high, 150-yard-long dam across the valley in order to make a lake of approximately 15 acres. The reservoir cost £400 and the dam £945 – more than the cost of a modest steam engine.[8] When the new mill at Ebley was built in 1818, the large part of an adjoining meadow was flooded to provide a reservoir which may well have equalled the Longfords one in cost. Halmer Mill on the Ewelme had a reservoir covering nearly 4 acres, which 'never fail[s] in supplying water to drive the mill power to its full extent'.[9] But in the crowded middle sections of valleys few mill owners had sufficient land on which to construct a reservoir.

WATER RIGHTS

Water rights were a tender topic and one which could take mill owners to the courts. Three issues predominated: raising the mill dam which could hold back water from the mill below; raising the mill dam so as to flood back into the tailwater of the mill above such that their waterwheel(s) ran in 'backwater'; creating a leat, or diverting part of the river or a spring and thus potentially depriving the mill downstream. In 1608 the owner of the lower of the two mills at Lodgemore (Lower Latemore) brought a court action against the owner and tenant of the upper mill (Upper Latemore/Merretts), accusing them of robbing his mill of water by raising the height of a sill between the two mills so that, in summer, the mill could 'yeld but small profit'.[10] Ebley Mill and mills downstream in the parishes of King's Stanley, Stonehouse and Eastington were the subject of a suit in the Court of Chancery in 1653. Feelings between mill owners ran high, twenty-three fulling and corn mills being affected. The issue concerned the alleged diversion of some of the River Frome to King's Stanley. No objection was raised providing sufficient water was left, if 'theyr cloth spoyled for want of fulling', the sill was removed.[11] A weir between Ebley and Dudbridge mills was the subject of an agreement in 1685, the owner of Ebley being permitted to insert boards in it when the water was 'low and starte', the condition being that the boards be removed when the water 'shall annoy recoyle upon, prejudice or injure the said mills called Dudbridge Mills'.[12] The weir again became the subject of a lawsuit in 1832 when boards had been put into the weir to raise it, resulting in the tailwater rising at Dudbridge.[13] When Humphrey Austin leased Langford Mill at Wotton to Thomas Mercer in 1801, a clause in the lease specified that Mercer should not 'dam or pen up the water' and obstruct 'the working of another mill … belonging to Humphrey Austin called New Mills … which are situated upon the same stream but below the mill called Langfords Mill'.

The steep gradients of some Cotswold scarp valleys, such as the Painswick and Slad brooks, enabled a high density of mills to be built with each having an adequate head – if not volume – of water. In any section of a valley where the gradient was steeper, there was often a nucleus of mills. Even the River Frome falls 125ft between Twissell's Mill at Chalford and Stafford Mill at Bowbridge. And clusters of mills would develop where there was a change in gradient, as at Chalford (see Map 6).

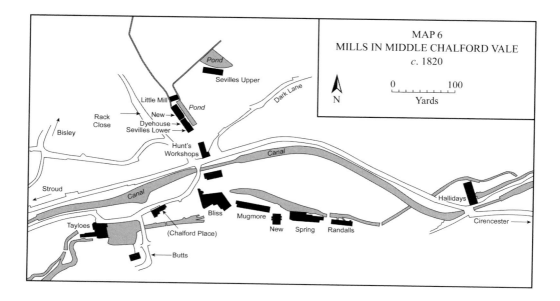

MAP 6
MILLS IN MIDDLE CHALFORD VALE
c. 1820

ANATOMY OF MILL SITES

Not only were the valleys most suitable for water-power development chosen by mill builders, but where possible, watercourses were constructed to optimise the power potential of the site. When a mill was constructed the builder basically had two alternatives. Either the stream could be utilised in its normal channel with dams, ponds and overflows constructed from this; or part of the stream could be diverted as a mill race or leat, enabling a greater fall to be achieved at the mill and more freedom in the layout of the buildings. The mill driven by the main stream was the earlier type of site. A greater head of water could be obtained if the main stream was dammed and a pond created upstream of the dam. This type of site is more often found in the upper reaches of valleys where a high fall could be obtained for an overshot wheel. In some narrow valleys the dam could be constructed right across the valley with less banking-up at the sides being required (for example, Upper Kilcot Mill).

Sites with mill races were probably fairly common by the sixteenth century. The principle involved the water supply for the mill being taken out of the stream higher up a valley and conducted in a near horizontal channel to the mill, the leat being several feet above the level of the stream at the mill. Some leats were half a mile long. Nearly all the mills on the Little Avon in Kingswood parish, as well as those on the lower Ewelme and the lower Frome, were constructed on this principle. The wheels at Dunkirk Mill, Nailsworth, were fed a half-mile leat.

At some sites a pond was created by diverting water from the stream with a weir – a kind of enlarged leat. Bayliss's Upper Mill, Painswick, is an example. At other sites a pond was created partway along a leat, while at others a leat took water to a large embanked pond and the mill was directly sited below the dam, as at Park Mills and New Mills, Wotton-under-Edge. In order to store water some mill owners built parallel ponds fed by a leat, water levels in each being controlled by sluices. This was the arrangement at Day's Mill, Nailsworth, and Broadbridge Mill, Alderley. A variant was the twin-leated type to be seen at Egypt Mill, Nailsworth, where separate channels fed each wheel. An ingenious scheme was adopted at Monks Mill, Alderley, whereby one leat was fed by the main stream and the twin leat by a spring.

Ingenuity was deployed to augment water-power potential at fulling mills, sometimes resulting in a hybrid, the resulting watercourses combining features of both stream and leated sites, as power requirements increased. A stream could be dammed, for example, and instead of the mill being sited immediately downstream of the dam, a leat took water from the pond to the mill which was sited parallel with the valley, as at Lower Steanbridge Mill, Painswick. Damsell's Mill, Painswick, was originally designed to be powered by the main stream but a leat was subsequently constructed to provide an additional head of water.

THE POWER OF WATER

The potentially destructive power of water is evidenced by the burst dam at Upper Dorey's (Painswick) Mill in c.1830 and when Wyatt's Mill dam burst in 1914 the mill was entirely demolished. The Playne brothers did not have complete confidence in their new reservoir at Longfords and built Lake Mill parallel to the valley rather than across it, water being taken to the mill by a leat.

The River Frome and its tributaries provided a larger and more regular source of power than the streams of Wiltshire and Somerset, Gloucestershire clothiers obtaining anything from 17hp at Hope Mill to 130hp at Stanley Mill.[14] Most mills had two wheels and many had three or four – Stanley had five. Ham Mill had three wheels of 10hp each, Vatch Mill had two 12hp wheels; Charfield Mill wheels generated 50hp in winter; there were three waterwheels at Cam Mill with a fall of 18.5ft, and four at Ebley generating 80hp. Timothy Went of Uley had a 26ft diameter wheel and Edward Sheppard at Uley had

Vatch Mill, Slad Valley; a large steam-powered mill sited on a spring, late nineteenth century. (Howard Beard)

one of over 20ft.[15] Even in the tributary valleys clothiers managed to generate 10hp or more at each site. Vatch Mill burned down and was rebuilt in 1827 on an extensive scale, considering its tributary valley location, with two 12hp waterwheels and three steam engines together generating 66hp. A painting, dated 1805, of a mill on the Toadsmoor Valley, shows a narrow external wheel which must have been of the order of 25–30ft.[16]

Mills situated in the middle and lower reaches of tributary valleys suffered water shortages, particularly in summer; lower down the Painswick stream at Small's Mill the power varied from 14hp in winter to 7hp in summer. Charles Hooper of Eastington, owner of the lowest cloth mill on the River Frome in the 1830s, reported that water frequently did not reach his mill until 1 p.m. in summer. Nathaniel & Joseph Jones of Chalford told the factory inspectors that 'as water is so very very uncertain, we are obliged to avail ourselves of it as the stream serves or let it waste'. It was this evidence from water-driven mills that prompted the factory inspectors to report that 'great evil would result from any prohibition against adapting the hours of work to the irregular supply of water'. A few mill owners reported loss of working time from winter floods.[17]

When industrial premises were shared, arrangements were made for motive power sharing. In the sixteenth century Doynton fulling mill could only be worked providing there was sufficient water to work the corn mill, too. And, when Bowbridge Mills were divided, both fulling mill halves being in the hands of different members of the Partridge family,[18] a clause was inserted in the agreement that in times of water shortage both parties were to cease work until there was sufficient water for both to recommence work.[19]

There are no descriptions of the kinds of waterwheels which powered fulling mills until well into the Industrial Revolution, but it is likely that the majority of earlier mills were driven by simple wooden undershot or breast shot wheels. The dense crowding of mills in some valleys meant that what, in principle,

Waterwheel at Egypt Mill, 1963. (Jennifer Tann)

might have been the best engineering solution for water-power generation, could not, in practice, be implemented. An overshot wheel needed the least volume of water and, on the whole, generated more power, but it required a greater fall of water and did not work well in tailwater on account of its rotating away from the direction of the water intake. A high breast wheel took in water between the middle and upper part of the wheel and rotated in the direction of intake, as did low breast and undershot wheels. An undershot wheel was the least economical in terms of water utilisation, but, as the Wiltshire millwright T. Mines pointed out, it could work in tailwater more effectively than any of the others.[20] Local millwrights applied practical know-how, devising solutions in wheel design and watercourses which were appropriate to different sites. When iron wheels became more common from the mid-nineteenth century, further optimisation of the power potential at different sites was possible through broad diameter breast shot wheels with curved buckets. The point is well illustrated by the middle section of the Frome Valley where, due to the slight gradient, an ideal site would have had a longish leat and a high breast bucket wheel. In practice, the mills were crowded so closely together – twenty-three within 4 miles – that long leats were out of the question. Low breast or undershot wheels had to be installed at many of the sites but adequate power could be generated by the multiplication of wheels.

When cast and wrought iron began to replace wood in the construction of waterwheels, some considerable time after the engineer John Smeaton had recommended it, wheels were made with more precision and prices rose. The Haden brothers of Trowbridge were waterwheel builders, as well as Boulton & Watt engine erectors. In 1820 they installed a 16ft-diameter, 11ft-wide wheel and millwork for Hicks of Eastington for £250. The west waterwheel at Egypt Mill is a low breast shot wheel of 14ft 6in diameter and 6ft wide which has forty wooden floats set at right angles into cast-iron rims. The east waterwheel is 14ft 6in by 10ft wide, also with wooden floats. Three waterwheels survive at Dunkirk Mill, two forming a pair, although not mounted on a common shaft. They are 10ft diameter and 7ft 2in wide and, from the name on the lauder, may possibly have been made by Daniels of Stroud. They have sheet-iron buckets lined with wooden sole boards. The third, an all iron overshot waterwheel 13ft in diameter and 12ft 4in wide with rim gears, is of the type made by John Ferrabee of the Phoenix Ironworks at Thrupp. A Ferrabee-type breast wheel with slightly curved buckets is preserved at St Mary's Mill. Water-power engineering improvements, until the adoption of the turbine, focused on bucket design and power transmission.

The Frome Valley mill owners' dependence upon water power was a key reason for objections to successive plans to construct a canal from the Severn to Stroud (and thence to the Thames). Paradoxically and not surprisingly it was also the wealthier mill owners who were amongst the various canal schemes' promoters. First planned in the late 1720s, again in the mid-1750s, yet again in 1774, and finally, successfully, in 1776, the Stroudwater Navigation was opened to Wallbridge in 1779, the Thames and Severn Canal being opened ten years later.[21]

The continuing significance of an adequate water supply towards the end of the nineteenth century, when all the surviving manufacturers would have had steam engines, is demonstrated by the arrangements for 'mudding'. The major manufacturers and mill owners met to arrange for mudding the Frome and Nailsworth valley rivers in 1873, some 2,000–3,000 loads of mud being removed from the pond at Millbottom Mill, Horsley.[22] And in 1905 the Stroud Valley Millowners' Association had meetings with directors of the water company with regard to ensuring their supplies.[23]

ANIMAL POWER

Animal power was as important in the mechanisation in the woollen industry as it was for cotton manufacture. Richard Arkwright had used horses for his first cotton spinning mill in Nottingham[24] and, when rotary steam power began to be introduced into factories, it was the horse-power equivalent which became the universal means of measuring the power required by different machines. Fulling took from eight to over 12 hours – enough to defeat any team of horses – and animal power was not used. It was, however, used in scribbling and carding shops and, later, possibly for jennies and shearing. In the eighteenth and early nineteenth centuries horse power was an inexpensive way of entering the woollen industry, even in Gloucestershire, where water power was relatively plentiful. Daniel Marklove of Berkeley employed horses to turn his machinery in 1802 and sent his cloth to be fulled at a water mill several miles away. When the stock-in-trade and machinery belonging to Timothy Went (deceased) of Uley was offered for sale in 1807, a 26ft-diameter two-horse wheel was itemised;[25] there was a horse-powered workshop in Vicarage Street, Painswick, in 1804[26] and a 'capital 3-horse wheel' was advertised by Price the Stroud engineer in 1813.[27] Two Painswick clothiers used ox power in the 1820s.[28] A typical horse wheel was between 15ft and 24ft in diameter and would have had harnesses for between two and four animals. Horse wheels were simple, usually horizontal, wooden constructions which could have been made by a local wheelwright or millwright, and they were cheap, usually costing between £50 and £150. In the period before the adoption of power looms, the mechanisation of preparatory processes enabled the establishment of a proto-factory and the subsequent transfer to steam power in a larger mill or factory.

STEAM POWER

Wiltshire and Somerset clothiers had begun to experience power shortages by the 1790s and ten firms in these counties had written to Boulton & Watt enquiring about steam engines before the first Gloucestershire woollen manufacturer (Day, Smith & Alder of Nailsworth) enquired about the price of a 30hp engine in 1800. Between then and the first Gloucestershire order in 1802, three other firms (G. Adey of Dursley, W. Read of Ebley, G.S. Hornblower of Horsley) made enquiries, none resulting in an order. Then, in 1802, H. & G. Austin of Wotton-under-Edge ordered a 6hp engine. From 1802 to 1825, twenty-five years after James Watt's patent had expired and by which time there were a number of reputable alternative engine makers, thirty-one new and one second-hand Boulton & Watt engines had been ordered by Gloucestershire woollen manufacturers (see Appendix).[29] Only four further Boulton & Watt engines were ordered by Gloucestershire woollen manufacturers after 1825 – a 30hp crank engine for Henry Hicks of Eastington in 1826, a 6hp A-frame engine for Playne & Co., a 20hp side lever engine for W. Long of Charfield in 1827 and a 30hp engine for Stanton & Sons of Stafford Mill, ordered in 1839. At the beginning of the nineteenth century clothiers made tentative enquiries after small engines, only two clothiers enquiring about the price of an engine over 14hp and neither of them ordering one. The highest horsepower ordered before 1820 was 14hp, although a Bradford on Avon, Wiltshire, clothier ordered a 32hp engine in 1816. The demonstration effect is noticeable with H. & G. Austin's engine appearing to set a precedent at Wotton-under-Edge, a flurry of enquiries from other clothiers in Wotton following their order. Similarly, Edward Sheppard of Uley's order of a 14hp engine in 1805 was followed by enquiries from two other Uley manufacturers within two years. From 1820 onwards the number of engines ordered per year increased, as did the size. Apart from J. Cripps of Cirencester's 6hp engine, all the engines ordered were of 10hp or more, R.P. & G.A. Smith of Wallbridge and H. Hicks of Eastington ordering 24hp engines in 1821 and 1822, and Harris Stephens & Co. of King's Stanley and N.S. Marling of The Vatch ordering 40hp engines in 1822 and 1825 respectively.

By the 1820s some woollen manufacturers were ordering engines without having first enquired about the price or availability of a specific sized engine. This may have been due to the fact that, by this time, the Haden brothers were well established both as millwrights and as agents for Boulton & Watt, and initial enquiries as to price may have been made through them. The existence of Boulton & Watt representatives in the West of England wool textile region probably accounts for their engines appearing to have had a large share of the post-1800[30] steam engine business of the area.

The majority of engines were various kinds of beam engine – rotative crank (eleven new plus one second hand) and independent six-column beam engines (eleven). In addition, one bell crank engine was supplied to Strange's of Wotton in 1805. This was a design developed in the early days of marine engines but was supplanted by the more compact side lever engine, a boat-type engine used on land – seven of which were sold to Gloucestershire woollen manufacturers, together with three independent engines. These side lever engines were more compact than beam engines and were probably promoted by the Haden brothers of Trowbridge who seemed to favour this design.

It is clear that a number of clothiers used steam to supplement water power rather than to supplant it. Samuel Clutterbuck of St Mary's Mill told the factory inspectors in 1834 that 'in very dry seasons' he used 'steam to supply the deficiency' of water power. Barnard & Co. of Nailsworth reported that they used water for about half the year and water and steam the other half. William Lewis of Brimscombe estimated his water power as equal to 60–80hp and his steam power as 80hp, 'of which 40 only is in use', suggesting that steam was an insurance against particularly dry periods in the year. N.S. Marling of Vatch Mill used both water and steam power on a daily basis.[31]

Some engines were claimed in factory sales bills to have been made by Boulton & Watt but are not recorded as having been supplied by the firm. They may have been second-hand engines, or they may have been made by other engineering firms on Boulton & Watt's principles.[32] St Mary's Mill, Chalford, for example, had what was claimed to be a Boulton & Watt engine, but there is no record of an enquiry or a sale by Boulton & Watt. There were certainly some non-Boulton & Watt engines in Gloucestershire woollen mills by the 1820s; among the mills which may have had engines of other makes by the early 1820s are Cam, Rockstowes (Uley) and Jeen's (Uley), although there is no record of their make.[33] One mill for which the alternative engine maker is known is the Sevillowes site which, by 1821, had a 12hp high-pressure engine by Woolf.[34]

The factory returns for 1839 record the proportion of water and steam power and the amount generated in each manufacturing parish. These figures confirm the continued dependence upon water power. Water was the sole source of power in Dursley and the major source in Kingswood, Painswick, Stonehouse, Rodborough, Minchinhampton and Horsley. In Stroud, however, the sources were about equal, as they were in Standish and Eastington. In Charfield and Wotton-under-Edge steam power predominated and in Leonard Stanley only steam power was employed (see Map 7 overleaf). By 1850 the total steam power used in the Gloucestershire woollen industry was 806hp, compared with 1,495hp generated by water power, i.e. 54 per cent. By comparison, nearly 70 per cent of the power in the Wiltshire

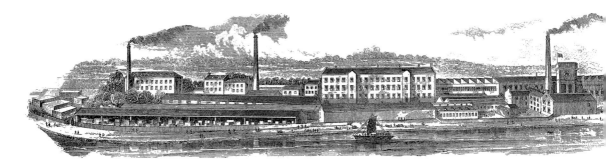

Bliss Mill estate, showing (right to left) Bliss, Mugmore, New, Spring, and Randalls Mills. (*Industrial Gloucestershire*, 1904)

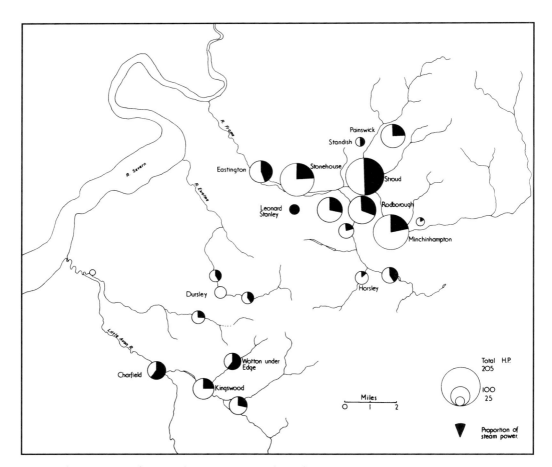

MAP 7: The proportion of water and steam power in each parish in 1839.

woollen industry was steam and the Yorkshire figure was 76 per cent.[35] The comparison with Wiltshire can, however, generate misleading conclusions,[36] since there were far fewer woollen mills there by this time and opportunities for generating water power were poor in Trowbridge, one of the main centres of the Wiltshire industry.

The apparent reluctance of Gloucestershire manufacturers to adopt steam power cannot be attributed solely to the price of coal. Coal was available from the Forest of Dean and the Stroudwater/Thames & Severn canals (opened in 1779 and 1789 respectively) facilitated its carriage to the Frome Valley mills. Some of the larger mills had wharves for unloading coal. The peak years for coal carried on the Thames & Severn Canal were around 1850, after which the tonnage carried declined.[37] Steam-driven mills in the Painswick Valley were disadvantaged and had to rely on coal being transported by wagon. After the opening of the railway link to Stroud in 1845 some mills, such as Stanley Mill, had dedicated spurs or, like Cam Mill, a nearby rail link at Coaley Junction. Pithead prices in the South-West remained greater than those in Lancashire/Cheshire or Yorkshire, being 7s (35p) in 1882, compared with 5s 10d (29p) in Lancashire/Cheshire and 6s 6d (33p) in Yorkshire; however, it has been suggested that 'the cost of coal was only a very marginal expense for manufacturers'.[38]

John Ferrabee and, later, his son James manufactured steam engines at Phoenix Ironworks, Thrupp – little is known of the type of engines or who their customers were, although it is likely that at least some of the engines mentioned in sale notices in the 1830s were theirs. In 1875 H.J.H. King established his engineering business in Nailsworth. A prolific inventor, he registered a patent every year from 1869 until his death, with the exception of 1874. His first major steam engine patent was taken out in 1885 and horizontal engines were supplied to Playnes at Longfords Mill, P.C. Evans at Brimscombe and Apperley Curtis & Co. of Dudbridge.[39] It is significant, however, that the greater number of known engine

Lightpill Mill, a large woollen manufacturing site, comprising several separate mills, now an industrial estate. (Mike Mills)

customers came from the newer industries which had moved into vacated cloth mills: Dangerfields (walking sticks) at Bliss Mill, Critchley Bros at Wimberley (Pin) Mill, Tubbs Lewis (elastic fabrics) at New Mill Wotton and Merretts Mill (shoddy) in Nailsworth.[40] Finally, dependence on steam was reduced as the surviving businesses installed gas or oil engines and, later, employed electric power.

MILLWRIGHTS

The task of constructing a wooden waterwheel or horse wheel and of conveying the power to the machinery was that of the millwright. Millwrighting was not an occupation created by the Industrial Revolution but a survival from an earlier age. The pre-Industrial Revolution millwright worked in wood.[41] The millwright's work involved the creation of simple power systems or the installation of more complex ones and the linking of the prime mover via shafts, gears and rope or leather belts to the machines. The millwright was called upon to assess the power requirements of particular machines, to decide on the optimum speed for operating them, to estimate the loss of power by friction and to calculate power ratios. In some cases he planned the mill building and the layout of machinery in relation to motive power. The early Industrial Revolution posed few technological problems; most millwrights were not theoreticians and simple traditional methods were all that was required for fulling, scribbling and carding. Even when a country millwright was aware of the more theoretically advanced power systems, such as the overshot compared with the undershot waterwheel, this was not necessarily an appropriate solution in practice, given circumstances of overcrowding.

By the 1760s and '70s some specialist millwrights/engineers had emerged – Brindley, Smeaton, Jessop, and Telford all undertook millwrighting besides their better-known canal engineering. In Yorkshire and Gloucestershire the general millwright continued to cater for country mills, although it is likely that some specialisation had occurred by the 1770s. The 'famous' Richard Remington of Woodchester installed an early knapping mill in Stanley Mill in 1735[42] and 'The famous Mr Chinn' of Tewkesbury installed 'and so completely finished' an overshot waterwheel such that 'in the driest summers there can be no want of water' at Stanley.[43] The millwright's activities ranged from making and repairing uncomplicated horse wheels to complex production engineering.[44]

The more sophisticated millwright engineer emerged in the 1790s when mechanisation was advancing in the wool textile industry, when cast iron was being introduced in machinery and millwork, and when boom conditions prevailed. By the 1820s the skills limitations of the country millwright became clear as cloth manufacturers began to investigate the adoption of steam power. With the nearest coalfield being the Forest of Dean across the Severn, and no local iron or heavy engineering industries, the western slopes of the Cotswolds lacked a pool of skilled engineering labour on which to draw. Nearly every pre-1815 enquiry to Boulton & Watt raised the question of the supervision and maintenance of steam engines.[45] The Haden brothers of Trowbridge undertook millwrighting, repaired gigs, cutters and drying machines and made waterwheels. They introduced iron cloth racks into the region. When Harris Stephens & Co. of Stanley Mill ordered a Boulton & Watt steam engine in 1822 they specifically requested the Hadens not to erect their engine, as they had 'machinery that is not wished to be seen but by their own people and you, being engaged in such work, without directly intending it, might copy these inventions'. Nevertheless, as their order books show, the Haden brothers repaired and erected new machinery in nearly every West of England woollen mill between 1824 and the mid-1840s.[46]

In 1822 Henry Hicks employed Thomas Hewes, the eminent Lancashire millwright, perhaps at the suggestion of Boulton & Watt, for work at Churchend Mill, Eastington – it is not known if he undertook any other Gloucestershire work. Twelve millwrights are recorded in Gell & Bradshaw's 1820 Gloucestershire

Directory. A number were members of well-known cloth-making families, including Packer, Clutterbuck, Gardner, Savory and Underwood. John Price of Stroud was textile millwright to one of the largest Gloucestershire clothiers, Edward Sheppard of Uley, and also made textile machinery. A member of the Gardner family of Stroud was appointed sole manufacturer of J.C. Daniell's patent shearing frame.

Some built larger engineering workshops, while others, for example Ferrabee, branched into iron-founding and machine making. Some of these new millwrights had been trained as instrument makers, others had been educated in nonconformist academies or had been apprenticed to a well-known millwright/engineer. They were able to evaluate the practical realities of a given factory location against the theoretical ideal. The West of England dyers who rejected new fuel-saving methods of heating dyeing furnaces, for example, would have done so on the advice of millwrights who pointed out that the theoretical savings in fuel costs would not equal the interest on the capital outlay in a medium-sized enterprise.[47] This was also the case with factory gas lighting, which was adopted by some of the larger Lancashire and Derbyshire cotton spinners but not by Gloucestershire woollen manufacturers. The millwright in Gloucestershire, as elsewhere, is tantalisingly elusive.

The adoption of power systems technologies reflects the overall shift of technical expertise from the West to the North of England from c.1825 onwards. Woollen manufacturers who had been locally supplied with spinning and weaving equipment in the early nineteenth century had to look to the north for mules and looms from the 1830s onwards.

Endnotes

1. T. Rudge, *The History of the County of Gloucester*, 2, 1803, pp.216–17
2. The Case of the Stroudwater Navigation, 1775, JX14.21gs
3. D1388, 109; PP 1834, XX, pp.279–80; RV 216.3gs; RX 180.3(1)gs
4. The Case of the Stroudwater Navigation
5. Jennifer Tann, 'Horse Power', in F.M.L. Thompson, 'Horses in European Economic History, a Preliminary Canter', *British Agricultural History Society*, n.d., pp.21–30
6. PP 1833, XX, p.262
7. PP 1834, XX, p.256
8. A.T. Playne, *A History of Playne of Longfords Mills*, 1959, pp.40–1
9. RX180.3(1)gs; RF65.30gs
10. D149/L1
11. RF 289.6gs
12. 289 (63)gs
13. D873B
14. PP 1834, XX, pp.254–6. Some PP sources say 200hp but this is unlikely
15. PP 1834, XX, pp.254–5, 276–80, 284–5; RX 319.1(19)gs
16. Copy in the Museum in the Park, Stroud
17. PP 1834, XX, p.285. Water mills were still allowed to make up for lost time up to 60 hours per week after the passage of the Factory Act and in 1836 young people of twelve to eighteen years of age were found, in some cases, to have worked 72 hours. PP 1836, Rept. Inspec. Factories, p.14
18. D947/M2
19. Box 69gs
20. Jennifer Tann, 'The Employment of Power in the West of England Wool Textile Industry', in N.B. Harte & K.G. Ponting (eds), *Textile History and Economic History: Essays in Honour of Miss Julia de Lacy Mann*, Manchester University Press, 1973

21 Michael Handford, *The Stroudwater Canal*, Alan Sutton, 1979; Humphrey Household, *The Thames & Severn Canal*, new edn, Alan Sutton, 1983; Michael Handford & David Viner, *Stroudwater and Thames & Severn Canals: Towpath Guide*, Alan Sutton, 1988
22 D4542/1/2
23 D4542/2/2
24 Jennifer Tann, 'Richard Arkwright and Technology', *History*, 58, 1973, pp.29–44
25 GJ, 8 February 1807
26 GJ, 1 October 1804
27 GJ, 13 December 1813
28 Hammond MSS Box 8
29 Several writers have produced (incomplete) lists of Boulton & Watt engines in Gloucestershire, e.g. Nicholas Kingsley, 'Boulton & Watt Engines Supplied to Gloucestershire: A Preliminary List', *GSIA Journal*, 1990, pp.49–53, or have discussed the introduction of steam power: Stephen Mills, 'Coal and Steam – The Arrival of Steam Power in Stroud's Woollen Mills', *GSIA Journal*, 2004, pp.42–52. The list in the appendix is derived from an ESRC-funded three-year study on the enquiry:order ratio for B & W engines
30 James Watt's patent having expired by the beginning of 1800 and the market being, therefore, open.
31 PP 1834, XX, pp.258, 259–60, 279–80
32 GJ, 23 April 1813
33 See Part II
34 Sale Bill, Hammond MSS
35 J. de L. Mann, *The Cloth Industry in the West of England from 1640 to 1880*, p.221
36 Mann implies on p.202 that Gloucestershire was backward in the adoption of steam power
37 Handford & Viner, p.20
38 Mann, pp.155–6; D.T. Jenkins & K.G. Ponting, *The British Wool Textile Industry, 1770–1914*, 1987, p.172
39 Tony Youles, 'The Nailsworth Engineers, Part I', *GSIA Journal*, 1989, pp.22–31
40 Ibid.
41 Jennifer Tann, 'The Textile Millwright in the Early Industrial Revolution', *Textile History*, 5, 1979
42 GJ, 24 June 1735
43 GJ, 12 September 1735
44 Jennifer Tann, *The Development of the Factory*, pp.95–108
45 Jennifer Tann, 'The Employment of Power in the West of England Wool Textile Industry'
46 These are in the Wiltshire County Record Office
47 Jennifer Tann, 'Fuel Saving in the Process Industries during the Industrial Revolution: A Study in Technological Diffusion', *Business History*, 15, 1973

CHAPTER SEVEN

INDUSTRIAL REVOLUTION 1790–1835: LABOUR AND ITS DISPLACEMENT

At the beginning of the Industrial Revolution the finishing processes (fulling, raising and cutting) were largely undertaken by men working in mills or workshops, the raising operation requiring the assistance of a boy. The preparatory processes (spinning both warp and weft) was women's work, assisted by a child. With the introduction of jenny spinning some women's work was lost, as a man with two children would sometimes spin the weft yarn. Slubbing, a new process created by the introduction of the billy, was undertaken by a man with two children; machine scribbling and carding being done by a child. Weaving was cottage-based and undertaken by men with the assistance of a child. Unlike at the early cotton and silk mills, many of which depended upon the labour of pauper apprentices from workhouses, there is little evidence of the use of pauper apprentice labour in the wool textile industry (although there was some in worsted spinning).[1] The reasons were partly technological and partly social. In the cotton industry the jenny was largely superseded by Arkwright's water frame, which was water-powered and could be operated by children,[2] and this led to the rapid emergence of dedicated cotton-spinning mills with mechanised carding and, with it, the demand for child labour on a large scale. In the woollen industry finishing had been undertaken in mills for centuries. The mill, thus, became the focus for machinery as other processes were mechanised and woollen manufacture became more vertically organised; most processes (with the exception of weaving) were located at the mill by 1830. Child labour was part of the family economy in wool textile manufacture and probably came into the mills as part of a family unit, children not being required in anything like the same numbers as in cotton or worsted spinning.

TRADE FLUCTUATIONS

Short-term trade fluctuations in the 1770s and 1780s caused considerable difficulties in the woollen industry. Prices fell in 1771–72 and there was a general shortage of money; the East India Company, which bought a great deal of Spanish stripes from Chalford, delayed its annual sale by two months. The domestic market was very slack in 1773 and the starving condition of Gloucestershire workers was described in the *Bath Chronicle*.[3] Depression continued in 1774, both West of England and Yorkshire clothiers suffering from a loss in demand for British cloth, but largely for different reasons. The East India Company's exports fell to under 10,000 pieces a year, Rudder commenting on industrial decline in Chalford.[4] In times of slack trade, with domestic weavers unemployed, cloth theft became common.[5] Indeed, it became so common that watchmen were appointed to guard tenters,[6] and if a thief was caught and convicted he could be punished by death: 'At our Assizes last week, no less than 16 criminals received sentence of death, among others Thomas Rudder for stealing cloth from Fromehall Mill.'[7] A writer in 1775 remarked that of the 100 mills in the area, 'some of them from the great decay of the coarse trade, are now turned in part, or wholly, to other purposes. Some lie totally idle; others are little or not half employed.' He continued with the observation: 'One third of that number fully employed would be more than sufficient for the whole clothing trade of the county; it is also to be feared they ever will be sufficient unless more hands could be employed in fine spinning.'[8]

Trade recovered somewhat between 1778 and 1782, exports to India compensating for the depression which began with war in France and Spain and later Holland. By 1781 the industry was said to be flourishing in Gloucestershire and its neighbouring clothing counties. But this was short-lived; a crisis in 1783, combined with a severe winter and a further slump in East India Company exports, contributed to one of the most severe depressions experienced in Gloucestershire in the eighteenth century. Relief committees were established in Stroud and it was estimated that there were 15,000 destitute persons in the fifteen Stroudwater parishes in January 1784.[9]

Despite trade fluctuations of the 1780s and 1790s, wealth was acquired by a number of clothing families. Members of the Hicks family, for example, had been at Damery Mill, Upper Cam, and Nind mills in the seventeenth century, but it was Henry Hicks who transformed the woollen industry in Eastington in the late eighteenth and early nineteenth centuries. He first leased Millend Mill with Edward Sheppard of Uley, in 1799, buying Churchend Mill from the trustees of John James. Churchend was transformed from a small fulling mill and by 1806 it had been rebuilt and enlarged. In this year Henry Stephens sold his estate to Hicks who became Lord of the Manor of Eastington. By 1820, he had been joined by his two sons and the family was also operating Bonds Mill. As the family wealth increased, Hicks in c.1815 built a mansion house, The Leaze (now Eastington Park), but he had the reputation of being remote from his workpeople; perhaps seduced by the life of a gentleman, his business failed in 1835.[10]

While some of the largest Gloucestershire firms undertook their own direct selling to foreign merchants, the majority of Gloucestershire woollen manufacturers continued to depend upon selling through London merchants who lacked specialist knowledge and did not involve themselves in the complexities of overseas markets. By comparison, many Yorkshire merchants (who were more closely concerned with the Yorkshire industry) had been sailing to America in large numbers since the end of the War of Independence in 1782 and sold great consignments of cloth there.[11] And, even if West of England manufacturers had been more fleet-of-foot, there was only a limited potential American market for high-quality broadcloth. The American War of Independence was disruptive to trade beyond America too, but the years between the peace of 1782 and 1796 were buoyant for the West of England. Clothiers complained of rising costs, particularly the price of English wool, and at a general meeting in March 1792 they proposed to increase the price of cloth by 3*d* per yard on cheaper cloth and 6*d* a yard on finer cloth.[12] Exports particularly increased between 1794 and 1796, but 1797 was a depressed year, although Spanish cloth sales made up some of the shortfall and coarse cloth for army and navy uniforms was in demand during the French wars. Clothiers were still complaining of losses on their cloth sales and, in 1798, a circular letter was sent to the chief manufacturers in the county setting out the cost of making a piece of cloth from 60lb of wool. The total cost amounted to £25 18*s* 11½*d* but the cloth was only sold for £24 6*s* 6*d*.[13]

Although there were the beginnings of competition from Yorkshire broadcloth, there was, as yet, little challenge to woollen cloth's supremacy over worsted for men's clothing and 'the country squire still wore Uley blue on Sunday and Stroudwater scarlet on Monday', but 'the more one follows the fortunes of the West of England broadcloth trade at the end of the eighteenth century the clearer it becomes that it was no longer a growth industry'.[14] Demand slumped with the declaration of peace in 1802, while trade with Russia suffered with the outbreak of war in 1807, cassimere and Spanish cloth manufacturers suffering particularly badly – H. & G. Austin of Wotton-under-Edge lost heavily. European trade was dislocated between 1805 and 1808, while 1811 was a disastrous year. To some extent losses suffered by Gloucestershire clothiers were offset by the East India Company which changed its practice of buying through factors to dealing directly with the manufacturers. Some of the larger manufacturers such as Edward Shepherd, Humphrey Austin, Wallington, R. Cooper and Daniel Lloyd opposed this practice when the Company's charter came up for renewal in 1812, since they preferred to export through private merchants or independently. Other manufacturers, however, including S. Wathen, W. & P. Playne, J. Innell, C. Ballinger, N. Wathen, W. Bayliss, S. Wood, W. Gardiner, J. Iles and D. Cox were just as anxious for the practice to continue.[15]

The market for broadcloth was changing, less being exported while home demand increased. Overall, it was clothiers who were prepared to diversify into new, lighter fabrics that benefited most from the growth in demand in the later 1780s, although Yorkshire competition was increasing, not least because spinning machinery had been more rapidly adopted there. Gloucestershire, perhaps due to the supremacy of its dyed broadcloth, was slower to take up the manufacture of cassimeres on such a large scale as some Wiltshire and Somerset manufacturers. However, the technical adjustments necessary for transferring from broadcloth to cassimere production may have been down-played, yarn for cassimere being about three times as fine as for broadcloth, and many breaks could occur during weaving.

Home demand maintained the Gloucestershire industry in the years immediately following the peace of 1815, although the fall in demand for military cloth hit Bisley and Chalford badly. The years 1816 and 1819 were particularly difficult and it was alleged that half the workers in Gloucestershire were out of work.[16] The Spanish stripes trade, upon which Chalford depended, fluctuated between 1820 and 1829 – the boom year of 1827 being followed by the worst in the decade the following year.

There were six bankruptcies in 1819, even those manufacturers who had a larger capital base to tide them over making heavy losses; the Austins of Wotton-under-Edge made a loss of over £15,000 in 1819 and over £22,000 in 1820.[17] Edward Jackson of Uley was the most notable bankrupt. Nonetheless, new factory building and extensions to existing ones went on apace between 1815 and 1825 and the wealthier manufacturers seemed confident of the future – 140 woollen manufacturers are listed in Gell & Bradshaw's 1820 directory, a number of smaller businesses being omitted[18] (see Maps 4 and 5).

It is clear that, to survive, a capability for business was essential and even this could not guarantee success or continuance. Humphrey Austin had been a dynamic clothier entrepreneur. When he retired in 1811 he built himself a fine house (Warren House) on land purchased some years previously from his friend Lord Berkeley. The new Austin partnership, George and his cousin Edward, a London merchant in the trade, did not fare so well. It was an optimistic but poor business decision to build Waterloo Mill (1815) and Old Town Mill (1817) in Wotton-under-Edge, for neither proved to be profitable. Nor was it wise to extend New Mills. George Austin withdrew more than he should from the partnership and when he died in 1815 he was insolvent, owing £30,000 to Edward and nearly £10,000 to Humphrey. A loss of £20,000 was made in 1820 and when the business accounts were investigated a total deficit of £45,000 was reported. Much of the indebtedness was to family members and the insolvency was covered up for ten years while Edward and his sons tried, and failed, to turn the business around.[19] Both large and small manufacturers were badly hit; emigration was a solution, one small manufacturer, possibly H. Wyatt, recorded in his Bill Book in 1831, 'Lent Thos Ellay £14 to take him to America'.[20]

With the prosperity of 1825 came the weavers' strikes. Edward Sheppard acceded to most of the weavers' demands and encouraged his fellow manufacturers to do the same, but the strike dragged on into the autumn in Stroudwater. In December 'the panic' occurred in which a number of country banks failed and the industry plunged into deep depression. William Playne Senior of Longfords was in London at the time and, anticipating that there might be a run on Tetbury Bank, collected every sovereign that he could acquire and, taking a chaise and four horses, galloped to Tetbury where he found an angry mob surrounding the bank clamouring for their money and the cash drawer nearly empty. At the sight of Playne, followed by a servant carrying a bag of gold, the run was stopped and the bank saved.[21] The distress was exacerbated by an increase in American import tariffs and the collapse of the South American markets. The panic of 1825 sealed the fate of many of the smaller clothiers; they had a smaller capital base and depended more on credit from the London factors. If this did not put them out of business directly, they lingered on using out-of-date technologies, having no capital for repairs or extensions. The first to fail were, on the whole, businesses located in the smaller mills in tributary valleys. The year 1826 was a severe low point; sixteen Gloucestershire firms went bankrupt, nine of them from Uley and Wotton-under-Edge; a further four were insolvent, three being from Uley and Wotton.[22] By 1828 the West of England woollen industry was said to be on the decline, notwithstanding the assertion that 'The Yorkshire make very fine cloth but the West of England are preferred'.[23] There was some improvement in 1827 and by 1828 trade was fairly brisk. But Gloucestershire, together with Wiltshire and Somerset, was feeling intense competition from Yorkshire; men there, with little capital and hard work, were willing to take risks and respond more quickly to demand by innovating different fabric designs.[24]

The crisis led to a steep decline in prices as well as profits. John Lewis of Oil Mills, Ebley, recalled that his father 'made as much profit on one cloth as I do on twenty';[25] Millman & Long of Charfield and Kingswood estimated that more profit was to be made on five ends of cloth formerly than on fifteen by 1838.[26] Edward Sheppard, who chaired the Gloucestershire manufacturers' committee, sought to assure ministers that the distress was universal within the clothing area of Gloucestershire.[27] But some cloth manufacturers showed optimism: William Marling of Ham Mill took his son Thomas into partnership in 1826, when the firm's capital was £6,000, and in 1832 another son, Samual Stephens Marling, by which time the capital was £22,000. Of this amount Thomas Marling's share was £9,000, built up from zero in the year he joined.[28] 1832 was an exceptional year for East India Company exports to China but in the following year the Company's monopoly expired and many suppliers, including those in Chalford Vale, were badly hit. William Playne of Longfords Mill, probably the largest stripe manufacturer in Gloucestershire, gave up making them after his weavers struck for an additional 10s per piece.[29]

When the factory commissioners visited the southern area they commented on the large number of mills which had ceased production and the fact that twelve of the nineteen names given them had either failed or retired from business within the previous three years:

> In the parish of Uley, in the town of Dursley and in the town and immediate neighbourhood of Wotton-under-Edge, we found many mills which have been shut up for a long time, some as much as three years, such has been the decay of trade in those neighbourhoods.[30]

Only two of this number, the smaller ones, had been declared bankrupt; whilst Austins had made a composition with its creditors, others had retired while they were still relatively well-off. The trade difficulties in Dursley, Wotton and Uley may have been connected with a fall in the export of cassimeres (largely, but not exclusively, to America),[31] overall exports of which had fallen from 15,000 pieces in 1827 to just over 2,000 in 1830.[32] While in 1833 J.W. Partridge of Bowbridge stated that there had been fewer unemployed in the previous year than for a long time previously,[33] there was destitution in parishes where no woollen manufacturers and few wealthier people lived, such as Coaley and Randwick.[34] More cloth was manufactured in Gloucestershire in 1832 than in any year since 1823 (Table 5) but the figure dipped significantly the following year and remained low for the rest of the decade.

Table 5 – Total yardage manufactured in Gloucestershire 1823–38

Year	Yards	Year	Yards	Year	Yards	Year	Yards
1823	1,769,762	1827	1,434,175	1831	1,976,622	1835	1,864,526
1824	1,741,100	1828	1,659,329	1832	2,169,340	1836	1,916,518
1825	1,750,243	1829	1,517,252	1833	1,426,689	1837	1,749,208
1826	1,499,661	1830	1,602,913	1834	1,966,846	1838	1,592,594

Source: Mann, p.339

TIME TAKEN TO PRODUCE A BROADCLOTH

In 1840 a full-scale Parliamentary inquiry was published into the woollen manufacturing process which had, until the late 1830s, remained largely a cottage trade – weaving. The Commissioner for Gloucestershire considered the effect of mechanisation on each of the processes of cloth manufacture separately. The Commissioner for Wiltshire and Somerset, however, produced a series of tables showing the labour requirements and cost of making a broadcloth. These data give an invaluable insight into the progressive displacement of labour by machinery over the period 1781–1838. Scribbling wool for a piece of broadcloth by hand in the early 1780s took one man 96 hours. When the scribbling machine was introduced, the equivalent work was completed by one child in 14 hours and another child undertaking the carding in 13 hours 21 minutes. Spinning warp yarn by hand took one woman and a child 260 hours and abb/weft took a woman and two children 352 hours. Slubbing by machine took one man and two children 7 hours 4 minutes, spinning warp on the jenny took one woman 38 hours 17 minutes and abb spinning on the jenny took a man and two children 34 hours 17 minutes. What had formerly taken 708 hours took just over 106.5 hours by 1805. The adoption of the flying shuttle (spring loom) led to a reduction from 364 hours of two men and one child to 252 hours' labour of one man and a child. Whereas raising by hand took one man 88 hours, raising by gig took a man and a boy 12 hours.[35] Table 6 shows that, at first, women were particularly hard hit; this was due to the spinning jenny which displaced women's labour in agricultural villages distant from the clothing villages. Thereafter, labour requirements for women and children were relatively unaltered until the 1820s. The requirements for adult males declined at each of the five periods.

Table 6 – Proportional labour requirements for the production of one superfine broadcloth

Labour	Period 1 1781–96	Period 2 1796–1805	Period 3 1805–20	Period 4 1820–27	Period 5 1828
Men	100	41	36	35	31
Women	100	25	25	25	18
Children	100	30	30	30	25
Total	100	32	31	30	25

Source: Adrian Randall, Before the Luddites, CUP, 1991, p.54

As a result of these technical innovations the labour cost of manufacturing one broadcloth fell from £8 18s 2¾d to £6 3s 5d, a fall of over 30 per cent.[36] The only other major technical innovation introduced between 1805 and 1827 was the shearing frame. This reduced the task of one man taking 72 hours to that of one man taking 18 hours. When, between 1820 and 1827, the rotary cutter was introduced this further reduced the time taken to 6 hours. The extent to which the Wiltshire/Somerset figures are accurate for those counties and the extent to which they can be relied upon for Gloucestershire is debateable. The Wiltshire Commissioner allowed four weeks for weaving a broadcloth when other evidence indicates that this could be accomplished in three.[37] He also suggested that the spring loom could complete a cloth in 70 per cent of the time taken by a double loom, using a traditional shuttle. Although the spring loom was slower when it was first introduced, experience and the use of better quality jenny-spun yarn led to time savings, although weavers were also working longer hours from c.1800, thereby completing cloths in a shorter time. The Gloucestershire Commissioner stated that it took about two days to dress a cloth in a gig mill, while the Wiltshire Commissioner gave 12 hours. The figure given for cutting is probably about right, William Partridge noting that a rotary cutter could cut a 20-yard-long broadcloth in 15 minutes. A 38-yard broadcloth would therefore take 28 minutes per cut by this criterion, equivalent to some twelve cuts in the 6 hours indicated by the Commissioner.[38] Whether these figures are accurate in fine detail or not, they conclusively demonstrate a high level of labour displacement throughout the Gloucestershire woollen industry in the period 1780–1820.

Mechanisation and the labour surplus led to wage reductions over the early nineteenth century. Wages of women and children were reduced in each category of work, female jenny spinners experiencing the greatest single reduction at 57 per cent; men's wages varied. Some, such as sorters, rowers and enginemen experienced no reductions; others, such as millmen and brushers, experienced small reductions, while master weavers experienced a reduction of 37.5 per cent between 1805–15 and 1838.

The productivity of a middle-range clothier may be estimated at twelve cloths per week, 600 per year in the period 1781–1828. Larger clothiers produced a great deal more and smaller clothiers very much less. Factory employees worked around 72 hours per week, domestic handloom weavers probably working longer than this after 1800 and, since almost all of the Gloucestershire mills were water-powered before 1820, many factories worked irregular hours. Randall estimates that whereas the labour of 659 men, women and children was required to produce twelve superfine broadcloths per week in 1781–96, this figure had fallen to 165 by 1796–1805 and to 128 by 1828.[39] What is certain is that the fall in employment in the Gloucestershire woollen industry during the early Industrial Revolution was significant and had profound social consequences. In 1834 a list of 'Prices for Weaving as paid by the Principal Manufacturers' in Gloucestershire was published. Prices were set out for casserene, cassimere, felt or broadcloth (coloured and white), and Spanish and super stripes. The most highly paid work was coloured felt or broadcloth for which 1s 9d per ell was paid. The cheapest was super stripes for which 9d per ell was paid for an 11 hundred fabric.[40]

LABOUR UNREST

The Gloucestershire textile worker was more vulnerable to mechanisation than his/her Yorkshire counterpart on account of the specialisation of different crafts. The introduction of machinery in any single branch could threaten a specialist trade, and the worker might not easily be able to transfer to another branch of manufacture. Moreover, technological innovation had the potential to create a knock-on effect in other processes of manufacture. Whether the introduction of machinery elicited opposition from workpeople or not was closely tied to the extent to which a particular machine was perceived as being likely to cause unemployment. And the level of opposition to a specific machine was not necessarily an indicator of its strategic importance to the industry. The slubbing billy, an essential machine in the three-fold sequence of mechanised yarn production, induced no opposition for it was a new trade. And local custom was a significant issue. In Yorkshire the scribbling machine and jenny were readily adopted. It was not until the introduction of the gig mill and shearing machine that Luddism amongst the croppers became a significant movement, with the creation of a trades union which had links with the West of England – and the Wiltshire Outrages. Since gig mills had been in Gloucestershire for some hundreds of years for raising coarse cloth, the application of the machine to fine cloth in the eighteenth century, while causing alarm, did not lead to widespread disturbances.[41]

Clothiers were quick to respond to a riot in Shepton Mallet in 1776, issuing a declaration to the effect that spinning machines would benefit the cloth industry. A second resolution applauded the 'spirited conduct' of the Shepton clothiers, suggesting that the jenny was 'conceived to be of general utility' to the industry.[42] An argument particularly emphasised by Gloucestershire clothiers and, on occasion, by workmen too was that the introduction of the jenny (and, by inference, other machinery) would not push up parish poor rates.[43] There were mixed reports of the reception of the jenny in Gloucestershire but there was no rioting.[44]

Scribbling, as with other wool textile processes, was a separate trade in Gloucestershire. It had been carried out since the later seventeenth century with the development of medley cloth production, scribblers usually working at the clothier's mill or other premises. A Gloucestershire clothier described scribblers as 'a poor, sickly, decrepit race of beings'.[45] It was stated in 1794 that a man and a boy using a scribbling machine could do the work of twenty hand scribblers.[46] Scribbling engines met with fierce opposition in Wiltshire in 1791 and 1792, riots occurred in Somerset in 1794 but there were only minor actions in Gloucestershire, the scribblers choosing, in 1794, to petition Parliament for limits to be set on the adoption of the machine in order that there be sufficient work for all.[47] No further direct action was taken by Gloucestershire scribblers. Further disturbance in Wiltshire in 1795 required troops, once more. Why was there less protest in Gloucestershire? Randall draws attention to the response to any given machine varying widely, notwithstanding the identical effect on labour, the reaction to machinery being not only economically determined, but also socially.[48] The authority exerted by gentleman clothiers differed between the south-western counties. In Gloucestershire fewer magistrates came from the clothing parishes and they were not drawn from amongst the clothiers, whereas in Wiltshire clothiers sat on the bench and the clothworkers, probably rightly, doubted a fair hearing. This may explain why the Gloucestershire scribblers took the lawful route of petitioning Parliament rather than rioting. The failure of the scribblers' petition seems to have ended resistance to wool preparation machinery in Gloucestershire. By 1803, in addition to scribbling machines at clothiers' mills, there were some public scribbling mills for smaller clothiers in Gloucestershire, as in Yorkshire.[49]

Gloucestershire textile workers did riot in times of high food prices, as in 1766 and 1795 when wheat crops failed. But there is no evidence of accompanying industrial conflict. The presence of the military, whilst considered necessary, was deliberately kept low-key.[50] When, in 1797, there was a shortage of circulating coinage and this coincided with a depression in the broadcloth trade, it was a fear of riots caused by clothiers' inability to pay their workers, rather than anti-machinery riots, which concerned the masters.[51]

The gig mill and shearing frame triggered violent industrial action in Wiltshire and Somerset.[52] It was certainly Gloucestershire's relative peace over cloth dressing and the Wiltshire clothiers' need to have cloths dressed in order to remain competitive, which served to endorse Gloucestershire's pre-eminence in the fine cloth trade.

Gloucestershire shearmen, unlike their Wiltshire and Yorkshire counterparts who resorted to violence, took the legal route. Having long accepted that the gig mill was permitted by law for coarse cloths, they sought to legitimise this and prohibit its use for fine cloth. The lawyer who acted for the scribblers was engaged to fight their case.⁵³ An association was formed – its membership was later said to number 500 – and subscriptions raised. The clothiers met and, allegedly, blacklisted shearmen who were members, claiming that an illegal 'combination' had been established. The shearmen responded alleging that they sought a hearing and relief only in ways 'consistent with the law'.⁵⁴ The petition was presented too late in the Parliamentary session for any action to be taken and no further attempts were made by Gloucestershire shearmen to seek prohibition of the use of the gig mill in the fine cloth trade. The legal and peaceful methods adopted by Gloucestershire shearmen enabled the relatively peaceful introduction of shearing frames from 1799 onwards. There was the odd skirmish, as occurred in Paul & Nathaniel Wathen's Woodchester business, when cloth belonging to them was cut down from their racks. The Gloucestershire clothiers met to discuss the matter and issued a conciliatory statement acknowledging their desire not to cause harm to workpeople. One man was subsequently caught, tried and hanged for the crime.⁵⁵ By 1802/3, it was claimed, the shearing frame had put many Gloucestershire shearmen out of work. Paul Wathen received a threatening letter in 1802 but it was not accompanied by violence:⁵⁶

> Wee hear in Formed that you got Shear in mee sheens and if you Don't Pull them Down in a Forght Nights Time Wee will pull them Down for you ... And Bee four Almighty God we will pull down all the Mills that heave Heany Shearing me Shjens in.

The flying shuttle was designed for the narrow loom, which explains its earlier adoption in Yorkshire than the West of England. Its introduction to Gloucestershire reflected the earlier pattern of the adoption of the scribbling engine, deputations being sent to the Stonehouse clothier who adopted it in 1793, rather than the weavers resorting to violence. Clothiers, however, were sufficiently anxious to meet and discuss the general situation in the industry – and they received a delegation from the weavers. The clothiers agreed to sell the shuttles to weavers whose anxieties, whilst not entirely removed, were greatly lessened.

It was the Gloucestershire handloom weavers who took up the cudgels on behalf of craftsmen in the woollen industry, seeking Parliamentary endorsement of the old regulatory statutes which in times of good trade were ignored, but in trade downturns were a means of drawing attention to technological innovations that threatened their livelihoods. While, by 1793, Gloucestershire weavers had accepted the spring loom, it was largely used by independent narrow weavers rather than weavers of broadcloth. However, around this time some clothiers in and around Dursley and Uley began to install looms in purpose-built shops for weaving cassimeres. Technically, if located beyond the bounds of a city, borough, market or corporate town, the clothier in question was breaking the Act of 2 & 3 Philip and Mary c.11, which forbade them to own more than one loom or to let out looms for hire, and in June 1792 a mob of around fifty weavers threatened a weaver accused of taking work below the price paid to out-weavers. While forty-four were subsequently charged with riot, those found guilty received light sentences.⁵⁷ It was this incident, coupled with the 1793 trade depression, which led to the establishment of the Gloucestershire Society of Broad and Narrow Cloth Weavers, following a meeting in Stroud in November 1793.⁵⁸ There were some disturbances in the middle of November, troops being sent to restore order, but by the end of the month calm was restored. The peace of 1802 brought a cessation in the need for military uniforms and with it a severe reduction in demand for Gloucestershire cloth. Many clothworkers were thrown on the parish – the overseers' accounts for Bisley peaking from 1801 to mid-1802.⁵⁹ Broadcloth demand revived in 1803 but the pre-1800 prosperity for outworkers was not regained as a weaver stated in 1806: 'I knew many before ... [the peace] who could have good beer in their houses and a sack of flour, who cannot have anything of the kind now.'⁶⁰

While there were undoubtedly other issues on which the weavers had grievances, it was the employment of 'illegal' workmen – those who had not served a full apprenticeship – as well as the spread of loom shops, on which they clashed with the clothiers. While in good times, such as 1798, non-apprenticed workers counteracted labour deficits, it was during downturns that journeyman weavers, in particular, suffered. One hundred and fifty men, probably mainly employed in making cheaper army and navy cloth, were served notices threatening prosecution in 1802 and these were, allegedly, only the better-known offenders.[61] It was claimed that around 100 did give up weaving. As in earlier clashes, Gloucestershire weavers resorted to legal means rather than riot, basing their case upon the Elizabethan statute prohibiting those who had not served an apprenticeship from weaving cloth (5 Elizabeth c.4 and 2 & 3 Philip and Mary c.11) and limiting the number of looms which could be kept by one person. Legal action was begun by an association of weavers at Bisley who resisted the employment of non-apprenticed weavers and the establishment of loom shops by clothiers in Chalford. A solicitor was hired to prosecute a Chalford clothier named Webb and the case was settled out of court, Webb undertaking not to employ any illegal weavers (he subsequently broke the agreement). The initial success of this action led directly to the establishment of the (Gloucestershire) Woollen Cloth Weavers' Society. The weavers sought to ensure that their organisation was legally constituted, modelling it on benefit societies.

This did not have the desired effect of pacifying the clothiers who, in October 1802, commenced the actions which were to lead, in 1809, to the repeal of the old regulatory statutes. In the years from 1802 onwards the clothiers' organisation became an efficient lobbying body. John Wall, George Austin, Richard Cooper and Daniel Lloyd reached a draft agreement on the desired revision of the woollen industry statutes and a committee of twenty-one manufacturers was formed, seventy-eight manufacturers signing the agreement.[62] In order to defray expenses, the clothiers levied a subscription, placing themselves in three categories according to the extent of their trade. Eleven firms with the most extensive business subscribed £20, the twenty-one in the second group £12 and the forty-six in the third group paid £6 each.[63] The joint committee of the clothiers of Gloucestershire, Wiltshire and Somerset usually met at Bath and Edward Sheppard frequently took the chair.[64] The clothiers claimed that, if the statute concerning apprenticeship were to be enforced, much of the Gloucestershire trade would be lost 'as a very great proportion of the weavers at present employed have not been apprenticed'. Sheppard reported that he employed seventy-five weavers, of whom only five had served apprenticeships, and that there was no surplus of competent weavers, claiming that 'the coarse woollen trade has already been in a great degree lost ... principally for want of weavers to carry on that branch of manufacture'.[65] Thomas King, a smaller clothier of Stroud, supported apprenticeship on the grounds that the introduction of the spring loom had caused over-production and apprenticeship would reduce the number of weavers.[66] The clothiers established a London base at Henderson's Hotel for a month in February 1803 in order to lobby more effectively. The weavers had, by 1803, come round to the idea that some alteration in the statutes was necessary, but they remained adamant in their opposition to loom factories.[67] Thomas Plomer, speaking on behalf of the clothiers in 1804 suggested that there would be no demand for cloth permitted to be made under the various statutes and to prohibit practices would be 'to prohibit altogether carrying on the trade'.[68] Equally vehement arguments were made on behalf of the clothworkers, the proceedings having an unintended lighter side when Randall Jackson described a brigade of Russian soldiers who, newly clothed in English cloth, were exposed to a drenching shower, whereupon their overcoats shrank to waistcoats.[69]

But first, in 1803, following an inquiry into the woollen industry, the statutes were suspended. A minor strike in Wotton in 1802, from which the Stroud weavers disassociated themselves, further hardened the positions of both sides; estimates of the number of weavers in the society ranged from between 1,200 and 1,300 to 3,000. A flurry of petition and counter-petition from both sides ensued and the House of Commons set up a large Select Committee to hear the evidence from all sides (excepting Yorkshire clothiers). The Committee recommended repeal which was rejected by the Lords. The Suspension Act was renewed every year until 1809, the clothiers growing despondent both as to the time and cost involved. A new Select Committee was established in 1806 and finally, in 1809, after three further Suspension Acts, the old statutes were repealed. Selective memory was clearly invoked by the witness who, in 1833, said 'during the burnings and destruction of property going on in other districts, here was no attempt of the sort, neither as regards the destruction of the machinery of the manufacturers, nor the property of the agriculturalist'[70] – although the damage was little compared with Wiltshire.

The high food prices of 1812 prompted weavers (allegedly of the whole region) to petition employers for an increase in their rates of pay, which they claimed had not been increased for twenty-five years, despite the soaring cost of living. They claimed that they deprecated the notion of combinations, riots, or 'tumultuously turning out from their labour';[71] whether there was any concerted response from the clothiers is unknown but no strikes ensued. When Luddites were burning some Yorkshire mills in 1812, the only sign of tension in Gloucestershire was the sending of a letter by one 'E. Lud' to John Lewis of Brimscombe threatening to burn down his mills if the mill workers, particularly shearmen, were not better paid. The letter was written by a 15-year-old boy, possibly at the instigation of older men, and he was transported for fifteen years as a consequence.[72] No violence followed. A small disturbance was reported at King's Stanley in 1819, but riots were 'frequent' in Wiltshire.[73]

There were strikes in Wiltshire and Somerset in 1822 but nothing serious occurred in Gloucestershire until 1825. In that year, despite trade being good, the weavers struck for a rise in pay rates to compensate for the harder and longer work required in weaving finely spun yarn and the weaving of more weft into the piece. It was a well-organised strike, each weaver in the region being required to surrender his shuttles and 'in about forty-eight hours all the shuttles were laid in the silent grave', the membership of the Stroud Valley Weavers' Union increasing in a few days from 400 to 5,000.[74] The weavers downed tools on 29 April and a large number from Wotton and Dursley met on Stinchcombe and Breakheart Hills; the Stroudwater weavers congregated on Selsley Hill on 2 May to hear whether the manufacturers had acceded to their demands. A Nibley magistrate recorded that Timothy Exell (who gained the title King of the Weavers in the strike) 'and three or four others of the same rank in life possess superior ability and indeed enviable talents for the management of a large assembly'. While several of the employers agreed to the prices asked, the weavers considered that too few clothiers agreed and they declined to return to work, the *Gloucester Journal* commenting that the weavers' conduct had been orderly and respectful.[75] Towards the end of the month relations became strained. On 31 May a weaver who had taken a chain from Ham Mill, agreeing to weave it under price, was 'abused' by around 200 weavers who told him to return the chain. Two days later around 500 weavers went to the weaver's house and took away the chain, harness, slay, and beam and proceeded to Ham Mill where they threw one of the mill employees into the brook and attempted to throw another in but he escaped to the counting house. On 3 June around 200 men went to Vatch Mill and some of them held sticks over the proprietors' heads, threatening to knock their brains out and destroy the mill. The following day Peter Wyatt, went to issue a complaint at the magistrates' office and was assaulted again. That same day around 3,000 men gathered at Vatch Mill and returned various unfinished pieces of cloth which they had taken from the weavers and the mob in Stroud seized several men and ducked them in Holbrow's fish pond. On 6 June several weavers were fetched from their houses and thrown into the brook at Nailsworth. More weavers were ducked in Holbrow's pond and on 7 June 'many acts of violence' were committed.[76]

In an endeavour to restore order, twenty-nine special constables, many of them clothiers, were sworn in and the Secretary of State was requested to allow a squadron of the military to go to Stroud. A squadron of 10th Hussars arrived in the town on 6 June and were stationed between Bliss and Tayloe's Mills in

Chalford, the horses being stabled by the Coffee tavern and Company's Arms. A number of the rioters were committed to Gloucester gaol.[77] Edward Sheppard of Uley agreed to pay the weavers two-thirds of the rate that they asked for, until circumstances changed. Most of the other employers in and around Uley followed suit. But the Stroudwater employers were indignant and William Playne retorted that these rates could not be paid by manufacturers under contract to the East India Company. This was not unreasonable since the weavers' demand had not made any acknowledgement of different types of cloth made in the region. There were riots in Stroud and Chalford in June and striking continued into the autumn with an outburst of violence at Wotton-under-Edge. One November night a party of weavers, led by Timothy Exell and an ex-soldier, marched to seize the partly woven cloth and beam of a weaver who had agreed to accept a lower price. They went on to Neal's Britannia Mill where stones were thrown and the Neals fired on the crowd with shotguns, wounding twelve men and one woman. Thirty-one weavers and the Neal brothers were sent for trial in Gloucester. The Neals were acquitted, two weavers were sentenced to three years' imprisonment and ten others for lesser terms. Timothy Exell had made himself scarce and was not caught.[78] Violence waned with the onset of depression in 1826, the weavers waiting until conditions improved.

Out of a total population of around 6,000 in Bisley, only 658 people were in full work in 1826 (the average earnings of adults and children together being 7s per week) and 460 in half-work; '2,026 [are] totally out of employment who are able and willing to work'. When asked what prospects there were for employment in the future the reply was 'nothing is in prospect but the Harvest … From the ruin of the trade the manufacturers property is so much reduced in value [that] several Clothing Mills [are] quite shut up'.[79]

The situation was as serious in the south, a Benevolent Committee in Wotton reporting on the distressing destitution there in December 1826.

In 1829 the weavers' organisation sought to attract membership from the middling classes and other tradespeople. The objective was to strike against one clothier at a time and the first Gloucestershire master to be attacked was a man called Haigh, who was forced to sign an agreement covering hours and wages at his factory, promising to abolish shop looms and to pay only in cash (rather than truck). This may have been the strike, referred to ten years later, which lasted nine weeks, after which spinning mules were installed and forty-one spinners dismissed.[80] Some magistrates met without the clothiers and, being persuaded that the sole purpose of the union was to abolish truck, which they agreed was an infamous practice, saw little harm in the organisation. The Stroud magistrates were nearly all clothiers and were sufficiently alarmed to attract the attention of the Home Office which sent a spy, one Francis Fagin, a Bow Street runner, to inveigle himself into one of the weavers' lodges and report back. He observed a theatrical ceremony of induction for new members involving swords, masks, turbans and skulls and an oath of loyalty and secrecy to the union. Timothy Exell and other leaders were quietly warned of the danger of transgressing the Conspiracy Act (under which the Tolpuddle Martyrs had been transported) and were warned to abandon secrecy and oath taking.[81] It was clear that the majority of union members desired to abolish shop handlooms and to prevent the adoption of power looms, but there was anxiety amongst some members and the Stroud Club's rules were revised to confine its objects to the suppression of truck and for payment of wages in cash.

No legal enforcement of measures to prevent truck were introduced and smaller and less successful manufacturers continued to pay in kind. Nathaniel Jones of Chalford, who was making stripes for China ten years later, told the Assistant Handloom Weavers Commissioners that he saw no future in the trade without truck. The minor depression of 1834 prompted some strikes against individual factory owners, one being against Edward Sheppard who attempted to reduce prices. Sixty to seventy weavers and other workers struck for six weeks, Sheppard estimating his loss to be in the region of £100,000. Action against William

Playne at Longfords and Playne & Smith at Dunkirk Mill was triggered by the price for weaving stripes, but there was another motive too, namely to raise the prices paid for other goods to the level of those paid by Stephens at Stanley Mill; 400 weavers and 200 other workers struck for seven weeks at Longfords.[82] The final settlement was a considerable decrease over that which had been paid in 1825.

The factory inspector Howell summed up the situation: 'I have been informed … that the manufacturers are diminishing … whilst they are increased in the north.' Mechanisation had provided the means by which fewer mills could fulfil the demand for broadcloth. To keep the former woollen industry workforce in employment would have required a major increase in demand for high-cost superfine cloth, or diversification into new products, which Gloucestershire manufacturers were slow to consider.

Endnotes

1. Katrina Honeyman, *Child Workers in England, 1780–1820*, Ashgate, 2007, pp.91–111; the wool textile industry is not directly addressed but the detailed use of parish and workhouse records shows all pauper children apprenticed to factories being sent to cotton mills, with a few to the much smaller number of silk and worsted mills
2. Arkwright used horses first at his Nottingham mill but subsequently used water power and his water frame licences per 1,000 spindles ensured that most early cotton mills on Arkwright's principle were remarkably similar in scale. Jennifer Tann, 'Richard Arkwright and Technology', *History*, 58, 1973, pp.29–44
3. *Bath Chronicle*, 25 February 1773, quoted by J. de L. Mann, *The Cloth Industry in the West of England from 1640 to 1880*, 1971, p.55
4. S. Rudder, *A New History of Gloucestershire*, 1779, p.289
5. GJ, 5 October 1772
6. GJ, 30 January 1775
7. GJ, 4 April 1774
8. The Case of the Stroudwater Navigation, JX14.21gs
9. GJ, 12, 26 January 1784; Mann, p.56
10. S. Mills, 'The Rise and Fall of Henry Hicks, Clothier of Eastington', *GSIA Journal*, 2002, pp.19–28
11. R.G. Wilson in N.B. Harte & K.G. Ponting (eds), *Textile History and Economic History: Essays in Honour of Miss Julia de Lacy Mann*, 1973, p.240
12. Stroud Museum MSS
13. D1759
14. D.T. Jenkins & K.G. Ponting, *The British Wool Textile Industry, 1770–1914*, 1987, pp.70–1
15. Mann, p.153, quoting GJ, 13, 20, 27 April, 4 May 1812
16. JF 13.27 fol 119gs
17. Mann, p.158
18. Only five businesses are listed for Painswick yet there were approximately thirty mills at work in the parish in the early 1820s
19. Mann, pp.183–4; R. Perry, *Wotton under Edge: Times Past – Time Present*, 1986, p.105
20. D1241 Box 68 bdl 3
21. A.T. Playne & L. Long, *A History of Playne of Longfords Mills*, 1959, p.19
22. Mann, p.168
23. PP 1828, H. of Lords CCXXXVI, pp.78, 314, 580, 582, 583, 593
24. Asst HLW Comms 1840, V, p.362
25. Suppl. Rept Fact. Comm. 1834, ii, B.1.40

26 Asst HLW Comms 1840, V, p.367
27 GJ, 6 March 1830
28 D873 B
29 Playne, p.151
30 1st Rept Fact. Comm. 1833, B1, 52; ibid., XX, p.950
31 Probably connected with the raising of the American tariff at the end of 1828
32 Mann, p.173
33 1st Rept Fact. Comm. 1833, E.10
34 RQ.246.1gs
35 Adrian Randall, *Before the Luddites*, pp.54–5
36 Ibid., p.55
37 PP 1802–03, 7, pp.8, 13
38 Randall, pp.62–3; Partridge, *A Practical Treatise on Dyeing*, pp.82–3
39 Randall, p.65
40 Stroudwater Textile Trust MSS
41 The gig mill may well have been used in the fine trade earlier, although no distinction is usually made between a mosing mill and a gig mill in property deeds
42 GJ, 16 September 1776
43 GJ, 14 October 1776
44 Mann, p.129, citing *Stroud Journal*; Partridge, p.71
45 PP 1802/3, 7, p.308
46 JHC 49, p.600, cited Randall, p.78
47 JHC 49, pp.599–600
48 Randall, p.86
49 Mann, pp.129–30
50 Randall, p.104
51 Ibid., p.97
52 K.H. Rogers, *Wiltshire Woollen Mills*, Pasold, pp.173, 175
53 Randall, p.124
54 GJ, 19 May 1794
55 GJ, 27 January; 4, 18 August 1800
56 D.M. Hunter, *The West of England Woollen Industry under Protection and Free Trade*, 1910, p.21
57 RV 319-6gs
58 GJ, 11 November 1793
59 Randall, p.198
60 PP 1806, 3, 338
61 JF 13.9(1)gs
62 JF 13.6 gs
63 JF 13.27 gs
64 JF 13.25 gs
65 PP 1802–3, V, p.245
66 PP 1802–3, VII, p.8
67 JF 13.27 (i) gs; JF13.9 (2) gs
68 JF 13.24 p.9 gs
69 11787 gs; HC 12.3
70 PP 1833, XX, p.943

71 GJ, 29 June 1812, cited in Mann, p.232
72 GJ, 29 July 1812
73 Mann, p.232
74 J. Loosley, *The Stroudwater Riots of 1825*, Stroud Museum Assoc., 1993, p.7
75 GJ, 16 May 1825, cited Loosley, p.19
76 Loosley, pp.15–16
77 D4693/14, cited by Loosley, p.17
78 Mann, p.236
79 Loosley, pp.57, 60
80 Asst HLWC 1840, II, p.459; Mann
81 R. Perry, *Wotton under Edge: Times Past – Time Present*, p.107
82 PP 1840, XXIV, p.473

CHAPTER EIGHT

CONCENTRATION AND DECLINE 1836–80

INNOVATION: PRODUCT DIVERSIFICATION; MANUFACTURING TECHNOLOGY; POWER

From the earliest years of the Age of Machinery the larger Gloucestershire clothiers had taken a lead in adopting the inventions of others, and of experimenting with different makes of machine. The Austin brothers attributed their success in the Russian market at the beginning of the nineteenth century to 'machinery; the machines have been the means of ameliorating the cloth and keeping down the price'.[1] This readiness, on the part of some, to innovate promoted the development of machinery manufacture in the area.

Innovation in wool preparation and finishing technologies – the processes for which there had been a number of Gloucestershire inventors and innovators earlier in the nineteenth century – continued in the mid-nineteenth century and a stream of patented inventions came out of the Phoenix Ironworks, first under John Ferrabee and later by his son James. In 1830 John Ferrabee patented a machine for preparing the pile of a length of cloth (no.6058). In 1838 William Lewis, clothier, and John Ferrabee patented certain improvements in the dressing and finishing of cloth (no.7584) in which 'the dressing cylinder and the cloth [move] in the same direction continuously and regularly delivering the cloth to the cylinder instead of allowing it to be dragged through or between tension bars or rollers, or from having a break [sic] attached to it'. In 1857 James Ferrabee and Charles Whitmore (engineer and mechanic of Stroud) patented improvements in carding, scribbling and condensing; James Ferrabee patented further improvements to machinery for carding, scribbling and condensing wool and other fibrous substances in 1858 and an improved feed for carding in the following year. Improvements in fulling machines were made in 1861 and in 1869 for machinery for felting.

Carding was the key to the production of good yarn. While, in principle, mechanised carding was the same process in the mid-nineteenth as in the late eighteenth century, there were important developments in practice. What emerged was a series of rollers which fed each other in an integrated process. After carding the slubbings were hand-pieced to produce a continuous slubbing ready for spinning. Several types of intermediate feed were tried, the Apperly being invented in Stroud; in the 1870s Cook Vick were local textile machinery manufacturers making Apperly and Clissold machines under licence.

It was at the end of the carding machine that the most important developments took place. The doffing comb knocked off pieces of slubbing ready for the billy boy to join them on the slubbing billy. A number of different machines were developed for piecing the slubbings but none was completely successful and none was widely adopted.[2] A solution, the condenser, was developed in the USA, and although British manufacturers were slow to adopt it, condensers certainly existed in the West of England by the 1850s where it was reported that 'a system of carding is carried on almost identical with that of our foreign competitors'.[3]

There were, however, few developments in either spinning or weaving technologies in the West. Until the mule became self-acting using Richard Roberts' 1820s improvement (designed for cotton), the carriage had to be returned and the yarn wound onto cops by the operative. It is highly likely that the mules in Gloucestershire in the 1830s and 1840s were not self-acting but were what were called 'hand mules'. Nevertheless, they had many more spindles than the largest jenny and were, therefore, more productive. The self-acting mule, while having been adopted by some of the more advanced firms in Yorkshire by the 1850s, was adopted more slowly in Gloucestershire and other West of England woollen manufacturing counties, although there was one at Cam Mill by 1867.[4]

Power looms were first employed in the Yorkshire worsted industry and adopted by leading Yorkshire woollen manufacturers in the late 1820s and early 1830s. It was a much slower machine in woollen weaving than worsted, making forty-two to forty-eight picks per minute in broadcloth, compared with 160 in worsted. The main difficulty was the width of the material, being 108in before milling. It was this relative slowness, little different from a handloom, which deterred many manufacturers from making the large capital investment in buildings and machinery. Four power looms were recorded in Gloucestershire in 1835, the factory inspector remarking that they had been introduced experimentally and 'if it succeeds many more will be used'.[5] William Hunt of Dyehouse Mill in the Nailsworth Valley installed power looms in 1836; he was followed by Charles Stanton of Stafford Mill who purchased four looms from Rochdale, and Samuel Marling of Ham Mill who purchased several later that year.[6] Stanton told the commissioners that he had installed the looms because they were so prevalent in the north and 'there is certainty, regularity and quick return in favour of power looms'.[7] Several manufacturers considered the power loom to be of doubtful benefit. N.S. Marling wondered whether the estimated two days' saving in the labour of weaving a cloth was offset by the requirement for better wool and more careful preparation of the chain, lest it snap during weaving; Peter Playne estimated there to be a marginal difference in favour of the power loom: 'men are racing against power and any disturbance would cause the general introduction of power looms'. There was more widespread adoption during the later 1830s, 101 being recorded in Gloucestershire in 1838. In all, they provided work for eighteen men, seventy-two women and eleven children, the latter chiefly as quillers.[8] John Ferrabee and Richard Clyburn patented certain improvements in power looms in 1836 (no.6986) but no record has been found of their adoption. Of the forty-three Gloucestershire manufacturers who completed factory returns for 1840, thirty had installed looms in their factories, in all some 1,054 looms. But the majority of these were handlooms. There were fifty handlooms in Dunkirk and Iron mills, sixty at Lodgemore and Stafford mills, and fifty in Hooper's Eastington mills. Stroudwater was the only area in which power looms had been installed, there being none in mills in the Little Avon or Ewelme valleys (see Map 8). Bonds Mill and Wallbridge Mill both had a small number of power looms and no handlooms, while Ham Mill had more than 50 per cent of its loom capacity as power looms – forty-five power looms and twenty-nine handlooms.[9] Stanton had increased his power looms to twenty-eight and William Hunt to eight. In all the other mills of which there is record, power looms comprised less than half of their loom capacity. However, two years later, the factory inspector reported: 'the power loom is gradually extending the factory system by superseding the old handloom weavers and collecting within the walls of the factory the power looms and their attendants.'[10] It is noteworthy that the mills which survived the longest in the lower Frome Valley – those in Stonehouse, Eastington and King's Stanley – had not adopted power looms by 1840, although there were handlooms in the factories which implies that being an early adopter of this particular innovation was perceived to carry high risk. The relative scale of factory-based

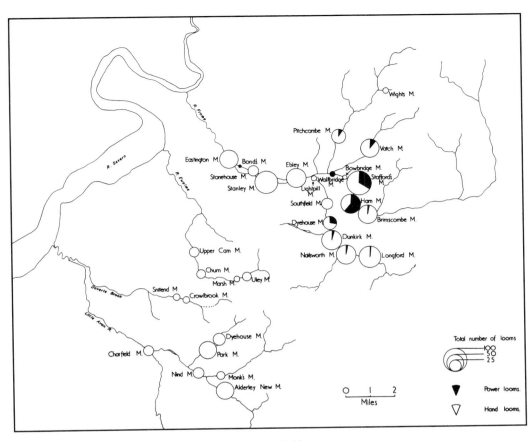

MAP 8: Distribution of mills in which looms had been installed by 1840.

operations in the lower manufacturing area is also clear. Where factories were still (or largely) dependent upon water power which was irregular, the perceived benefits of power loom weaving were less for, if a factory power loom stood idle for want of water power, a handloom weaver could work.[11] The factory inspector noted in 1841 that 'nearly twenty years have passed since the application of the power loom to wool, and though constantly extending, it is still much less employed for that purpose than the hand loom.'[12] However, power looms had reached the lower region by 1844, Lower Mills, Dursley, containing 'six capital power looms',[13] and by 1850 there were 224 power looms in Gloucestershire.[14] It was not until the capital and operating costs of steam power fell in the 1850s and 1860s that power looms were more widely adopted and, as Table 7 shows overleaf, it was from the 1860s that power loom adoption increased markedly. There are no separate figures for Gloucestershire power looms between 1870 and 1904, the figures being subsumed in the category 'Southwest'. The highest point was 1878 when 3,042 were recorded, the subsequent diminution (2,630 in 1885) reflecting the decline in the number of mills. Technical improvements enabled power looms to work at greater speed from the 1870s, the number of picks per minute increased and new fast looms were introduced in the 1880s.[15]

Table 7 – Spindles and power looms in Gloucestershire, 1850–70

Year	Spindles	Power looms
1850	61,896	224
1856	63,256	421
1861	59,986	618
1867	65,094	1,231
1870	48,241	878

Source: Mann, pp.220–1; Jenkins & Ponting, The British Wool Textile Industry, 1770–1914, p.112

Fulling in stocks was gradually replaced by the rotary milling machine which was developed in Wiltshire by John Dyer and patented by him in 1834. The cloth, with its ends sewn together, passed through a mouthpiece, between rollers and then between boards. The squeezing action produced the desired degree of felting more quickly and in a more controlled way.[16] In 1847 a Hunslet (Yorkshire) firm began advertising in Gloucestershire and John Ferrabee was manufacturing milling machines from the 1850s. Fulling by stocks continued, however, into the early twentieth century in Gloucestershire, stocks and milling machines sometimes being used successively in the felting process.

In the early nineteenth century cloth-drying houses, sometimes called air stoves or tenter houses, were built to protect drying cloth from the vagaries of the weather, substituting for drying out-of-doors on racks or tenters. Sometimes they were heated by steam pipes. During the latter part of the nineteenth century the process, at least for larger manufacturers, was accelerated by the tentering machine in which cloth was fixed to a travelling lattice by hooks and the whole placed in a steam-heated chamber. A favourable report on the machine in the 1862 Exhibition drew attention to its labour-saving potential – one man and a boy being able to dry 2,500 yards of cloth a day, regardless of the weather, whereas it previously took eight to ten people to accomplish the equivalent by older methods. Wool-drying machines, in which wool was conveyed on a moving surface through a hot chamber, were also available from the 1850s but the considerable amount of steam heat required deterred many manufacturers from adopting them – they continued to use the old circular stoves heated by open grates or buildings heated by steam pipes.

Peter Playne, whose business was centred on Dunkirk Mill, calculated the water horsepower of his four mills in 1848:

Dunkirk Mill	28hp
Egypt Mill	12hp
Horsley Mill	6hp
Inchbrook Mill	5hp[17]

And, in the same year, he estimated the power requirements of the machinery used at Dunkirk, from which it can be seen that the finishing processes of milling and raising made by far the largest demands on power and that water power alone could not provide this:

Willies	3hp
Washers	3hp
Roughing	11hp
Scribbling	8hp

Milling   20hp
Cutting   4hp
Spinning   2hp

Figures given for power utilisation in Gloucestershire between 1850 and 1871 show a slow but steady increase in steam and a decrease in the use of water power (see Table 8), the proportion of steam exceeding water power by 1861 and emphatically so by 1871. As earlier in the nineteenth century, comparison with Wiltshire and Somerset is misleading, for the Wiltshire industry in particular was poorly supplied with water power. Nonetheless, Gloucestershire's better water power endowment may have encouraged some manufacturers to delay adopting steam power and, as the demand for power increased with larger and faster machinery, the delay could have proved critical.

Table 8 – Power utilisation in the West of England woollen industry

Date	Gloucestershire Steam hp	Gloucestershire Water hp	Wiltshire/Somerset Steam hp	Wiltshire/Somerset Water hp
1850	806	1,485	850	487
1861	1,079	1,045	1,264	430
1871	1,022	591	1,136	148

Note: 1850 figures are for nominal hp; 1861 and 1871 are for indicated hp.
Source: Jenkins & Ponting, The British Wool Textile Industry, 1770–1914, p.122

CAPACITY AND CONCENTRATION

There were seasonal and cyclical trade fluctuations in the Gloucestershire woollen industry in the period 1835–80, as in earlier periods, but with the significant difference that they were accompanied by enduring structural changes characterised by concentration and eventual decline. During these years the woollen industry in Yorkshire and Scotland expanded. Two sets of figures can be set alongside the official factory returns for the decade 1831–41: those of 1839 compiled by Miles, the commissioner investigating the situation of the handloom weavers, and those compiled by Samuel Stephens Marling in connection with the anti-Corn Law meeting of 1842. These are shown in Table 9, the 1838 figure being adjusted (plus 5) for Kingswood which was, at the time, in Wiltshire. The reduction in the number of mills over the decade is clear.

Table 9 – Estimates of woollen mills at work in Gloucestershire, 1831–41

1831	1835	1836	(June) 1838	(March) 1839	1841
133*	118+	118+	101 + 5+	(79)	77*

Key:
* Marling's figures
+ Factory returns
( ) Miles's figures
Source: Mann, The Cloth Industry in the West of England from 1640 to 1880, 1971, p.180

Two hundred and five woollen mills were recorded in the three West of England woollen counties in 1835, the figure for Yorkshire being 406, with 90 in Scotland. Output of woollen cloth rose in the West between 1834 and 1836; American demand for British cloth boomed in 1836 and there was a great deal of speculation. But in the following year exports fell and in 1838 poor harvests led to a fall in home demand. With the East India Company's monopoly of the China trade being broken in 1833, orders to Chalford manufacturers suffered badly and by 1839 only fifteen of its former forty-one mills survived as woollen factories.[18]

The depression of 1837–41 probably accounted for the majority of failures. The stripe market was dead and Yorkshire competition was keenly felt. There were thirteen failures in Yorkshire in 1839 and seventeen in 1840 but the situation in Gloucestershire was far worse with fifty manufacturers failing, the number in business falling from about eighty to forty-two in the period 1832–42.[19] Ebley Mill changed hands several times between 1827 and 1840, the Clissold family having sold out to a London-based firm of textile merchants. John Figgins Marling was at Ebley by 1834 but by 1840 he had got into financial trouble and immigrated to America, the mills being taken over by two other members of the Marling family, Thomas and Samuel Stephens.[20] The most well-known bankrupts were Hicks of Eastington in 1835 and Edward Sheppard of Uley in 1837. These clothiers took too much capital out of the trade in bad, as well as good, times to support conspicuous levels of consumption; Sheppard lived in some splendour in his large clothier's mansion, The Ridge, at Uley, and may have been struggling financially for some time. Sir Paul Baghot (formerly Wathen) of Lypiatt Park failed three times: in 1821 as a banker, in 1826 as a merchant and in 1837 as a clothier.[21] However, what was referred to as 'large establishments and expensive habits of living'[22] should not be over-played; many clothiers who failed had smaller businesses and modest standards of living. Others, such as the Playnes and Marlings, built up their firms by employing good business practices and living within their means. Some manufacturers came to agreements with their creditors and yet others retired before all their capital was lost. If S.S. Marling's estimate of fifty business failures between 1831 and 1841 is correct, many must have taken the former route.[23] Production fell in 1837 and 1838, a small increase earlier in 1839 being the calm before the serious downturn later in the year when the factory inspector reported that he had never known the area to be more depressed. By the following year working hours had been reduced due to the depressed state of the woollen industry.[24]

Table 10 – Mills, employment and power 1838

County	Mills	People employed	Steam hp	Steam percentage of total	Water hp available
Gloucestershire	101+5	5,740	873.5	34	1,720.5
Wiltshire	48-5	2,993	688	68	320
Somerset	30	2,133	260	41	372
West Riding of Yorkshire	543	26,180	7,492	78	2,067

Source: Return of Mills and Factories, year ending 30 June 1838, cited in Mann, p.186

There are discrepancies within the factory returns and between these and the figures presented by the Assistant Handloom Weavers Commissioner, Miles, in 1840, most of which data had been collected in 1838. Whichever Gloucestershire figure is the more accurate, the trend is clear, and what can be in little doubt is the comparison between the West and Yorkshire. With 543 mills and 78 per cent of the power requirements being met by steam, the Yorkshire industry was growing at the expense of the West.

The continuing concentration of the woollen industry in Stroudwater is clear: fourteen mills in Stroud, twelve in Minchinhampton and nine each in Horsley (including Nailsworth) and Painswick. Nevertheless, the mill at Berkeley was still in operation. The utilisation of steam power had increased to almost 50 per cent of the total power available, the single engine in Stanley Mill generating 50hp; the average engine size in Stroud and Cam was around 25hp, and in Eastington 22.6hp. The average engine size was smaller in the tributary valleys such as Painswick. The significance of young employees is clear at this date, comprising something of the order of 50 per cent – more in some parishes – and notably more than the number of adults in Alderley, Horsley and Avening.

A directory of 1840 provides what is likely to have been an under-representation of woollen manufacturers in Gloucestershire. Nevertheless, the woollen industry was still in evidence in Uley, Dursley and Cam on the Ewelme, as well as in Wotton-under-Edge, North Nibley, Kingswood and Charfield on the Little Avon and its tributaries (Map 9). In Stroudwater four mills were recorded in the Painswick Valley, while the middle Frome and the Nailsworth valleys still had many working mills. The puzzlement of Gloucestershire contemporaries was expressed by Reuben Hill who observed in 1840: 'It appears to be a strange fact that the masters are breaking and the men are in rags, yet there is as much cloth

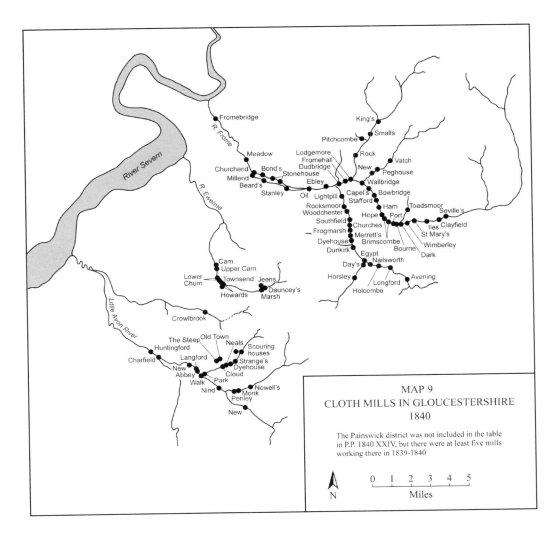

MAP 9
CLOTH MILLS IN GLOUCESTERSHIRE
1840

The Painswick district was not included in the table in P.P. 1840 XXIV, but there were at least five mills working there in 1839-1840

made as ever.'[25] The depression continued for Yorkshire woollens until the end of 1843 when the tide turned in both the North and the West, although the market for fine woollens was slower to recover. T. & S. Marling sent goods to Canton in 1841 and 1842, and increased demand in the China market was reported for cloth made in Bisley (probably Chalford) and Nailsworth.[26]

The years between 1836 and 1848 were ones in which William Playne & Co.'s profits at Longfords Mill declined.[27] Their order book for the period 1836–49 shows a number of small to middling orders for the home market; gone were the days when most of the cloth was dispatched to London merchants. Instead, the Duke of Beaufort ordered 'I end wool dyed scarlet 15/- or 16/-'; James Price of Fairford ordered 'I black double mill'd not too blue'. Cloth was sent to Manchester and Liverpool tailors via Pickfords Boat or Deacon's Waggon to Doncaster or 'steam via Hull to Leeds'. Cloth samples were sent to regular customers. Overseas customers were sent larger quantities either as ordered, such as the East India Company which ordered eighty-five ends of scarlet stripe in 1837, or sent on commission in the hope of a sale. In 1836 a Mr North was an intermediary regarding cloth that had been dispatched and here the letter harks back to the days when Blackwell Hall factors held the whip hand:

> we do not wish positively to limit our friends Messrs FRH to any particular price but this statement will serve to shew them whether or not they can dispose of them profitably for us. We would certainly prefer that they should hold them for a time for the chance of a better market rather than sell them under the cost price.[28]

The Gloucestershire industry was at a low ebb in 1842. Forty-seven manufacturers were named in a directory of that year, although there were several notable omissions, including Charles Hooper at Eastington and Yeats of Monks Mill, Alderley. By this date the industry had lost many of the great clothing family names which had been represented in the petition for the repeal of restrictive statutes in 1802. At least twenty-four of the eighty-three signatories had failed, plus many other woollen manufacturers, besides associated businesses such as dyeing and wool stapling.

Trade began to improve in 1843; the railway was eventually opened through the Golden Valley, with a station at Stroud in 1845, and new factory buildings and extensions were being erected. The process of industrial concentration continued with smaller businesses ceasing and the larger both increasing their output and the number of workpeople employed. Gloucestershire was less badly affected than Wiltshire and Somerset in the depression of 1847–48 caused by the collapse of railway mania and the high price of grain (although profits were down). A change in fashion in 1848, however, introduced what A.T. Playne called 'the palmy days' when mills worked to capacity producing broadcloth. Demand was so great that William Playne & Co. frequently sold their entire output to large wholesale houses in London, Manchester and Scotland as soon as it was finished.[29] This new demand for broadcloth lasted approximately forty years, proving to be a two-edged sword by cushioning industry leaders from paying sufficient attention to possibilities of future fashion changes. A directory of 1849, which named manufacturers rather than premises, identified two cloth manufacturers along the Ewelme at Cam and Dursley (Townsend). Crowlbrook was the only remaining cloth mill in North Nibley, while twelve mills remained in and around Wotton-under-Edge and Kingswood. There was further concentration in Stroudwater, the Painswick, Slad and middle/upper Frome valleys being particularly affected; mills in the middle to lower Frome and Nailsworth valleys, having weathered earlier storms, benefited from the demand for broadcloth. The list appears to have been an underestimate, however, since the factory returns for 1850 record (but do not name) eighty mills and factories, although some manufacturers operated more than one mill, and the factory inspector noted the prosperity in the area. But, while 147 mills were recorded altogether in Gloucestershire, Wiltshire and Somerset (Gloucestershire having more than twice as many as either of the other two counties), the woollen industry in Yorkshire had expanded to 880 mills.[30]

Despite intense trade fluctuations and falling profits, one woollen cloth manufacturer owned or leased a number of mills, presumably as an intended investment for the longer term. Nathaniel Samuel Marling either co-owned, owned, or leased (and sublet) nineteen mills in the mid-nineteenth century, some for a few years, others for longer. They ranged from large mills, such as Stanley, Lodgemore, Fromehall, Ham, Port, Griffin and Vatch, to small mills such as Freames and Pitts in the Nailsworth Valley, and Peghouse, Upper and Lower Doreys in the Painswick Valley; Steanbridge and the Jenny Mill in the Slad Valley. While some Painswick clothiers owned or leased two or three mills, none invested on anything like the scale of Marling and no other member of the Marling family was similarly inclined.[31]

By 1856 the number of woollen mills in Gloucestershire had fallen dramatically, sixty-four mills being recorded in the factory returns. There was a further fall in Gloucestershire to forty-nine (as well as falls in the other western cloth-making counties) by 1861 and by 1870 there were 956 mills in Yorkshire and over 200 in Scotland, while the Gloucestershire figure had fallen to twenty-eight (see Table 11).

Table 11 – Number of woollen mills and employment in Gloucestershire 1850–70

Year	No. of Mills	Persons Employed
1850	80	6,043
1856	64	5,409
1861	49	4,687
1867	62	6,368
1870	28	3,848

Source: Factory returns, cited in Mann, pp.220–1

Surviving Gloucestershire manufacturers made significant additions to their mills and factories between 1850 and 1861, besides adopting further new machinery, resulting in a decrease in employment of 22 per cent, although the average number of employees per mill rose from seventy-five to ninety-five.[32]

At the Great Exhibition of 1851 seven Gloucestershire woollen manufacturers exhibited. The *Stroud Journal* commented on the lack of space which prevened 'the number or quantity of specimens originally intended being sent, or the proper display of materials exhibited'. The displays were said to be representative of the firms' usual goods, rather than being 'got up expressly for the occasion'.[33] One shoddy manufacturer, M. Grist of Capel's Mill, was 'the only exhibitor of this manufacture' – some ten years previously a shoddy manufacturer exhibiting would have been unthinkable.[34] Shoddy enabled some manufacturers to remain competitive into the twentieth century. The jury believed that the Stroudwater manufacturers had misunderstood the object of the Exhibition, which was to show variety of production, and had instead only exhibited cloth at the top of the range. This begs the question as to what extent the exhibiting manufacturers actually produced cloth across a range of weights and qualities. It was later suggested that the whole woollen cloth production of Gloucestershire at this time amounted to £900,000 a year, a figure which, it has been suggested, was some £300,000 less by value than the production of the prosperous year of 1836.[35] Notwithstanding continuing mill closures, Gloucestershire, together with the much contracted industry of Wiltshire and Somerset, maintained its position for the manufacture of the highest quality fabrics, the factory inspector noting in 1853 that 'During the last few years progress has been … energetic'.

Gloucestershire manufacturers seem to have comprehended the role of international exhibitions rather better by 1862 and eight woollen manufacturers were awarded medals in Class 21, Wool and Worsted, and for the first time, a shoddy manufacturer (Grist) was commended:

O. Bird, Southfield, Woodchester
R.S. Davis & Co., Stonehouse
C. Hooper & Co., Eastington
Hunt & Co., Stroud
Hunt & Winterbotham, Cam
Marling, Strachan & Co., Ebley/Stanley
Roberts, Jowlings, Lightpill
Stanton & Sons, Stroud

In the Machinery Manufacturing Class 2 Gloucestershire manufacturers were awarded medals:

J. Apperley & Co., Dudbridge, feeding apparatus for carding machines. (They also exhibited their condenser for which they had patented four improvements between 1856 and 1858.)
J. Ferrabee & Co., Thrupp, apparatus for forming balls of fleece and for fulling cloth. Ferrabee, in addition, received Honourable Mention for his direct action vertical steam engine.

However, the 1862 jurors noted that the value of Gloucestershire's production had increased by only £100,000 since 1851, on account of 'the universal taste of late years being for fancy and undressed goods, a class which Stroud does not produce'.[36] Despite the fact that 'meltons', a softer fabric without a glossy finish, had first been made in the West, Gloucestershire manufacturers were slow to consider the possibility of changes in fashion and none exhibited 'meltons' at the 1862 Exhibition, although Marlings of Stanley Mills did in the Paris Exhibition of 1867.[37]

LODGEMORE & FROMEHALL MILLS IN 1862

Hunt & Co. of Lodgemore and Fromehall Mills, Stroud, received the accolade of a three-day visit in 1862, followed by a detailed description of the firm's business, in Henry Mayhew's *The Shops and Companies of London, and the Trades and Manufactories of Great Britain* published in the Exhibition year. The detailed account of manufacturing processes (besides the appearance of operatives) provides the best description of the manufacture of superfine cloth in the mid-nineteenth century. The medal awarded to Hunt & Co. was the justification for selecting these mills 'as the highest representative house connected with this branch of industry'.[38] The manufacture of a piece of superfine black or blue cloth at Lodgemore Mill took five months from sorting the wool to packing the finished cloth, the two mills together producing 180 ends, or almost 6,500 yards, of cloth each week.

The reporter eulogised about the setting: 'one of the prettiest and richest strips of meadow land, perhaps, in all England.' Between the two mills were the 'lake-like reservoirs, with the snow-white swans sailing proudly on their surface; so that, were it not for the many-windowed buildings and the tall, minaret-like chimneys rising above the roofs, the works themselves, skirted as they are with spacious lawn-like tenter grounds, would have more of the aspect of a park than a factory'.[39] Dark-coloured cloths were made at Lodgemore, the visitor noting that one part of the reservoirs was dyed black and another bright indigo. Fromehall Mill made the more delicate-coloured cloths – scarlet superfines for officers

Lodgemore & Fromehall Mills; an engraving from the 1862 Mayhew publication, *A Visit to Hunt & Co's Lodgemore and Frome Hall Mills*. (Museum in the Park, Stroud)

and hunters, besides orange, yellow, salmon, crimson and light blue. The separation of the processes was deliberate, for the delicacy of the colours was such that the smallest amount of dark-coloured wool finding its way into the light-coloured cloths would spoil them.

The Lodgemore wool-sorting room was a 'long, light, airy apartment' fitted with sorting boards against the windows. Some sorters were 'older women' who wore huge blue-and-white spotted aprons which covered their clothes, with a limp cotton headgear resembling sun bonnets. At other windows there were male sorters 'in slouched hats and huge white smocks, that look like night-gowns'. At one end of the room was a small boy in a very short coarse white smock whose task was to beat a bundle of fleeces on the floor with a bent wooden stock in readiness for the sorter. The women undertook the task of 'linting', picking out bits of twine, straw, oats and other bits and pieces in readiness for the sorters. The men's task was to separate the different kinds of wool from the fleece on a sieve-like grating.

The wool was sorted into six grades ranging from Electoral to 'Filthy'. As one of the sorters said, 'We're always a-learning … there aren't scarcely two flocks of sheep alike, and what's more, scarcely two manufacturers what has their wool sorted after the same manner.'[40]

After sorting, some wool was sent to the dyer to be dyed blue or black. On being returned to the mill it was taken to the picking room. Here, women in the same garb as in the sorting room stood in front of picking hurdles 'with their fingers and faces as blue as if they were in the last stage of cholera', pulling out the small bits of thread and twine, while boys in sack-cloth jackets 'all be-smudged with indigo, and with fingers and faces the colour of "blue mould" attended the blowing machines and the twilley'. The whole atmosphere was 'hazy … with the clouds of blue dust for ever flying about the place'.[41]

A continuous oiling process sprinkled 'a fine dew' over the wool which was spread on an endless travelling table, from which the fibres were automatically flung out some 20ft across the room. Scribbling, condensing and spinning took place in a designated room, the 'din of working machinery stuns the ears, and causes a sense of confusion in the brain'. The spinning frames 'keep moving in and out, as if by their own free will, like enormous harpsichords fitted with some hundreds of the finest strings'.[42] The workpeople's clothing was different here, the men wearing short jacket-like smocks and square paper caps, while the girls 'affect the elegance of hair-nets', coupled with the spotted bib-and-tucker aprons. The wool was passed through a scribbling engine to a condenser from which the slubbing was produced. And so on to a bobbin ready for spinning. The slubbing bobbins were carried by girls to the mules, each of which was attended by two girls wearing spotted pinafores and hairnets. Each mule had 250 spindles, warp yarn being spun more tightly twisted than the abb.[43]

The spun yarn was then taken to the wool loft to be weighed and from thence to the warping shop where the yarn was wound around the warping bars working 'like so many monster whirligig cages'. The warps were taken to the sizing shed to be immersed in a glue solution to give them sufficient strength and then dried. When dry, the chains were taken out to a strip of ground by the mill pond and laid over a series of hurdles and stretched 'almost to cracking point' by ropes attached to a windlass, when at their highest tension 'they appear not unlike the mass of fibres of some gigantic fiddle bow'.[44]

The stretched warps were taken to the weaving shop to be wound on to the beam and for the different threads to be passed through the harnesses that lifted the threads at different throws of the shuttle, depending on the nature of the fabric. The weaving shop was 'enormous … as large, indeed as two ordinary music halls'. Seventy-six power looms were at work: 'never was such a clatter heard in the world … one felt as if the drum of the ear would really split under the pressure of the noise.' Some seventy women and ten men minded the looms and were paid by the yard and made from 12s to 15s per week.[45]

The raw cloth picking room, to which the woven cloth was taken, was a well-lit workshop. Cloth was drawn over perches and the women sat in a row with their backs to the window, dragging the cloth down and picking out the various knots, slack threads, warp ends and broken shoots found in the fabric with a pair of tweezers.

The cloth was then taken to the washing house where the stones were 'all sloppy with the wet work continually going on'. There was a 'peculiar and not particularly agreeable odour' about the place. In the middle of the shed were two great undershot waterwheels 'as big as the paddles of the Great Eastern'. Two tanks were sunk into the ground at either end of the shed, above which two large wooden rollers revolved. Cloth was passed through the rollers and sprinkled with a mixture of sieved seg and pigs' dung, the source of the 'odour'. When the entire cloth had been wetted it was run in the trough for a couple of hours, then removed and thrown upon the stone floor where it remained for 12 or more hours, being frequently turned. It was then returned to the trough and the same seg/dung liquid and run for another 2 hours. Then the cloth was watered to cleanse the fabric: 'we were assured that nothing but the seg and dung would touch the cloth in this state … the object being to cleanse rather than thicken the fabric.'[46] The cloth was then spun in a centrifugal machine and taken to the stove to be thoroughly dried.

The next process was undertaken in the burling shop. The burling room at Lodgemore Mills was 'supposed to be the finest in England', being 120ft long and 27ft wide. 'It is cool, and at the same time light [sic] to perfection.' The object of burling was to detect any spills or bits of straw or lint left from the picking process, as well as any knots or anything which 'would serve to destroy its finish'. The cloth was drawn over perches, as in the picking room, to be examined by the female burlers.

In the fulling house the cloth was reduced by felting and shrinking such that a broadcloth of around 62 yards by 3 yards was reduced to 49 yards by 1¾ yards. Narrow goods of 62 yards in length and 42½in width were shrunk to 53 yards by 28in. Fulling was undertaken in 'boxes', the cloth being put in dry, and as it was carried up and down between the rollers it was sprinkled with a mixture of fine melted soap and water. As the cloth passed into the trough it was put under pressure from a small wooden wheel.

Soap not only cleansed the cloth but kept it soft in the milling process. The fulling process took from 20 to 30 hours depending on the nature of the cloth. Then the cloth was measured to see if sufficient milling had been done. It was, however, rare for a cloth to be returned to the mill 'for the fuller knows his work too well to make such a mistake'.[47]

After fulling, the cloth was taken to the gig or dressing house where the surface was raised. The largest teazles, called 'kings', were used for doeskins. The 'queens' were also used for doeskins, while the third size, 'middlings', were used for broadcloths or felts. Two hundred packs of teazles were used a year, each pack containing some 20,000 teazles and costing around £4 10s. Each handle had to be turned after half an hour's work as they became rapidly coated with fluff, and after another half an hour they all had to be removed for the flock to be brushed out of them. Then they were put to dry for 12 hours and were ready to be used again. The gig mills at Lodgemore had some seventy-two handles on the outer surface of the barrel containing 1,440 teazles. The damp cloth was wound on to a cylinder at the base of the machine (the 'bottom wind') and wound around a second cylinder, the 'top wind', above the teazle barrel, being stitched into a continuous piece. The gig barrel rotated at 160 times per minute, taking around 5 minutes for the cloth to be wound from one cylinder to the other. The process was then reversed. Altogether, it took between 20 and 30 hours for the cloth to be dressed, depending on its quality – fine broadcloth, cassimeres and doeskins requiring the most raising. Then the cloth was placed on a trestle to drip, after which it was taken to the stove house to be dried in folds on tenterhooks in a heat of around 120 degrees, 'the atmosphere of the place being like that of a Turkish bath'.[48]

When dry, the cloth was carried to the brushing & cutting shop. First the nap was raised by brushing in a contrary direction to that in which it had been worked on the gig mill. The length of cloth was continuously brushed for around an hour. Both the brushing machines and the twenty-eight cutting machines were operated by men and boys. The cloth was carried to the cutting machine for the back of the cloth to be cut, after which it was returned for brushing to have the fibres brushed up the wrong way on the right side, following which it was returned to the cutting machine for the face of the cloth to be cut. The finest cloths were cut from forty to fifty times, being steamed and brushed between each cutting, 'and even the commonest cloths made at the mills of Messrs Hunt & Co are thus treated at least half that number of times'.[49] The cloth was then taken to the passing or examining shop where it was measured and minutely examined for faults or tears. 'Such are the honourable principles upon which these works are carried on', that women stitched coloured silk by a fault or tear so as to identify it.

Having been measured and passed, the cloth was once more scoured with finely strained fuller's earth and water in a washer, then allowed to drip before being mosed on the dressing gig, the teazles having only been used once previously and therefore had a fine sharp hook. The cloth was taken to the stove-loft and then to the cutting shop to be brushed and cut and re-examined before being taken to the press shop for hot-pressing.

From the hot press the cloth was taken to the scalding or steaming house, wound upon rollers and covered with a piece of fine linen cloth 'till it has the appearance of an enormous roly-poly pudding' with the ends tied for boiling. It is then plunged into the 'dippers' – large iron tanks – in which the water was nearly 190 degrees and 'cooked' for around 5 hours. On being removed the cloth was left to cool, taken again to the washers for the fuller's earth and water procedure, then rinsed in cold water and mosed. All black and blue cloths were treated like this for eight successive times and following the final scalding they were once more scoured in earth and water. White and grey doeskins were scalded and mosed five or six times and 'black singles' six to seven. Then they were sent to the stove rooms to be tentered and dried. They were taken once more to the cutting shop to be steamed, brushed and cut, and finally re-examined for burrs and lint. They were scrutinised against light in a darkened room to check for any small holes, hot-pressed once more, brushed one last time and then cold-pressed before being wrapped for dispatch.

'When finished the sleekness and skin-like texture of the fabrics produced at the mills ... seem well to repay all the trouble and care bestowed upon them; for we saw cloths there, compared with which the satin-like coat of the most thorough-bred racer would have been coarse and harsh to the touch.'[50]

All the cloth made at Fromehall Mill was woven and dressed 'perfectly white'. The manufacturing processes up to the first roughing stage were the same as at Lodgemore, but after the first roughing and cutting, the cloth was sent to the dyehouse to have any stains removed 'by some secret process' and subsequently dyed. The key to excellence lay in making the sorted wool 'as white as mountain snow' by bleaching. Then, after being scalded and rolled a sufficient number of times the cloth was sent to the dyehouse. Scarlets were dyed twice; the first time to be 'grounded', after which they were returned to the mills to be 'linted' or examined for imperfections. Then they were returned to the dyehouse for final colouring. Billiard cloths, liveries and all fancy cloths were similarly treated.

At the warehouse the visitors were shown 'the finest products of these Mills, which are allowed to be among the finest in the world'; as the sheets of scarlet were spread out their eyes 'positively ached with the glare', the cloths feeling 'as soft and homogeneous in their texture as a baby's skin'. The visitors, having spent three days at the mills, bowed their heads in respect 'to the great manufacturing artists, like Messrs Hunt & Co., who think it worth their while to devote some five months' continued labour to the production of a single piece of *perfect* broadcloth'.[51]

ORGANISATION OF THE GLOUCESTERSHIRE WOOLLEN INDUSTRY 1860s–1880

In 1865 there was a short-lived attempt to run Stanley and Ebley mills separately – the architect G.F. Bodley's chateau-like addition to Ebley dates from that year. Strachan had retired in 1862 and the business had become Marling & Co. once more, Samuel Stephens Marling running Ebley and William Henry Marling running Stanley. But in 1869 the businesses were reunited under Marling ownership until 1920.

1867–68 were depressed years. A directory of 1867 recorded forty-four woollen manufacturers. There had been further concentration in all three areas, only Cam Mill being left on the Ewelme, Crowlbrook Mill at North Nibley and Monk Mill in Alderley (although this was empty for a year after the lease was advertised in 1868, at which point the machinery was removed to a mill in Stonehouse). Strange's and Park Mill at Wotton were still operating, while in Stroudwater two mills were recorded in Painswick, and three in the Slad Valley. A number of mills in the Nailsworth Valley were no longer recorded (Nailsworth, Holcombe, Dyehouse, Merretts and Rooksmoor), while Hope Mill in the middle Frome Valley was no longer mentioned. While this list probably omits some manufacturers, it is indicative of the continuing process of concentration. As in earlier times, the outbreak of the Franco-Prussian war in 1870 favoured the Gloucestershire woollen industry with orders for military cloth, although markets in France and Germany were disrupted; home demand was buoyant following good harvests.

Although the 'palmy days' were over for good by the later 1870s, some firms with established reputations continued, by further investment, to make profits into the late 1880s. However, many mills were mortgaged, some for a very long time, and this could serve to obscure a firm's financial state of affairs. William Barnard, a partner in Hunt & Barnard of Lodgemore Mills, admitted to having been on the edge of bankruptcy for years before it finally happened in 1860.[52] He appears to have engaged in some dubious financial practices involving using his Lodgemore partnership to prop up his partnership with T. Sampson of Ham Mill, without his partner Hunt's knowledge. Park Mills, Kingswood, was for sale in 1867, as was Millend Mill, Eastington, just before Charles Hooper's death in 1869, although his son continued the business at their other mills. And in 1868 Apperleys put Dudbridge Lower Mill, recently

Monks Mill, North Nibley; a nineteenth-century photograph. (Jennifer Tann)

built and filled with modern machinery, up for auction after they had been unable to find a tenant. By the 1870s the site of Hawker's Dyeworks was in the hands of the Marling family who leased the knapping mill and dyehouses to Adolphos Charles Smith, dyer. A new high-pressure steam engine had been installed to pump water to the indigo houses.[53]

How much capital was required for a new entrant to the industry is difficult to conjecture. And there were new entrants. John Libby, a partner in a cloth factor business in Stroud, commenced as a cloth manufacturer at New Mills, Stroud, and was required to invest not less than £2,000 in the business, although he already owned the mill. J.G. Strachan, formerly a successful cloth factor, left his partnership with N.S. Marling at Ebley Mill in 1861 (although the partnership still had some years left to run), buying Hunt out at Lodgemore and Fromehall mills for £30,000. Marling, Strachan & Co. of Stanley Mills had a capital of over £160,000 before Strachan left the partnership and, by 1869 when Stanley and Ebley mills were united, the capital was £176,000. Vatch Mill, New Mill and Ham Mill, all near Stroud, as well as Cam Mill were either acquired by new owners or the existing manufacturers found new partners.

There was some revival in 1871 after the end of the Franco-Prussian war and 'commercial prosperity and industrial activity' was noted in Stroud,[54] although James Ferrabee noted that 'fine cloth is gradually declining', predicting that 'another thirty years will probably see it confined to two or three large manufacturers or altogether vanished from the neighbourhood'.[55] Superfine broadcloth manufacturers continued to win more medals at exhibitions than those from other woollen manufacturing areas; Marling's black and blue broadcloth was the best one writer at the 1867 Paris Exhibition had ever seen, 'holding a middle place between the extreme glossiness of the French and Belgian cloths and

Ebley Mill, a mid-nineteenth-century painting; an older clothier's residence dominated by the New Mill. (Museum in the Park, Stroud)

Ebley Mill courtyard. (Roger Tann)

the dullness of the Saxon'. The reporter, however, noted with some surprise that there was far less that was new or remarkable than had been expected in the English section.[56] West of England cloths were characterised as being superior to others and *The Times* commented that there was nothing to compare for finesses, softness and firmness of texture, with the rich lustre of the best Stroud blacks and blues.[57] In the Paris Exhibition of 1878 four Gloucestershire firms exhibited, Hooper's elastic cloth and a diagonal coating woven under a new patent being reported 'by far the most remarkable in the place'.[58] However, firms seemed unable to respond creatively to changes in the marketplace, of which there had been many clear signals: 'The productions of the West of England rank second to none in the world, though few but Englishmen can afford to wear them.' When the demand for fine cloth in America was transferred from the West of England to Germany, 'it was a matter rather more of price and finish than quality and make'.[59] The American tariff was not prohibitive until 1880, but Gloucestershire manufacturers were slow to acknowledge and act upon changes in demand. There were, by then, many who would rather have had a fashionable new coat or suit every two years than one which lasted for ten.

Gloucestershire was included in the West Midlands' factories return for 1874, but since the other counties included in that category had so little cloth manufacture at the time, the figures probably refer almost wholly to Gloucestershire. If this is the case, there had been a considerable increase in spindles in the county, 4,978 being recorded, compared with 3,957 in 1870.[60] Decline commenced at the end of 1875, although even four years later the number of spindles had not fallen dramatically. From the late 1870s to 1880 there was a rise in the number of mills on the market, including some belonging to larger, well-established manufacturers: Southfield and Churches mills in 1875, Woodchester in 1875, Vatch Mill in 1877, Nailsworth Mill in 1878, while Charfield Mill was let to elastic cloth manufacturers.[61] In all, six mills closed between 1870 and 1874 and three more between 1874 and 1879. Trade was said to be 'in a very depressed state' in 1878,[62] and by 1879 Rowland Smith of Stonehouse had drawn attention to Yorkshire's supremacy in lighter fancy coatings made of a woollen/worsted mix which cost 25 per cent less than West of England broadcloth,[63] while Scottish competition in fancy tweeds had gained market share.[64]

The unwillingness of Gloucestershire firms to learn from the evidence of market changes is clear. The large rise in demand for woollens was at the cheap end of the market, while the middle classes were far more influenced by changes in fashion than earlier in the nineteenth century. If manufacturers had altered their machinery to accommodate different market needs, might they have lost the reputation for high-quality fabrics upon which their businesses had grown? And, perhaps, the difficulties of making machinery alterations have been underplayed. The rise of the ready-made clothing industry, particularly in Leeds and its environs, brought fashionable suitings within the pockets of the less well-off. Stroud developed a ready-made clothing industry but it was not supplied with locally made fabrics.

The Scottish tweed industry flourished, fashion dictating that any well-dressed man should have a Scottish

Toadsmoor Mills, a complex site involving two mills, united in the nineteenth century. (Mike Mills)

tweed suit. Gloucestershire firms did begin to make tweeds but they were too dull and too late. Scottish firms experimented with a variety of wools, buying the finest of the Australian clips when required. And, as demand for both high quality and variety increased, Scottish mills produced yarns and fabrics from cashmere and vicuna. The reason usually given for the West's failure when Scottish firms triumphed was the quality of Scottish design. However, it was not sufficient to attract one or more textile designers from Scotland; the culture of traditional ways of doing things in West of England firms also needed to change. Scotland also developed a successful knitting industry alongside the woven textile trade which, as knitting began to challenge woven fabrics at the turn of the century, enabled Scotland to hold on to a wool textile trade.[65]

LABOUR

As the factory system advanced so the number of persons employed in factories increased. The first available figures are for 1834 when 1,229 males and 777 females worked in Gloucestershire woollen factories.[66] By 1839 the figures had increased to a total of 5,058.[67] The figures for factory-based employees were highest in Stroudwater. Despite the fact that there were factory looms in all the major mills by 1838, there was still a large number of handloom weavers whose situation had become desperate. In 1834 the outdoor weaver earned on average just over 8s per week. The factory weaver's average earnings were 11s 9d although rates varied from mill to mill (14s 1d at Eastington; 13s 5d at Lodgemore; 11s 9d at Dunkirk).[68] Trade improved in 1836 and the factory inspector noted complaints by both manufacturers and parents about the factory legislation which prohibited children between the ages of 11 and 12 from working more than 48 hours per week.[69] The condition of the factory weaver was reported to be 'far above that of the out-door workman' in 1838.[70] With the greater concentration of weavers in mills, cottages containing loom shops fell empty from being too large. Of sixty cottages built in Cam for handloom weavers, around one-third were empty by 1838.[71] The depressed state of trade in 1838, fuelled by high wheat prices and the shortage of money in the previous two years, besides the panic of three large business failures in 1837, led to a reduction in working hours and the undertaking of a major Parliamentary report on the condition of the handloom weavers. The labour market was overstocked with weavers and while, in a good year, few might be unemployed, in a recession the unemployed could be many hundreds: 'the manufacturers take men on and turn them off according to the sudden orders that might come into the district'.[72]

'The out-door weavers are undoubtedly in deep and severe distress,' wrote Miles, the Handloom Weavers Commissioner for Gloucestershire, while the Governor of Horsley Prison noted that weavers were grateful for their daily food and left prison with regret, not knowing where the next meal would come from. 'The only furniture in many [weavers' cottages] being an old stool, a broken table, and a few cracked and broken cups and pans. In one or two cases I found them sitting on the stair steps, having no chair or stool. In others a log of wood is a substitute.'[73] Sir William Marling recollected: 'I always felt very sorry for the handloom weaver as he brought his piece to the wool loft for his earnings which, even in the fifties, I fancy did not average much above 10 shillings a week.'[74] The diet of many weavers consisted of bread crusts, a little fat and water. Many had not tasted meat for years. In Randwick 'starvation threatened the village'.[75] Most weavers were in debt to small shops, 'the consequence of this is ruin to the proprietors, inasmuch as a weaver is not worth suing'.[76] One witness to the Commission estimated that seven-tenths of the weavers were obliged to seek occasional relief from their parish and the number of paupers relieved both in and out of the workhouse increased in nearly all the cloth-making parishes between 1837 and 1838. Minchinhampton parish relieved a total of 2,017 in 1838; Bisley, the worst hit, relieving 3,501.

A further burden many outdoor weavers suffered was the truck system. The passing of the (anti-) Truck Act in 1831 (prohibiting payment of wages in tokens redeemable at the manufacturer's shop) had been greeted with great celebration in Gloucestershire. But it was a short-lived victory. Employees were afraid to complain for fear of being put out of work. By 1838 there was only one money-paying master left in Chalford and it was payment in truck which precipitated the 1839 struggle there. In Chalford, the truck system was open, since it was almost universal, but in Nailsworth, while practised, it was 'carried on with too much subtlety to be evidenced'.

Living under these degrees of hardship, it is not surprising that weavers were prone to embezzle wool or yarn; and the gains so made might be spent in one of the many beer shops. There was one exception to this otherwise almost universal image of manufacturers' failures and workers' distress. This was Stonehouse. The masters were said to be 'benevolent and kind' and each cottage well furnished.

Several measures were proposed for the provision of relief. Some manufacturers attempted to establish benefit societies for their employees but levels of distrust were too high and they came to nothing.[77] Individual initiative was taken in Randwick where the men were set to work on the roads and women in making clothes in return for clothing, furniture and food. Allotments were created in Uley, Horsley and Nailsworth, although in the former parish they were blamed for keeping families in the parish who might otherwise have moved on. Plans were made for enclosure of parts of the common land in Minchinhampton and Bisley.

The most strongly favoured remedy was migration or emigration. Employment was found in Yorkshire for eighteen people from Bisley and for a further sixty-six in a flax factory in Shrewsbury. Emigration took place on a larger scale; seventy-eight left Uley for Canada in 1835 and 200 went to Australia from Horsley, Kingscote, Stanley, Uley, Minchinhampton, Owlpen and Kingswood. In 1838 around forty-five left Minchinhampton for Australia and in 1839 sixty-eight left Bisley for America and Australia. While the immediate effect was a reduction in the levels of poor relief, the villages must have appeared somewhat ghost-like – there were nearly 200 empty houses in Bisley, for instance. Uley was judged to have been transformed: 'A parish once full of factory people is now an agricultural district and happier than ever it was.' But Charles Stanton of Stafford Mill was insightful: 'Locally it might be beneficial; as a general principle it is like bleeding a person, only a temporary expedient.'[78]

The depression of the early 1840s was reflected in the recorded population losses in clothing villages. These communities, which had gained overall from 8 per cent to 9 per cent between 1821 and 1831, lost all the gains by 1841. Uley lost over 900 inhabitants and Wotton-under-Edge 780. The lower cloth-making region lost 16 per cent overall, while Stonehouse and Eastington gained. The population decline in the Ewelme Valley was severe; it declined in almost all the Little Avon parishes and in Stroudwater it was most marked in Bisley and Painswick. Only Stroud, Stonehouse and Eastington were exceptions, the growth being due, in part, to the continuance of the wool textile industry in fewer but larger units, besides industrial diversification. The population of Avening more or less held up between 1831 and 1871 due, in large measure, to Playne's Longfords Mill.

The Gloucestershire woollen industry had suffered severe setbacks in the past but, from the mid-1830s, the pace of decline increased and, notwithstanding the 'palmy days', the demand for superfine high-cost cloth decreased and with it the number of mills. By 1880 the surviving businesses were leaner, more professional and highly mechanised, and were beginning to diversify their product ranges. There were still profits to be made but by fewer firms.

Endnotes

1. PP 1802–03, VII, p.347
2. D.T. Jenkins & K.G. Ponting, *The British Wool Textile Industry, 1770–1914*, 1987, p.104
3. Report on the Paris Universal Exhibition, 1867, iii, 67. Much of it printed in SJ, 12 October 1867
4. SJ, 30 November 1867, cited Mann, p.289
5. PP 1835, XLV, p.153
6. PP 1840, XXIV, pp.396–7
7. Ibid., p.455
8. Ibid., pp.388, 396–7, 456
9. PP 1840, XXIV, pp.396–7, 454
10. PP 1842, XXII, p.345
11. Asst HLW Comms, XXIV, p.553
12. PP 1841, X, p.302
13. RX 115.13 (32) gs
14. PP 1850, XLII, p.455
15. Jenkins & Ponting, pp.113–14
16. K.H. Rogers, *Woollen Industry Processes: II, The Factory Industry*, Trowbridge, 2008, pp.39–40
17. Ian Mackintosh, 'Power and the Playnes', STT typescript
18. J. de L. Mann, *The Cloth Industry in the West of England 1640 to 1880*, p.175; the Company's monopoly of the India trade had been lost when its charter was renewed after 1812
19. GJ, 15 January 1842; quoted in Mann, p.179
20. E.A.L. Moir, 'Marling & Evans, King's Stanley and Ebley Mills, Gloucestershire', *Textile History*, 2, 1971
21. E.A.L. Moir, 'The Gentlemen Clothiers', in H.P.R. Finberg (ed.), *Glos Studies*, p.243; Mann, p.181
22. Asst HLWs Comms 1840, V, p.455
23. Mann, pp.194–5
24. PP 1838, XXVIII, p.85
25. PP 1840, XXIV, p.414
26. GJ, 17 December 1842
27. A.T. Playne & A.L. Long, *A History of Playne of Longfords Mills*, 1959, p.45
28. D4644/3/1
29. A.T. Playne & A.L. Long, p.45
30. Mann, p.220
31. See Part II
32. Mann, p.220
33. SJ, 6 July 1951
34. Ian Mackintosh, 'Shoddy Work, Stroud. Well done!', *Warp & Weft*, series 2, issue 4, 2011/12. In 1871 there was some confusion in the railway company with five separate shoddy businesses all named Grist in Stroudwater
35. Repts by Juries on the Int. Ex. of 1863, class XXI, p.11; Mann, p.198
36. Ibid., p.11
37. Rifleman and Hussar, p.3, cited in Mann, p.204
38. Henry Mayhew, *Lodgemore & Fromehall Mills*, 1862, p.4
39. Ibid.
40. Ibid., p.10
41. Ibid., p.11

42 Ibid., p.13
43 Ibid., p.15
44 Ibid., p.17
45 Ibid., p.18
46 Ibid., p.19
47 Ibid., p.22
48 Ibid., p.24
49 Ibid., pp.25–6
50 Ibid., p.28
51 Ibid., p.32
52 SJ, 25 August 1860
53 R. Perry, *The Woollen Industry in Gloucestershire to 1914*, 2003, pp.135, 144
54 SJ, 30 December 1871
55 SJ, 19 August 1871
56 SJ, 29 June 1867
57 SJ, 5 August 1871
58 *Leeds Mercury*, quoted in SJ, 10 August 1878
59 Ibid.
60 Mann, p.222
61 Ibid., p.212
62 Stroud Gaslight & Coke Co., RQ 293.34gs
63 SJ, 8 February 1879
64 Mann, p.214
65 Jenkins & Ponting, p.175
66 PP 1834, XIX, p.266
67 PP 1939, XLII, p.151
68 PP 1834, XX, pp.324, 403, 458
69 PP 1836, XLV, p.167
70 PP 1840, XXIV, p.377
71 PP 1840, XXIV, p.441
72 PP 1840, XXIV, p.419
73 PP 1840, XXIV, pp.429, 433
74 W. Marling, Trans. BGAS, XXXVI, p.329
75 RQ 246.1gs
76 PP 1840, XXIV, p.444
77 PP 1844, XXIV, pp.421, 446, 452, 454, 495, 526
78 PP 1834, XXIV, pp.546–9

CHAPTER NINE

DECLINE 1880–1914

Concentration, stagnation and decline accelerated from the 1880s onwards. Between 1878 and 1894 six mills closed – New Mills, Wotton, closing in 1898 – and employment in the Gloucestershire woollen industry fell from 4,801 persons in 1881 to 3,898 by 1891. 'Manufacturers who were unable, or unwilling, to meet the new conditions imposed by science and invention have been forced to suspend operations. The result is the survival of the fittest.'[1] Machinery had been speeded up – it was claimed in 1910 that a woollen spinner could spin twice as much yarn as in the 1880s,[2] although the increase in loom speeds was less significant.[3] Nonetheless, in a non-expanding or declining market, fewer manufacturing units were required to sustain production levels. Some, albeit incomplete, evidence of the scaling-up of production facilities at surviving firms is available for 1891[4] (see Table 12).

Table 12 – Capacity of some mills in 1891

Mill	Woollen spindles	Looms
Brimscombe	5,670	144
Cam	4,600	210
Stonehouse		120
Wallbridge		22
Peghouse	2,500	
Lodgemore	5,660	100
Fromehall	2,640	57
Ham	4,590	70

Source: Worralls Textile Directory, 1891

Only P.C. Evans of Brimscombe Mills had introduced worsted spinning.[5] Ritchie & Co. at Ham Mill failed in 1899; and, while Roberts Jowlings & Co. of Lightpill claimed to 'be doing a very wide and flourishing trade' in c.1900, employing 160 hands in making a range of traditional West of England cloth, the firm failed in 1909.[6]

By 1901 only 3,049 people were employed in the industry. A 1900 directory recorded seventeen cloth mills (see Map 10) of which only Cam Mill was identified in the lower region, Nind Mill being omitted. Longfords Mill, Avening, and Lightpill on the lower Nailsworth Valley remained. Southfield and

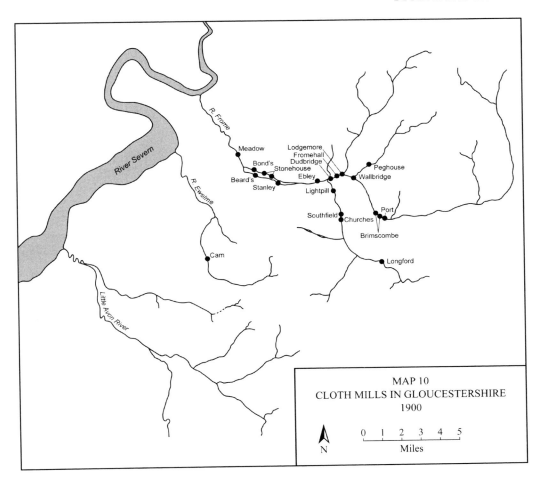

MAP 10
CLOTH MILLS IN GLOUCESTERSHIRE
1900

Churches mills, Woodchester, survived, while Peghouse Mill continued in the Slad Valley; no mills were recorded as surviving in the Painswick Valley. In the Frome Valley the surviving mills were in the lower and middle Frome areas. There were the Eastington enterprises of Beards, Bonds and Meadow mills, Stonehouse Upper and Lower mills, Stanley and Ebley, Lodgemore, Fromehall and Dudbridge mills below Stroud and no surviving mills above Brimscombe. W. Margetson of Strachans remarked in 1906: 'I have been forty years in this business and during that time I have seen forty businesses go to the wall while I have never yet, I think, seen a new business opened.'[7] While business directories often under-represent, they are useful in showing trends, eleven firms occupying around fifteen mill premises as recorded in Kelly's Directory of 1906. By 1910 the figure was eight firms, seven of which were still in business in 1931. Six firms remained in 1935:

Marling & Evans – Stanley and Ebley mills
Millman Hunt & Co. – Wallbridge
Howard & Powell – Wallbridge Mill
Hunt & Winterbotham – Cam
Strachan – Lodgemore and Fromehall mills
Playne & Co. – Longfords Mill[8]

It may have been overly optimistic for Chance and Bland of Gloucester to claim in 1904, 'If the manufacturers in Gloucestershire will keep abreast of the times, there is no reason why they should not at least hold their own',[9] and it was in this spirit that they published *Industrial Gloucestershire* to show that 'in the larger life of today Gloucestershire is still distinguished by the resourcefulness and adaptability of its people'.[10] Eight leading wool textile firms were featured, the businesses being located at Cam, Nind, Dudbridge, Brimscombe, Woodlands (Stroud), Lightpill, Eastington and Stonehouse. It is curious that Playnes of Longfords were not included, for they weathered the twentieth century longer than many.

Of the eight woollen firms featured in *Industrial Gloucestershire*, four mills (Lightpill, Brimscombe, Cam and Stonehouse mills) were using some worsted yarns, but Brimscombe Mills was still the only mill with a worsted spinning facility. Some of the mills were making vicuna and llama fabrics (Lightpill, Cam and Stonehouse), the latter mill also manufacturing waterproof coatings, while P.C. Evans & Co. of Brimscombe Mills had introduced lightweight indigo serges for tropical markets.[11] But, by comparison with the Yorkshire worsted industry where first cotton warps and subsequently exotic fibres such as mohair, alpaca and vicuna were adopted earlier, Gloucestershire woollen manufacturers were slow to respond to new demands and opportunities.[12]

The rise of the knitting industry alongside tweed manufacture in Scotland served as some kind of cushion to woollen manufacturers.[13] It is interesting in this context to note a rather short-lived attempt to replicate this in Gloucestershire, for Dunkirk Mill, Nailsworth, was acquired by W. Walker & Sons Ltd, hosiery manufacturers of Nottinghamshire, in 1891. The mill was completely remodelled, electric lighting installed, a new steam engine bought to replace the old beam engine, and a policy introduced for continuous technical improvement. By this date part of the mill had already been converted to walking stick manufacture by another firm which Walkers took over.[14] It was not very long before stick manufacture proved to be more profitable than hosiery.

Dunkirk Mill, Nailsworth; the mill extended along the valley. It is now housing. (Howard Beard)

The manufacture of elastic fabrics had been introduced by Charles Hooper of Eastington as an innovative product alongside the manufacture of more traditional cloths from the 1860s. And towards the end of the nineteenth century Tubbs Lewis, which had acquired Abbey and Langford mills, Kingswood, and New Mills, Wotton, introduced the manufacture of elastic fabrics at the latter, 'for which looms of special pattern have been devised'. Within a few years they had begun to make shoddy, in addition to a range of other products.[15] This was an important, yet distinct, branch of the textile industry in Gloucestershire by the early twentieth century, in particular for members of the Grist family, the largest enterprise being that at Merretts Mill in the Nailsworth Valley. In c.1900 Hunt & Winterbotham mixed mungo (top quality shoddy) in their products and, during the First World War, discarded soldiers uniforms became an important raw material.[16]

New investment took place in the mills owned by firms which had weathered the difficult times of the 1870s. At Cam, innovation was said to have been 'almost continuous' from 1859, when there had been a single mill building, to 1904 when the company owned a large industrial site. New wool sheds had been built in 1900; dyehouses and weaving sheds between the late 1990s and turn of the century; 120 power looms and a 500hp steam engine had been installed; part of the factory was lit by electricity and a private railway system had been built through the premises.[17] Charles Hooper had abandoned one of the older mills (Millend) in Eastington, the plant being taken to their other mills: Meadow, Churchend, Beard's and Bond's mills, which were lit by both gas and electricity. Hooper rationalised the work at the mills with spinning at Churchend, weaving at Meadow and Bond's, and dyeing at Beard's.[18] 'Scarcely a year has elapsed without witnessing additions to the mechanical equipment.' Amongst the firm's specialities were cricketing and tennis whites which were made in a separate department to avoid being stained. A regular sight in Eastington was 'two women with a basket between them walking from Churchend to the looms of Bonds and Meadow Mills', the gravelled path being wide enough for them to walk side

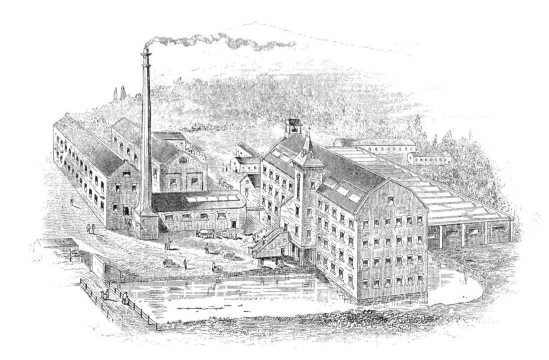

Stonehouse Upper Mill. (*Industrial Gloucestershire*, 1904)

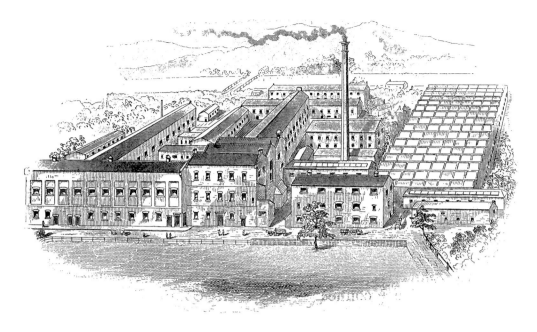

Stonehouse Lower Mill. (*Industrial Gloucestershire*, 1904)

by side.[19] Hooper was known as a good employer and had always depended heavily on the American market for superfine cloth. This may have been one of the reasons for the sudden closure of the mills in 1906.[20] Nind Mill was, by 1904, the only mill left in Wotton-under-Edge. It had become a limited liability company in 1890 and in 1893 took over Samuel Long & Co. of Charfield Mill. The Nind Mill property covered 16 acres and 'although old established, they have been kept thoroughly up-to-date by the addition of new buildings and the installation of the latest machinery', including a new weaving shed erected in 1890. Three waterwheels were still used in 1904 'supplemented' by steam and gas engines, the gas being generated on the premises. Besides apparel cloths the firm made piano, billiard and box cloths and also printers' blanketings.[21]

R.S. Davies & Sons of Stonehouse Upper and Lower mills were operating on a large scale in 1904 – in 1889 there were 5,000 spindles and 150 power looms in the mills. With direct access to the railway, the firm was in a good position for coal supplies, nonetheless the firm closed in 1906. Apperly, Curtis & Co. of Dudbridge claimed that their progress had been rapid following the arrival of Alfred Apperly in 1872.[22] A fire in 1891 had enabled extensive alterations to be made 'in accordance with modern needs and a largely increased trade' and, in 1902, the firm had acquired an adjoining factory in order to extend production. Motive power was provided by two 500hp engines made by King & Co. of Nailsworth, supplemented by a 50hp turbine. Apperlys made worsteds and lighter fancy goods, travelling rugs being another speciality. The firm was commended in *Industrial Gloucestershire* for its technical innovations which 'saves hand labour or facilitates production', and also for the attention paid to the health and well-being of the operatives. Washrooms, supplied with hot water, were provided on every floor and the mills were well ventilated; ovens were provided for dinners to be warmed.[23]

Bowbridge Dyeworks continued to specialise in reds. Ralph Bassett recollected: 'My father used to relate how he was dyeing madder reds "in the Bell" at midnight for the army, during the Crimean War, when news of the victory of the battle of the Alma came through to Stroud.' At the beginning of the twentieth century the ground floor of the mill was used for cochineal, indigo and argol or tarters. The first floor was the cloth room grinding shop and girling room. (The girling process consisted of binding the lists of fine Superfines Refines, Mediums and Venetians to be dyed scarlet, resulting in a broken white edging to the piece which was 'recognised in the trade as a guarantee that the dye was pure cochineal'.) But during the 1914–18 war the dyeing of scarlet virtually ceased.[24]

Several of the larger firms became limited liability companies between the 1880s and 1890s, amongst them Hunt & Winterbotham of Cam in 1887 and Apperly, Curtis & Co. in 1896. The latter company was valued as follows:

	£	s	d
Freehold land and buildings	30,189	0	0
Machinery and plant	18,030	0	0
Stock in trade at cost	48,181	9	6
Book debts (good) + cash	43,732	9	9
Goodwill	19,278	7	11
Total	159,411	7	2[25]

Notwithstanding the large industrial site, there was less fixed capital investment than working capital in the form of raw materials and cloth unsold, together with money due for cloth supplied. Apperly, Curtis & Co. was amongst the first to introduce pensions for employees. Marling & Evans became a limited liability company in 1920.

The cloth manufacturers that survived the post-First World War depression continued in business through a variety of means which usually involved innovation – in product, process, technology or work organisation. One quirky innovation had a domestic origin from the 1870s and contributed to saving one Gloucestershire woollen firm for 100 years, besides contributing to another's prosperity: the tennis ball. In the early 1870s J.M. Heathcote, the MCC's real tennis champion, found that a rubber ball was awkward on a wet court. So he asked his wife to cover one with flannel. This was the beginning of a non-apparel innovation at Playnes' Longfords Mill and one which continues at Lodgemore and Cam mills. By the 1970s well over 14,000 balls were used at Wimbledon alone.[26] Assisted, on occasion, by the Wool Industries Research Association (WIRA), which was founded after the First World War, Playnes sought advice on bleaching cloth for tennis ball coverings and the rubberising of tennis ball cloth.[27] On rubberising WIRA noted that rubber could be applied to yarn or 'with less advantage' to the finished cloth. Playnes were told that rubberising at the yarn stage would not prevent the subsequent milling of cloth but the end result would be different.[28] Experiments were undertaken with rubberised cloth but, by the 1970s, the covering of a tennis ball was melton – a cotton warp with a wool and nylon mix for the weft. This was felted and gigged to achieve a fluffy surface.[29] A further innovation in tennis ball coverings was discovered by chance. When A.L. Long was mill manager, one of his tasks was to inspect and pass cloth at all stages of manufacture. During the cropping of a batch of tennis ball coverings one piece was not cut as much as specified and the customer complained, Mr Long being sharply rebuked for the error. Yet, as he later recounted:

> had we but known, I and those culpable, were twenty years ahead of our time. Today we use strong coarse wool and do not crop the cloth at all – it is only common sense to let the player wear the ball out.[30]

By 1910 it had become clear that the Old Mill at Longfords was not safe for modern heavy and faster-running machinery. Moreover, the workshops, being saturated with oil, were a fire risk. Playnes were faced with the by-no-means-straightforward question of whether to cease manufacture or to rebuild the greater part of the mill and rearrange the machinery. The company decided to invest but was faced with the constraints of an already crowded site in a narrow valley. It was decided to build over the middle pond using ferro-concrete piles which were driven into the bed of the pond and upon which reinforced concrete beams were installed which supported the 160ft by 85ft new loom shed. A new spinning mill was built to house self-acting mules of 420 spindles each, the whole being powered by electricity from a central power plant, electric current going from the dynamo to the motors in the various workshops. A.T. Playne commented that four mules required the attention of one girl to mend any yarn which broke, without the necessity of stopping the machines.[31] The machinery was 'capable of turning out more cloth than at any time during the last 50 years'.[32]

In 1920 a merger established the group to be known as Winterbotham, Strachan & Playne, thus bringing together the Longfords business with those at Cam and Lodgemore/Fromehall mills, the individual mills continuing to operate separately. At the 200th anniversary of Playne & Co. in 1959 the then managing director, A.L. Long, attributed the company's survival to the 'prestige and good will of the Playne family business … [and] the utmost use of their 160 years of traditional skill' allied with 'new and better methods' – no reference being made to the Group.[33]

Cam Mill had been owned by Thomas Hunt of Nailsworth in the mid-nineteenth century but the business lacked capital and Hunt went into partnership with A.B. Winterbotham in 1859, the firm becoming Hunt & Winterbotham. Arthur's two sons, H.B. ('Mr Bertie') and his elder brother Arthur, succeeded him in the business, the firm becoming a limited liability company in 1889 with a capital of £100,000 in £100 shares, 97 per cent of which was held by the Winterbotham family. In 1907 £30,000 worth of debenture shares were offered for sale which contributed to funding new buildings. The brothers were very different characters; apparently their boardroom rows could be heard down the corridor. The mill was divided into two sections, one making traditional woollen cloth which was managed by Bertie, the other making cloth with worsted yarn which was managed by Arthur. Between 1899 and 1904 investment was made in new wool sheds, dyehouses, weaving sheds (containing 120 looms), soft water plant and offices, and by 1904 the premises covered almost 8 acres. It was considered sufficiently noteworthy of remark that water power had been abandoned, power then being supplied by a 500hp engine. Part of the works was lit by electricity and plans were in hand for lighting all the buildings. The firm was noted for its range of West of England cloths, besides billiard cloths 'of which they are the largest manufacturers in the district'.[34] Arthur Winterbotham and one of the employees ('Pa Wise' – father of sixteen children), during an evening's drinking, invented a new cloth which came to be known as Indian Whip:

> The design … defied all known text book standards … [being] of such a size of yarn and … with so many ends … and so many shoots … per inch it appeared physically impossible for them to be contained in a square inch … only a couple of drunks would have attempted it.

It was a 'magnificent cloth which defied all to copy it'. And in the First World War almost every officer was clothed in it. The largest single order was for 5,500 yards from a military tailor in Edinburgh, enough for more than 1,500 uniforms. However, the brothers continued to battle and in 1920 Arthur left the firm.[35]

Bertie Winterbotham had a reputation for admonishing employees and 'terrorised all and sundry', yet he had a shrewd understanding of human nature and a kind heart. The firm owned around seventy cottages for employees which were let at minimal rents and the family supported numerous charities in the district. Bertie Winterbotham was used to taking decisions without consultation and when, in 1915, the Board was informed that materials and a new electric motor had been ordered for the new scouring plant without its approval, he was given the nearest thing to a reprimand, being 'instructed' to make it clear to the suppliers that he expected to be treated fairly.

Excerpts from the Hunt & Winterbotham Board Minutes on various production issues between 1909 and 1915 show that the Board undertook its governance duties with care, and, when thought necessary, sought to rein in the managing director. In 1909 the Board discussed the condition of the steam engine which had been installed in 1881 at a cost of £1,150. Presented with an estimate of £1,250 for a new 350hp engine and the prospect of three weeks' stoppage of work, this was turned down and a modest repair recommended instead. In the following year the building of a new loom shed was authorised by the Board on the site of the Fuller's Earth Works and eight new 100in Knowles and five new Lee & Crabtree 120in box looms were authorised to be ordered. At an earlier Board meeting Cam Mill's loom tuner had been accused by Mr Crabtree of impeding the installation of the Lee & Crabtree looms, previously supplied, by preventing the suppliers' tuner from adequately instructing the weavers in the management of the looms. The Board took this seriously and informed the mill tuner that his services would be dispensed with if he failed effectively to run the new Lee & Crabtree looms.[36]

Bertie Winterbotham was one of the last of the 'Gentlemen Clothiers', living in a large house at Ashmead, Cam, with a staff of about sixteen, including a chauffeur, as well as outdoor staff to manage the grounds and to look after his hunters (for he hunted on two or three days of the week in the season). In May he departed for Scotland for a month's salmon fishing and usually spent some time in Wales in the autumn.[37]

Lodgemore and Fromehall mills seem to have employed a range of colourful characters in the early twentieth century, some of whom, including one particular manager, were more practical than cerebral and, as one former employee wrote, 'how he came to be appointed to such an exalted position was … a mystery for, apart from his knowledge of the ways of the mill, I would guess him to [have had] … a somewhat doubtful IQ'. Indeed, a foreman described him as 'the bist all hell and no notion'. In these mills, too, there were ragings and blusterings but when J.N. Tod, with 'reforming zeal', was appointed things changed.[38]

Bowbridge Dyeworks was part of Strachans and, as a commission dyer renowned for scarlets, it had undertaken dyeing for Hunt & Winterbotham (amongst other firms) before the merger which resulted in the creation of Winterbotham, Strachan & Playne. Cloth from Cam was taken to the Ship Inn at Stonehouse by horse and cart, whence it was collected and returned 'by a similar vehicle from Bowbridge'. There had been little investment in the works for a number of years and:

> at the first sight the whole set up looked like something straight out of Dickens or even earlier … the windows, such as they were, appeared seldom have been cleaned and dim flickering gas lights burnt most of the day. The dye vats had obviously been used for many a day and most were made of stone. The wonder was that from this archaic and dilapidated 'works' there could be produced cloths dyed with a purity of colour and fastness that was a model of its kind … and spoke volumes for the sheer 'craft' of those who wrought the wonder.[39]

The head dyer in the early twentieth century was Ralph Bassett, 'a product of his times', who followed his father and grandfather in the trade. 'There he was, a somewhat lonely soul who seldom saw his boss in this outback of the Stroud Valley and struggling ever with problems that only the sheer necessity of earning his daily bread could have kept him going.' There was a 'dungeon like place where the scarlet spirits were made'. Bassett's grandfather had been sent in to stir the spirits when only 10 years of age. There were ten dye tanks in the main dyehouse (with a date stone of 1775 above the door). The top floor of the mill served as a drying stove, two racks running the entire length of about 60 yards. Bassetts of

Bowbridge used to dye superfines intended for the tunics of the Russian Court in the days of the Tsar, for R. Davis & Sons of Stonehouse Mills.

Dyeing seems to have attracted pranksters. Bowbridge Dyeworks was still bleaching with sulphur in the late nineteenth century and the whitening house was 'right away in the far top corner of the premises' on account of the fumes. However, a lad dropped a handful of flowers of sulphur down the chimney of the dyehouse 'and then watched with great glee the spluttering choking crowd that came tumbling out of the door'. A 'fine new whitening house' was built in the yard opposite the mill in the 1890s, fitted up with tanks, washers and a hydro-extractor, and by the 1930s hydrogen or sodium peroxide were 'entirely superseding the sulphur stoving'. But Bassett remembered 'hanging up wet cloths – a very arm-aching job – and taking them down after a night in the burning sulphur fumes to the accompaniment of much coughing and choking'.

Colour matching was a highly skilled task, for no two people appear to see a particular colour in exactly the same way:

> what to one would appear to be a perfect match to the next man could equally appear too green, blue, too yellow ... so Ralph had not only to know his cloth but also his customers and sometimes his customers' customers ... and make allowance in his judgements accordingly.[40]

The men 'appeared to be a strange, ill assorted lot who tended the dye vats with an air of fatalism and general dejection'.[41] The War Office continued to require scarlet cloth for soldiers' coats to be dyed with cochineal long after other dyes had superseded it and Bowbridge was the last dyeworks to do this work. As with other colours there were different shades of scarlet. The Pytchley Hunt's 'pink' was dyed on a commission basis by Bowbridge for Howard & Powell of Wallbridge. On one occasion the Prince of Wales intimated that he would ride with the Hunt. This resulted in a sudden demand for the Pytchley's hunting pink and Bowbridge quickly dyed another length of cloth. Ralph Bassett decided to dye with a modern substitute rather than cochineal. The day of the hunt turned out to be very wet and 'ere the day was done everyone with a new coat was horrified to discover [that] the cloth bled and white breeches and saddlery were slowly but surely blossoming out in a perfectly pale, pretty, pink'. Moreover, the only grey polo pony, specially selected as the Prince's mount for the day, was dyed pale pink, too. Hunt & Winterbotham had to foot the bill for new coats as well as for cleaning the breeches and saddlery. But the pony had to remain pink until nature intervened with the annual moult.[42]

As in earlier times not all mills closed through the insolvency of their owners. Some mill owners decided that they had had enough and retired (or diversified) before it was too late. Alfred Stanton, whose grandfather William had bought Stafford Mill, Stroud, continued to innovate as his father and grandfather had done, but the mill ceased production as a woollen factory in the 1880s and became a paint factory. Stanton money was invested in a range of new industries occupying former textile mills in the valleys; and, like a number of other well-to-do woollen manufacturers, Alfred Stanton was active in political and social life in Gloucestershire, in due course becoming MP for Stroud. John Libby, a former cloth factor who became a partner in New Mills, Stroud, withdrew in 1898 for a comfortable retirement. Samuel Long of Charfield Mills retired in 1878 lending his son George and a cousin partner £10,000 each as working capital in the business, later giving George his share and leaving £61,000 at his death. George died only eight years after his father and the Charfield factories ceased production at that time.[43] Two former woollen mill-owning men continued their involvement with the woollen industry in other ways and at far less personal risk. Alfred Stanton became one of the trustees for the debenture holders when Apperly & Curtis became a limited liability company in 1896; and William Clissold, a descendant of a clothier family which had been in the woollen industry from the seventeenth century,[44] and had worked with James Apperly in the invention of the feeder for the billy and carder and on improvements in the condenser which replaced them, became a non-executive director of Apperly & Curtis in 1896.[45]

If the continuation of some firms in the early twentieth century and the failure of others was a case of the survival of the fittest, what constituted fitness? Sound leadership and management were essential by the early twentieth century, when the profits to be made on the sale of a single piece of cloth were much lower than formerly. In 1904 John Mackie published a small book entitled *How to Make a Woollen Mill Pay*, in which he discussed 'the necessity for looking after small things which so often escape notice in the busy rush of mill life'.[46] Interestingly, Mackie was from Trowbridge so would have had first-hand knowledge of the West of England woollen industry. The 'small things', rather than those which were sufficiently important to 'command the manufacturer's constant attention', were those which, Mackie claimed, 'too frequently hold the balance between solvency and insolvency'. Competition was keen and getting keener so that each item, however small, needed to be 'turned to account to cheapen the cost of production'.[47] In order to keep track of these 'small items' a sound system of bookkeeping both for the mill and the counting house was needed, for without this the manufacturer could not know whether any particular element of the business was being conducted at a profit or a loss. A firm of accountants, which had a number of clients amongst the woollen manufacturers in the North of England, reported that a large percentage of failures was due to a poor or inadequate system of bookkeeping, for it was 'impossible for the manufacturer to know, or be able to know, whether his business was a paying or a losing one until too late'. While Mackie supposed that double-entry bookkeeping was practised in most counting houses, he claimed that there was no system recognised as best practice for use in a woollen mill.[48]

Mackie recommended a series of departmental books, commencing with Wool Books, into which every bale of wool was to be entered, and concluding with the Make and Cost Book, which itemised how each piece of cloth was made. Altogether some twelve accounts books were recommended.[49] And these did not include each departmental foremen's notebooks.

The author of the notes on Cam and Lodgemore mills and Bowbridge Dyeworks was given the 'fearsome job of trying to build up a common costing system for the Combine'. Each mill had, until then, 'its own peculiar methods which they were reluctant to forego'. The office books at Bowbridge were kept by Ralph Bassett's daughter and it was not long before she was discharged. Shortly afterwards all the Bowbridge staff were transferred to the enlarged dyehouse at Cam. They were 'the rummiest collection of individuals'. A number of the dyehouse employees were men of faith and Biblical quotes would appear on the washers. On one occasion someone had written 'The wages of sin is death', to which a wag had added, 'And those of Cam Mill are not much better either'.[50]

Bowbridge Dyeworks may have appeared somewhat chaotic in the early twentieth century but, notwithstanding years of tacit expertise, they sought and acquired up-to-date knowledge on aniline dyeing processes, besides, on occasion, being consulted by dyestuffs manufacturers. Ralph Bassett approached the Bayer Co. of Manchester in 1913 regarding the dyeing of billiard cloths; the Clayton Aniline Co. Ltd on preferred methods of cochineal dyeing in 1926; and Bassett was consulted by a Huddersfield aniline dye company in 1929 regarding the dyeing of blue. It would appear that L.B. Holliday of Huddersfield, having asked Bassett to test their Fast Chrome Cyanine 2B, queried whether Bassett had used any indigo, to which Bassett firmly replied, 'there was no indigo used: that would have been at variance with your request'.[51] In 1929 government restrictions on importing chemicals required for dyeing necessitated an application to the Dyestuffs Advisory Licensing Committee for a licence to import Sulphonine Red G Conc from Sandoz Chemical Works. Bowbridge was required to submit conclusive evidence that three top British producers' equivalents were unsuitable. The three companies failed the 'fading test'.

To succeed, a twentieth-century woollen mill required an effective manager who needed 'all the energy, foresight, and tact he can command' in a role with 'a great weight of responsibility upon his shoulders'.[52] It is not clear whether the large family-owned mills which survived beyond 1900 were directly managed by the owners or not. Certainly the Hoopers of Eastington were closely involved in the leadership of the mills and exercised a paternalistic oversight of employees, paying (for the time) good wages and offering regular employment, but how closely they supervised the day-to-day business is open to question. Bertie Winterbotham of Cam was both a director and manager who operated in the manner of sole owner. Mackie emphasised the importance of a mill manager 'being fairly conversant with every department … from a practical standpoint'. And those manufacturers who had sons predestined to join the firm were advised to put them into the mill to learn how a piece of cloth was made.[53] Arthur Leslie Long became Managing Director of Longfords but only after he had served a period as mill manager. Moreover, before he joined the firm his father sent him to the USA for three months for what, today, might be called an internship, to study woollen manufacture there. Besides a knowledge of the manufacturing processes a mill manager needed to be just in all things, firm and not ostentatious, having no favourites and having one view, 'that of the success and honour of his firm'. The woollen mill manager 'should not trouble himself with routine work … The manager's work is to see that things are done, and not to do them himself.' One of the most important roles of the mill manager was to appoint departmental foremen, hiring the best people and paying the best wages. Mackie recommended that foremen be encouraged to submit ideas of their own for improving the efficiency of the business and, in the climate of more authoritarian leadership, this may have required some considerable attitudinal change. J.G. Strachan had brought another talent to woollen manufacture, namely experience as a cloth factor and a single-mindedness which prompted him to leave Ebley Mill, where he had been in partnership with Nathaniel Marling, to strike out on his own, buying Lodgemore and Fromehall mills for £30,000 in the 1860s. When he retired in 1890 the business was flourishing. Strachan was a capable businessman with wide knowledge of the trade which was put to good effect. While poor leadership and management might permit a business to survive for a while, it was unlikely to do so for long in the highly competitive environment of the late nineteenth and early twentieth centuries. And it may have been a recognition of this that led some woollen manufacturers to decide that they had had enough.

A secure financial footing for a woollen manufacturing business was essential. This had been recognised by William Playne in his rescue of Tetbury Bank in 1825 and in Arthur Winterbotham securing financial backing from his father's bank, Winterbothams of Stroud, before he joined Cam Mill. Attention to financial detail is evidenced in Arthur Winterbotham's notebooks where he calculated the costs of running the business and of making a length of cloth.[54] The Evans of Brimscombe do not seem to have lacked capital for expansion and, when Apperly & Curtis became a limited liability company, the need for working capital was clear. This must have been a necessity for much of the nineteenth century, if not earlier, since months could elapse between cloth being finished and payment being received, particularly if it was destined for export. While becoming a limited liability company did not ensure business success, there seems to have been a notable reluctance by firms to move in that direction. And this was not unique to wool textile firms in Gloucestershire.

Firms which survived the First World War were conscious of the importance of effective marketing, appointing their own staff to travel abroad and at home as company salesmen, besides appointing resident representatives in overseas markets such as Vienna, Paris, Milan, New York and Toronto, as well as Australia and New Zealand. Mackie emphasised the importance of direct selling through a firm's own salespeople rather than through agents: 'the representative of a woollen manufacturer should be one sent direct from the mill to the customer.' Residential agents were 'becoming out of date'. Mackie recommended that the manufacturer should accompany his representative to London or other large centres once a year to get to know the buyers' foibles and that the sales representative should be allowed a fairly free hand in engagement with customers, knowing where the line should be drawn. The sales representative, well briefed in the market, was a valuable resource to senior managers at the mill, but they were not necessarily consulted on their knowledge of customer preference and the competition. The successful Gloucestershire firms had London offices well stocked with cloth and sought to cut out middlemen in the domestic market by selling direct to tailors and ready-made clothing manufacturers. Mackie emphasised the importance of good display:

> a few patterns well mounted and well shown will have much more weight with the buyer, and will stand a much better chance of being well selected than three times their number untidily mounted … Merchants dread going through too much and they will become weary or inattentive before they have got half-way through an uninteresting … collection. A few patterns carried in the hand will often command more attention and gain better results than a whole case full of ranges.[55]

This was a lesson that many manufacturers failed to learn, judging by the large swatches of cloth – frequently 'variations on a theme' and often dull – which were being produced to entice buyers into the 1950s and beyond.

REASONS FOR DECLINE

Between 1900 and the eve of the First World War some large and apparently well-managed woollen manufacturing businesses failed. Some, it would seem, quite suddenly. There was no single reason for the decline of the woollen industry in Gloucestershire but a number of factors, some of which might have been surmountable singly but, when combined, were overpowering.

Vertical integration had been an advantage from the early Industrial Revolution to the mid-nineteenth century, enabling the clothier to have an overview of manufacturing processes and this may have accelerated technical innovation. But the potential costs of updating were higher in vertically than in horizontally integrated firms which specialised in, for example, carding and spinning or weaving and finishing. Moreover, the replacement of any machinery which sped up one particular process or was generally more efficient, had a disruptive knock-on effect both up and down the production line.

There were management issues. There are hints, albeit from Wiltshire, that members of some of the second- or third-generation clothier families, who had been used to selling through London merchants, were not at ease with the face-to-face selling which was a feature of the twentieth century; nor with the rapid response which was required when interest in a cloth sample was shown.[56] By the 1940s managers/ directors were invoicing beyond a due date which adversely affected liquidity. And the valley sites which had been so necessary for the generation of water power in an earlier age restricted spatial development, Longfords being a prime example; but also Wallbridge and Capels for a different reason, namely that the railway had cut through these sites – Wallbridge used the railway arches for storage and mending.

No manufacturer blamed the price of coal for the decline of the industry, although the Stroud Gas Light & Coke Co. reported in 1871 that it was 'almost an impossibility to procure any coal from which gas could be manufactured at any price at all'. Jenkins and Ponting dismiss coal price, one of the authors remarking that a (Wiltshire) woollen mill in which he had worked during the 1930s, with a turnover of around £100,000 p.a., had had an annual coal bill of £1,000.[57] Clothiers agitating for the opening of the railway line along the Frome Valley from Swindon had predicted the ruin of their industry without it, but this and the branch line along the Nailsworth Valley, while benefiting the larger concerns which had their own sidings, did not save mills in either valley from failing. The lack of railway communication to facilitate the delivery of coal for steam engines may have been a contributory factor to the decline of businesses in the Painswick Valley and at Kingswood and Wotton. Nevertheless, Charfield Mill with its rail communication failed to survive the Great Depression.

The so-called 'palmy days' for traditional Gloucestershire broadcloths, cassimeres and doeskins between the 1850s and 1880s may have lulled those firms that had survived the previous shorter trade fluctuations into a belief that the market promised stability and security. This was not to be. By keeping their eyes fixed on supplying the immediate needs for army and navy cloth, leisure pursuits such as billiards and hunting, as well as clothing the upper (and aspiring upper) classes, it is likely that some manufacturers failed to recognise that the incidence of war might be less frequent (although more devastating) in the future and that fashion would play an increasingly important role in the clothing of the upper and professional classes. As with the Staffordshire pottery industry in the mid-twentieth century, there seems to have been a failure to undertake research into new products, new designs and new markets. And, as has been demonstrated in the late twentieth and early twenty-first century, some of the most successful (non-textiles) businesses were ones which produced innovative goods that people did not even realise they wanted – or needed.[58] When fashionable, colourful cloth was produced elsewhere which was also cheaper, the fact that a (more expensive) broadcloth coat might last for ten or more years may have been an unacknowledged disadvantage to Gloucestershire.

The response to the demand for lighter fabrics was a case of too little too late and manufacturers were slow to diversify. Worsted yarns could have been adopted on a larger scale than was done – John Libby suggested that cloth made with mixed yarns was cheaper to make and the more superior mixed cloths could be sold at almost the same price as broadcloth. But this was to assume that there would always be a demand for broadcloth. And in this a correspondent to *The Times* missed the point:

> It is safe to say that if all the cloth which is sold as 'west' were woven and finished in the mills of that district there would be no cause for complaint of bad or unprofitable trade … there would not be a single vacant mill or idle loom in the district.[59]

Whilst different or modified machinery was required for worsted (or other fibre) spinning and the technical issues should not be underestimated, firms were, on the whole, unprepared to recognise the need for or to fund radical changes in design and manufacture.

Production costs for companies manufacturing a range of both dark and light-coloured cloths were high. In the manufacture of traditional superfine cloths which were dyed in the piece, light and dark colours had to be kept separate at all stages of manufacture, requiring a costly duplication of production facilities. Medley cloth, too, for which dyeing was undertaken in the wool, was separated from other cloth throughout its production. The more exotic fibres such as mohair and alpaca were relatively little used. Some West of England manufacturers began to make fancy cloths which were particularly complex to make. Fancy designs involved a longer time in setting up the loom and long runs were unusual; production costs were, therefore, higher. Moreover, the changeover from one kind of yarn or fibre to another made technical demands which have been underplayed by writers on the industry. However, Scottish wool textile firms,[60] which experienced a similar decline in demand for high-class woollens, introduced new fabrics which the upper classes and the better-off amongst the middle classes wanted. They replaced the West of England as the main producer of high-class, expensive cloth for both men and women and prevented Yorkshire from dominating that sector in the way it had in worsted. Jenkins and Ponting observe that native Scottish weavers had always used the 2/2 twill which was well adapted to colour effects. The West of England preferred the 2/1 twill which was not as good for colour patterns.

Scottish check designs expanded the appeal of tartans and Scottish textile designers, being amongst the best in the world, were headhunted to Yorkshire, the West of England and overseas. But it would have taken more than the importing of designers to change the manufacturing culture in the West of England.

Yorkshire firms were more fleet-of-foot in responding to new demands, the proximity of merchants in Halifax and Leeds, with their greater knowledge of markets, being an asset. West of England manufacturers were more often at arm's length from the tailors and mass clothing manufacturers, selling their cloth through London merchants 'in a very formal way typified in the annual spring and autumn showings of new ranges, followed by the production of card samples, lengths and finally, it was hoped, pieces'.[61] A large ready-made tailoring industry developed in the Leeds area to meet the needs of a mass market characterised by rising incomes and new attitudes to clothes buying; proximity to this market for cloth may have influenced some Yorkshire manufacturers to become more design-conscious and market-focused. Stroud was not far behind Leeds in the establishment of a ready-made clothing firm, Holloway Brothers being founded in 1849. In 1853 they purchased thirteen sewing machines from C.T. Judkins of Manchester and, in the following year, claimed to be the first company in the country to use steam power to drive sewing machines.[62] There were other ready-made clothing companies in Stroud: William Tratt & Co. and Hill Paul & Co. Ltd. However, whereas ready-made clothing manufacture in Leeds was a significant outlet for Yorkshire textiles, there was little link between Stroud mass tailoring and the local textile industry. By the end of the nineteenth century the managing director of one of the companies expressed scorn in being unable to find local cloth manufacturers who could supply him with cloth at the right price. Their argument, he claimed, was that they could not reduce their costs sufficiently. His response was 'keep trying'.

Yorkshire manufacturers were prepared to travel to potential overseas customers to learn of their needs at first hand. But the woollen industry as a whole suffered from protective tariffs in overseas markets. While Dorothy Hunter did not accept that tariffs were even a major reason for the decline of the woollen industry in the West, Jenkins & Ponting argue that the woollen industry was more adversely affected by tariffs than any other British manufacturing industry.[63] William Marling wrote of 'The disastrous effect ... [of] high tariffs ... the manufacturers of this neighbourhood know that our woollen exports to the continent and the United States of America continue to dwindle'.[64] William Davies, of Stonehouse Mills, who had formerly had a large trade with the USA, said that he found it no longer profitable to export there.[65] As A.T. Playne said, 'American markets are practically closed to us and those of Europe largely so.'[66] The tariff levels in the USA often exceeded 50 per cent by value. In addition, the level of bureaucracy in protected markets created an additional burden as manufacturers sought to alter their goods to take advantage of tariff categories and to keep in touch with the fine detail of legal requirements in foreign markets. Where there was no protection for British producers, as in the home market, British woollen manufacturers did well. But only if they were quick to respond to changes in demand or, even better, anticipated it or created it.

Gloucestershire was late in recognising the importance of technical education and was way behind Yorkshire where local Chambers of Commerce assisted in the development of training programmes. As John Libby commented:

> There is a go or earnestness about the Yorkshireman we would do well to imitate. We would urge upon manufacturers and managers the importance of establishing a system of technical education in Stroud.[67]

But there was by no means a unanimous verdict in favour and it seems clear that the provision of education for the wool textile industry was, as in other areas of technical education, seriously deficient in England compared with what was available elsewhere in Europe.

Until mergers and limited liability status created a larger capital base, firms which by nineteenth-century standards had been large, were undercapitalised for the twentieth century. Even families which had not withdrawn capital to promote gentlemen's lifestyles were hard-pressed. Wallbridge Mill, for example, was insufficiently capitalised to be able to replace obsolete carding and spinning equipment and its owners relied on Strachan & Co. for these processes.

Whereas at the beginning of the nineteenth century West of England woollen manufacturers could be supplied with some of the most up-to-date machinery by local machinery manufacturers and millwrights, this was not the case by mid-century. Notwithstanding the local invention and manufacture of milling and carding machines, besides, later in the nineteenth century, steam engines, the all-important spinning mules and power looms were supplied by Yorkshire and Lancashire firms. It has been suggested that the disappearance of machinery manufacturers was a result of the decline of the woollen industry, not a cause.[68] Nevertheless, the closer proximity of machinery manufacturers and their customers was, in earlier days (and later in the North), a benefit to both parties; machinery manufacturers had the opportunity for R&D to be undertaken, sometimes unwittingly, by their customers and were able to learn at first hand the most pressing technological needs of the industry, while the woollen manufacturers were able to see new possibilities, besides making their needs known and contributing ideas for modifications.

Could Gloucestershire manufacturers – or the British government – have done anything to prevent the collapse of the woollen industry? Hindsight is, of course, a wonderful thing. By adhering firmly to a policy of free trade during the mid-nineteenth century, at a time when growing economies overseas were raising tariff barriers, the British government contributed to the industry's difficulties. The woollen industry was not a direct parallel of the smaller British silk industry, which was rapidly killed off by a combination of European tariffs and overseas silk goods entering free-trade Britain, for British woollens did well in the open UK market. But high tariffs in selected European countries and the USA made it well-nigh impossible for high-cost West of England cloths to compete. By adhering to the top-of-the-range in quality and price, besides showing an apparent reluctance to adopt new yarns and fibres or to explore new woollen and mixed fibre products, the Gloucestershire woollen manufacturer showed insufficient vision for the future and did not, on the whole, appear to have the heart for the new business environment of the twentieth century.

Endnotes

1. Chance & Bland, *Industrial Gloucestershire*, Gloucester, 1904, p.26
2. D.M. Hunter, *The West of England Woollen Industry under Protection and Free Trade*, 1910, gs, pp.46–7
3. D.T. Jenkins & K.G. Ponting, *The British Wool Textile Industry, 1770–1914*, p.169
4. Worralls Textile Directory, 1991
5. Ibid.
6. RV 293.7gs
7. *Stroud News*, 12 January 1906
8. *Kelly's Directory of Gloucestershire*, 1935. Bonds Mill is recorded but it had ceased production the previous year
9. J. Libby, *Twenty Years History of Stroud 1870–1890*, 1890, p.82
10. *Industrial Gloucestershire*, p.3
11. Ibid., pp.24–9
12. Jenkins & Ponting, pp.178, 190–1
13. Ibid., pp.174–5
14. Dunkirk Mills leaflets, Stroudwater Textile Trust (STT)

15 *Industrial Gloucestershire*, p.23
16 Ian Mackintosh, 'Shoddy Work, Stroud. Well done!', *Warp & Weft*, series 2, issue 4, 2011/12
17 *Industrial Gloucestershire*, p.24
18 A.E. Keys, *A History of Eastington*, 1953, p.76
19 Ibid.
20 R. Perry, *The Woollen Industry in Gloucestershire to 1914*, 2003, p.133
21 Ibid., pp.24–8
22 *Industrial Gloucestershire*, p.28
23 Ibid.
24 Bassett, 'Fifty Years of Dyeing and History of the Yard', MSS in Stroud Museum
25 R. Perry, p.136
26 'Tennis Balls, my Liege', *The Field*, 1 July 1971
27 D4644/4/9
28 D4644/4/9, WIRA to Wm Playne & Co. Ltd, 3 February 1938
29 'Tennis Balls, my Liege', *The Field*, 1 July 1971
30 A.T. Playne & A.L. Long, *A History of Playne of Longfords Mills*, privately printed, 1959, p.60
31 Ibid., p.158
32 Ibid., pp.47–8; A.T. Playne, *Minchinhampton and Avening*, 1915, repr. 1978, p.158
33 Ibid., p.62
34 *Industrial Gloucestershire*, p.24
35 Herbert Winterbotham, typescript by a former office employee, nd, STT MSS. Mr C.W. Hill started as an office boy in 1911 and rose to be Company Secretary and a Director. His son Michael donated these recollections
36 Typescript excerpts of Hunt & Winterbotham Board Minutes, 1909–1915, STT MSS
37 Herbert Winterbotham
38 Strachans, typescript by C.W. Hill, nd, STT MSS
39 Stroudwater Textile Trust Archive, Bowbridge, typescript, by C.W. Hill, nd
40 Ibid.
41 Ibid.
42 Ibid.
43 The two Samuel Longs had been much involved in local affairs, both in their time being mayors of Wotton, besides contributing to charities including church, chapel and school
44 At Vatch Mill for much of the eighteenth century, then Ebley and, later on, at Paganhill Mill which closed in the 1850s
45 R. Perry, p.132
46 John Mackie, *How to Make a Woollen Mill Pay*, London, Scott Greenwood, 1904, p.v
47 Ibid., p.1
48 Ibid., p.2
49 Ibid., pp.16–39
50 Stroudwater Textile Trust Typescript, Bowbridge, nd
51 Stroud Museum MSS, Clayton Aniline Co. to Strachan & Co., Bowbridge, 24 April 1926
52 Ibid., p.40
53 Ibid.
54 R. Perry, pp.138–9
55 Mackie, p.62
56 Jenkins & Ponting, pp.275–81

57 Ibid., p.172
58 For example, Sony 'Walkman', Blackberry, iPad
59 Libby, p.63
60 Jenkins & Ponting, p.291
61 Ibid., p.172
62 D1405/20/1; D3944/3
63 D.M. Hunter; Jenkins & Ponting, p.293
64 *Stroud News*, 12 January 1906
65 Ibid.
66 Ibid.
67 Libby, p.83
68 Jenkins & Ponting, p.173

CHAPTER TEN

POSTSCRIPT

CLOTH MANUFACTURE

The future of woollen cloth manufacture in Gloucestershire for the large part of the twentieth century was vested in two firms – Winterbotham, Strachan & Playne and Marling & Evans.

There were some initial tentative moves towards rationalisation in both companies earlier in the century, but it was not until after the Second World War that horizontal integration was seriously addressed. Following the closure of Gyde & Bishop at Bowbridge, wool dyeing equipment was transferred to Lodgemore Mill, while piece dyeing equipment went to Cam Mill. Playnes had previously carried out all processes of manufacture except piece dyeing. Amalgamation presented opportunities for economies of scale; wool scouring and piece dyeing were thereafter undertaken at Cam and wool dyeing at Lodgemore, 'thus saving a considerable amount of steam and power'.[1] The Longfords fullers and washers were then transferred to the former dyeing and scouring plant. A new mending shop was adjoined by a new finishing shop. After the Second World War and for the duration of food rationing, Playnes installed a canteen, to enable staff to make their food coupons last longer. In 1947 the managing director presented a plan to the holding company for the erection of a new large building at the mill with a view to having a logical sequence of production processes. After what seemed to be an initial agreement, this was not approved – until 1950.[2] Manufacture ceased at Longfords in 1990.

Perhaps it was the sense of a passing age which prompted the Clayton Aniline Co. to write to Ralph Creed Bassett at Bowbridge Dyeworks in 1935 suggesting that he write a history of the dyeworks for *Dyer and Calico Printer*, since a history of dyeing in the West of England 'could only be compiled from the records of firms like Bowbridge Dyeworks, Gyde Bishop & Co and Moore Bros of Trowbridge'. Six years previously a message had been sent from the London office of Hunt & Winterbotham to Cam Mill, noting, 'It is certainly most interesting to see how other makers' scarlets fade'. Mr Margetson commented, 'It is dangerous to monkey about with other people's cheap dyes when we already have a good one.' In 1950 the Historical Records Committee of the Society of Dyers & Colourists wrote to Hunt & Winterbotham pointing out the need for records to be kept of buildings and techniques, soliciting the firm's support. E.V. Giles replied, 'Mr Bassett has a unique knowledge of the earlier methods of dyeing utilising the various vegetable products and it is believed that he has samples and a good deal of data in this connection.' Bassett had recently retired after fifty-eight years in piece dyeing, starting in 1898 as assistant dyer at £70 p.a., and finishing as Manager of Piece Dyeing at Cam. Bassett was subsequently contacted and invited to give or bequeath material to the Society.[3]

Hunt & Winterbotham opened a London shop in 1949: 'whatever happens after the devaluation of the pound … we must sell more of our goods to the dollar market.'[4] In that year Graham Carmichael, chair of the holding company, went on a promotional trade mission to the USA, representing the woollen industry of Great Britain, emphasising that it was essential to know what the market wanted.[5] Strachans had specialised in snooker, billiard and red hunting cloth at Cam; and tennis ball cloths, amongst other products, at Lodgemore. Strachans continued to make speciality military cloth, receiving an order in 1947 for cloth for naval uniforms for HM King George VI's South African tour[6] and ten years later a large order from Russia for 2,500 yards of civilian overcoating – a repeat order on an earlier one for 10,000 yards.[7] Strachan & Co. received an order for cloth for the Pope's robe which was 'a particular shade of white. There is nothing else quite like it.'[8] By 1951 all cloth was sold through ABL, a distribution organisation which focused particularly on the USA and retail business on ships.

The post-war sellers' market had ended by 1950; all three Winterbotham, Strachan & Playne mills were making similar cloth and more rationalisation was needed. There had been an experiment in the late 1940s to present a complete sales line with specialities from other manufacturers but this had not been wholly successful. In 1951 Graham Carmichael proposed to the Board of the holding company that complete horizontal integration take place in Winterbotham, Strachan & Playne. This met with a hostile reception by the managing directors of the subsidiary companies. The differences of opinion at Board level were acute – Carmichael was said to have had no technical training and to have had little practical experience in manufacturing (despite being selected to represent the UK woollen industry on a recent trade mission to the USA), and those who voted with him were reputed to have no experience of West of England cloth manufacture. An acrimonius general meeting was held, with Carmichael and one other director resigning. It could be suggested, with hindsight, that challenging views from outside the industry might have been of benefit to the company.

By 1958, with the reins of the holding company, Winterbotham, Strachan & Playne, in new hands, it was recognised that the demand for traditional, high-quality cloth was not sufficient for the combined capacity of the mills in the Group. The Group established links with Leeds University and began to undertake trials of new technologies, such as new Pirn winding machines, which were operated alongside older models in two of the mills. The Group was responding to the challenge of artificial fibres – nylon, orlon and terylene were all tried 'but none [was found to be] so good as wool'. Nonetheless, it was acknowledged that the attraction of synthetics was strong. It was recognised that the West of England 'must lead the fashion field with new colours and patterns knowing all the time that once the mass market takes a fancy to them Saville Row will cease to use them'.[9] The Group appointed a work study engineer in 1953 (he had no experience in the textile industry, neither was he local). It was acknowledged that the foremen in each department knew their materials and machines but that time reduction could be achieved in moving materials from one section to another. 'Only when the work study engineer was satisfied that he had established good relations with the foremen and the operatives was any stop watch used.' Spinning time was lost at Cam and Longfords in awaiting materials and there was a lack of oiling capacity at Longfords. Mules at Cam stopped frequently, awaiting spools, and time was lost in the movement of spools. At Longfords a creel fed the warping mill but when a red light came on with a thread breaking, the bobbins were so high 'no female can reach it'. The Group came to a decision to do what it believed it did best, at a level others found difficult to compete with: to 'stick to the materials for which we have become famous'.[10] Whether there were any realistic alternatives is a matter for conjecture. Perhaps there was a failure to recognise that a culture change was required, besides a preparedness to experiment for rather longer with new fibres and technologies. It took until the 1990s for a thoroughly radical solution to save the two remaining mills in the Group.

There were – and are – exceptions to the story of decline but the Gloucestershire textile businesses which survived the First World War, the post-war depression, the Second World War and, subsequently, a domestic market governed by clothing coupons, had a steep uphill climb. The firms which survived to the 1950s had received large orders for military cloth in the First and Second World Wars which, to some extent, served as a cushion for difficult post-war times.

Times were difficult in the later 1950s and in 1958 about one-third of the Cam labour force was on short time.[11] Winterbotham, Strachan & Playne was acquired in 1990 by Millikens, an innovative USA-based company, and was named Milliken Woollen Speciality Products Ltd, operating from Lodgemore and Cam mills between 1990 and 2011; the company is now called WSP Textiles Ltd after a management buyout towards the end of 2011 (the name reflecting the company's founding businesses: Winterbotham, Strachan and Playne). The business is British-owned, having been financed by Vespa Capital, and three senior managers from the Milliken days are now on the Board. The company employs 160 people at the two mills; the Strachan brand dominates the snooker cloth market (sporting successes made for good publicity: 'Mr Joe Davis achieved his 500th Break on the finest West of England Billiard Cloth – as he did most of his others'[12]) and Playnes' tennis ball cloth covers about a third of the 300 million tennis balls manufactured per year. Marlings Industrial Felts excepted, WSP Textiles is now the 'keeper of the flame' of the factory-based wool textile industry in Gloucestershire.

During the Second World War the Admiralty took over part of Stanley Mill, although blending, dyeing and finishing were still undertaken there by Marling & Evans. This partial horizontal separation was completed in 1946 when Marling & Evans moved all the weaving to Stanley. Washed wool was delivered to Stanley where it was dyed and then taken by lorry to Ebley for carding, combing and spinning, the spun yarn being collected daily and returned to Stanley for weaving and finishing. From 1949/50 the company began a new period of investment in machinery. Three new carders were installed on the ground floor of Ebley and the first pair of high-speed mules was purchased from J. Charlesworth of Yorkshire. However, they were not wholly effective until two employees at Ebley – Charlie King and Ken Stadden – devised a mechanism in the friction, which Charlesworths subsequently applied in their later models.[13]

The wool textile industry suffered keen competition from overseas manufacturers from the late 1960s. Some diversification, with the introduction of different fibres in apparel cloths, had taken place in the nineteenth century – A.T. Playne, having listed the main range of woollen cloths made at Longfords in 1915, remarked: 'No doubt there are other materials which will be made when the taste of the public demands them.'[14] Longfords were at full stretch producing khaki and navy cloth at the time, but a reactive stance did not auger well for peacetime competition. Marling & Evans experimented with the manufacture of different mixes of wool and synthetic fibres (the latter largely from Courtaulds), including Fibro in the early 1950s and acrylan (Rosemary Clooney starred in a commercial for this), which was both crease resistant and, where wanted, could be permanently creased. Acrylan was killed off by the introduction of terylene and Ebley produced terylene/wool mix yarns. Synthetic fibres were used in the manufacture of car seat belt fabric and later in industrial felts, particularly needle felting which, from 1993, was used for contract floor covering.

In the 1950s/60s Marling & Evans began to produce non-apparel cloths at Stanley Mill. One of their most innovative products was a fireproof cloth called Nomex, made under an exclusive licence from DuPonts, the US patentees. DuPonts sent Marling & Evans some synthetic fibre and enquired if it could be made into cloth. After much design and development – deciding on the optimum length of staple, whether the slubbing could be pieced, and the testing, with their suppliers of oil, to ascertain the best oil mixture for fine spinning[15] – the company succeeded in producing a pure white fabric; the first fireproof cloth ever made. It was used for racing drivers' and firemen's overalls, and in Concorde, and saved lives in racing car crashes. Despite being expensive there could have been many more applications for this fabric but, after changes in senior management and an apparent lack of technical knowledge at the top, other firms invaded the Nomex market.

In the early 1980s Marling & Evans took the decision to close Ebley Mill and to move all operations to Stanley Mill. This involved transferring only five (shortened) mules and three carding engines. In the mid-1980s there were a number of changes and financial difficulties which resulted in a bank nominee chairing the Board and in 1989 the business went into receivership. In December 1989 Marling & Evans (the apparel side of the business) closed, while Marling Industrial Felts continues.[16]

Stroudwater is now the home and work base for numerous artists, designers and craftspeople in contemporary textiles, some of whom are producing work which is acquired for international collections. Bobbie Cox is an eminent tapestry weaver whose abstract work is informed by landscape – she has travelled widely in India and this influence is present in her work; Matthew Harris was, until 2000, an artist working with paper, at which point he transferred to textiles using cutting, dyeing and hand stitching to create abstract imagery in textile artwork; Nick Ozanne has relocated to Stroud from London to produce her handwoven scarves and designs for silk upholstery; Liz Lippiatt creates cloth for costume and couture. She has a studio in Stroud and also works with six other craftspeople in the group known as Studio Seven which has devised Textiles in Performance as well as interactive displays with titles such as 'Cut 2 on fold' and 'Make do and Mend' at the Stroud International Textile Festival, Quenington and Hidcote. Tim-Parry Williams is a textile designer, weaver and academic who experiments with different fibres, combining hand weaving with collaborative design for industry. He has particular links with both the traditional and contemporary textile industries in Japan.

ALTERNATIVE INDUSTRIAL USES FOR FORMER WOOLLEN MILLS

Ever since the high point in the number of woollen mills in Gloucestershire – around 200 in 1815/20 – former mills have been adapted to alternative uses. A.T. Playne remarked in 1915, 'Though some of us may regret the decline of the old broadcloth, yet we must recognise that the new industries have done much to restore prosperity to the district.'[17] The smaller mills in the Painswick and Slad valleys, besides the tributaries of the Little Avon and Ewelme, were often converted to flour milling, saw milling, or bone grinding. Some larger mills became flock and shoddy mills from the mid-nineteenth century, for example Toadsmoor, Merretts, and Capels; Thrupp Mill was converted to the Phoenix Ironworks; and several mills were converted to walking stick manufacturers in the late nineteenth century, such as Bliss, Seville's, St Mary's and Dunkirk. Some twenty mills in the Stroud area were used for stick manufacture in the mid to late nineteenth century, although by the 1920s walking sticks and parasols were less fashionable and the industry declined. Pin manufacture was introduced at Frogmarsh and Inchbrook mills, as well as three mills in Painswick, including King's Mill. Pins, hairpins and knitting needles were made at Wimberley Mills, Brimscombe. Textile manufacture continued following the conversion of some woollen mills to silk throwing – there were eleven by 1856 but, following the 1860 treaty with France which opened the British market to French goods, whilst the French market was protected by tariffs, the number fell and there were four left by 1870, Warehouse Mill, Chalford, part of Seville's, and Hope Mill amongst them. A brief final flurry took place just before the end of the nineteenth century when Halliday's Mill, Chalford, and then part of Day's Mill, Nailsworth, were used for the manufacture of sewing threads.[18]

New engineering businesses were started up towards the end of the nineteenth century, such as Dudbridge Ironworks, established in 1891 for the manufacture of oil and gas engines. Attention was drawn in *Industrial Gloucestershire* to the fact that the parts of engines were standard and interchangeable. The company had a large overseas order book, as well as numerous UK customers. The Stroud Metal Co. was also located at Dudbridge, having been established in 1899 to save a brass founding business. Umbrella fittings, steam and water valves and gauges, as well as 'all kinds of castings' were made. Ebley Ironworks had been established in 1812 by the Shipway family, members of which were still running the business in 1904. As the character of manufacturing industry changed in Gloucestershire, so Shipways adjusted their product lines, becoming known, amongst other things, for their shoddy manufacturing equipment.[19] Perhaps due to the reputation of engineering businesses in Dudbridge and Ebley, the Hampton Car Co. relocated to Dudbridge from Hampton-in-Arden in 1919. However, over its short life the company struggled financially and, having been sold for the second time in 1924, went into receivership two years later.

It is surely not too fanciful to suggest that the growth of new industries in Stroudwater, Wotton and Dursley was assisted by the range of trained people in the labour market, for many skills were transferable. Those who understood how to run a condenser or tune looms would have been able to move into engineering, for example. The science of dyeing could be transferred to chemicals; fabric design to printing. And management skills could be transferred to almost any sector, as could sales and marketing.

In the twentieth century high technology businesses came to the valleys. Lightpill Mill was a major centre for the development of a new type of plastic called casein made from milk curd. Developed in Germany, a British company obtained a licence to manufacture and moved to Lightpill in 1911 (nearer to the supply of curd from Ireland than Enfield). Despite refrigeration, it was difficult to keep curd in good condition and a new dry process, using granules, was developed, being registered in 1913 as Erinoid. Lightpill became the main supply of this plastic outside Germany, the main use in Britain being for button manufacture. Erinoid ceased production at Lightpill in 1982, since when the large mill has had some innovative firms in different parts of the site. One of the buildings is now occupied by Cleeve Materials Handling Ltd which was established in Cheltenham in 1989 by a local engineer and his partner, selling screw conveyors for large industrial plants. Stroud Switchgear Ltd is also located at Lightpill. Founded in 1984, the company undertakes the design, development and production of custom-made electrical systems and switchboards. Brimscombe Mills was the home of Sperry Gyroscope from c.1938 until c.1969, making gyroscopic compasses for the Royal Navy. Ryecote was founded in 1969 'by an engineer with a problem – how to make audio recordings on location uninterrupted by the sound of the wind'; a family-owned company, the products are designed, made and hand-assembled on the former New Mills site in Stroud. With an enviable record of bringing out product innovations almost every year the company sells its windshields worldwide. Peter Sheppard Engineering is based at Charfield Mills.

A major employer in Gloucestershire is Renishaw, founded in 1973 by David (now Sir David) McMurtry and John Deer for the manufacture of inspection and measurement equipment, probes for machine tools, motion control laser encoders, dental CAD/CAM systems and now medical diagnostics. Renishaw is an R&D-intensive company whose head office is located at New Mills, Wotton-under-Edge (which has been very well restored). Renishaw also has a production facility at Inchbrook (where the old mill is well preserved), besides

New Mills, Wotton-under-Edge, in the early nineteenth century. (Renishaw plc)

Oil Mill, now Snow Mill, Ebley. (Roger Tann)

other locations. Western Windpower Ltd has been designing, manufacturing and installing wind-monitoring solutions for the wind energy industry since 1992 and exports around the world. Ebley Oil Mill became a flour mill in the mid-nineteenth century, run in conjunction with Stratford flour mill. It is now called Snow Mill for the very good reason that snow is made there – artificial snow. This is a classic kitchen table innovation story: in 1982 Darcy Crownshaw 'used ingenuity and excellent contacts' to make a high-quality biodegradable artificial snow from recycled materials. It could be sprayed on to plants, buildings and film sets; it made good snowballs and was actor friendly – you could even leave realistic-looking snow footprints in it. Snow Business's first market was the film world and television. But new markets came to them and these presented new challenges – the snows were required to meet different specifications. What in the marketing world is termed 'visual merchandising' provides a seemingly endless set of opportunities and challenges and the company recruits specialists to develop special products including the machine for the quietest falling snow in the world. Snow Business now makes 200 different kinds of snow and sells in over twenty countries.[20] Fourteen of the larger former woollen mill sites have become industrial estates, amongst them Bliss Mill (Chalford Industrial Estate), Lightpill (Bath Road), Nailsworth Mill and Frogmarsh Mill in Stroudwater, each with a number of different tenants. The former cloth mills continue to contribute to the regional economy by providing accommodation for a wide variety of industries.

MILLS AND THE LANDSCAPE

Industrial buildings provide visually engaging, challenging alternatives to new-builds when it comes to finding premises for small businesses, technology parks, professional services, health clubs, prestige offices, art galleries and theatres. The 1940s Baltic Flour Mills at Gateshead have been converted to an arts centre; Abraham Darby's works at Ironbridge are an industrial museum; Titus Salt's huge Saltaire Mill is now part of a world heritage site which has galleries, retail outlets and restaurants; Tate Modern is in a former power station; a dockside warehouse at Liverpool has been sensitively transformed into a hotel; the steel-making sheds at Templeborough near Rotherham have been converted into a science centre, winning the Stirling Prize in 2001:

> Cleaned of detritus and grime, odourless and dust free, its working parts left as sculptural objects in the space inside and out, the Templeborough Mill has become an object almost entirely out of its context – perhaps an apt enough metaphor in Britain, the birthplace of the Industrial Revolution.[21]

Langford Mill, Wotton-under-Edge, in 1963; a modest-sized mill, typical in scale to some in the Ewelme and Little Avon valleys, with potential for restoration but often allowed to fall into ruin. (Jennifer Tann)

Charfield Mill (called Bone Mill) – well restored but empty in 2012. (Roger Tann)

There are some iconic examples around the world, including an early 1900s flour mill in Spain converted to newspaper offices and Lowell, Massachusetts, a textile company town, now converted to an industrial museum. St John's Mill on the Isle of Man (complete with waterwheel and very well-preserved fulling stocks from 1815) has been converted into a conference centre. Historic industrial sites provide unique opportunities, indeed 'the fabric of industry is being transformed'.[22] Housing, too, is a possibility where the interior of a listed building is not compromised by lateral subdivision into small residential units and Dunkirk Mill, Nailsworth, is an excellent example.

The survival of former woollen mills in Gloucestershire is patchy and many have gone. In Wotton-under-Edge, New Mills excepted, little remains of the former water-driven mills, although some of the steam mills in the town have been converted to other uses. There is relatively little material evidence of Uley or Kingswood's former standing in the industry, although some splendid clothiers' houses remind us of the wealth that could be generated at the height of the industry. The group of three Charfield mills have been well conserved but one has been empty for several years.

But some mills have been saved and those which retain a contemporary industrial/commercial function have a particular integrity. Some of the new industries briefly described above are located in former woollen mills. Other examples include Stroud District Council's fine restoration of Ebley Mill as council offices; Belvedere Mill, Chalford (the former corn mill on an old fulling mill site), with its mill pond, complete with swans, is home to Heber Ltd which was established in 1984 to provide electronic design

Ebley New Mill, now Stroud District Council offices. (Roger Tann)

and manufacturing services to industrial, commercial and consumer markets. The company uses its market diversification to translate innovations across sectors and has an in-house product development service. Port Mill at Brimscombe has also been handsomely restored and is home to The History Press and other businesses. Bourne Mill is being sensitively restored for Noah's Ark cycle and snowboarding business. St Mary's Mill has been restored for contemporary businesses. One of the tenants is a computer software design company and the mill, with waterwheel, compound steam engine and other displays, is currently open three times a year. Egypt Mill, Nailsworth, retaining two internal waterwheels, has been converted to a hotel/restaurant and Ruskin (formerly Millbottom) Mill is a college for students with various disabilities. Gigg Mill has several tenants including the Stroudwater Textile Trust's (working) weaving machinery display.

Other mills have been converted to housing. Dunkirk Mill and associated buildings is now residential and also houses the Stroudwater Textile Trust's display of water-powered finishing equipment. Longfords and Iron mills have been converted to housing, as has Church Mill in Wotton-under-Edge, the engine house becoming a Roman Catholic chapel. There are plans to develop the unique, Grade I listed Stanley Mill for housing. Many of the smaller mills in the tributary valleys have been converted to single residences, while several are bed and breakfast businesses, for example in the Painswick Valley, although many have gone.

Gloucestershire's former cloth mills, most of them built of Cotswold stone, nestle in the river valleys; they and their predecessors are an established part of the landscape, having been there for some hundreds of years. Mill sites, like churches, were continuously occupied; when rebuilding or extension was necessary, the established site was used. The Frome Valley is so narrow in its middle and upper reaches that the Thames & Severn Canal runs closely alongside. The Five Valleys and their tributaries which constitute Stroudwater still have numerous attractive mills surviving. While some are sad looking, many are well maintained. Most of the surviving mills are Grade II listed; two – Ebley and Dunkirk – are Grade II★ and

Clayfield Mill, Chalford, in the early eighteenth century. It was reduced in size for road construction and is now a residence. (Roger Tann)

Stanley Mill is Grade I on the basis of its structural uniqueness. There is no other mill in the world with arcaded interior cast-iron work of such design or beauty. In addition to the mills there are elegant clothiers' houses in the valleys and along the valley sides. Stroudwater is now designated an Industrial Heritage Conservation Area with seven separately designated conservation areas within it. The Cotswold woollen industry landscape has its own special aesthetic. As such it makes an important contribution to tourism.

Landscape and townscape have a particular role in education. It is important for students of all ages – including adults – to learn about the many-layered environment in which they live and work; to hear the stories of those responsible for creating the human-made landscape; and the stories of those who worked in the mills and factories. By touching the heart, the mind is engaged, for this is his-story and herstory. Walking in the valleys to look at mills and clothiers' houses, visiting open mills, visiting Painswick churchyard to look at clothiers' tombs, studying nineteenth-century census materials, learning what it was like for an adult to work 70 hours a week in a mill, imagining being a 7-year-old child working at scribbling wool for 40 hours a week; all these (and many others) are threads which make up the story of Wool & Water in Gloucestershire.

Endnotes

1. A.T. Playne & A.L. Long, *A History of Playne of Longfords Mills*, 1959, p.51
2. Ibid., pp.54–6
3. The Museum in the Park, Stroud, Bassett MSS
4. *Citizen*, 23 September 1949
5. *Stroud News*, 21 November 1947

6   *Citizen*, 22 February 1947
7   *Citizen*, 30 January 1957
8   *Citizen*, 11 October 1958
9   J.H. Collins, 'West of England Cloth in Winterbotham Strachan & Playne Group', *Fibres International*, October 1958
10 Ibid.
11 *Citizen*, 19 September 1958
12 *Citizen*, 19 September 1953
13 This was an effective working partnership between King's unrivalled knowledge of spinning and Stadden's knowledge of textile engineering. Ken Stadden went to Yorkshire on several subsequent occasions to apply his practical working knowledge of the friction mechanism at Charlesworths
14 A.T. Playne, *Minchinhampton and Avening*, 1915, p.158
15 A typical finely spun West of England yarn was 58s; Nomex was 36s (the lower the number the higher the count)
16 'Marling invests heavily to meet widespread demand for its products', *Filtration & Separation*, Jan/Feb, 1988
17 Playne, p.159
18 Jennifer Tann, 'Worms to Riches', *Warp & Weft*, series 2, issue 4, 2011
19 Chance & Bland, *Industrial Gloucestershire*, 1904, pp.17–20
20 This paragraph is based on the websites of the respective companies
21 Gillian Darley, *Factory*, Reaktion Books, 2003, p.208
22 Ibid., p.200

APPENDIX

ENQUIRIES AND ORDERS FOR BOULTON & WATT STEAM ENGINES IN GLOUCESTERSHIRE WOOLLEN MILLS 1802–39

Date	Firm	Location	Enquiry hp	Order hp	Engine Type
1802	H. & G. Austin	Wotton		6	Small side lever
1803	Roberts	Painswick	2, 4, 8		
1805	E. Sheppard	Uley	12	12	Crank
1806	G.D. Harris	Uley	12, 14, 20		
1807	N.D. Lloyd	Uley	4, 14, 20		
1808	Tattersall	Wotton	30		
	G.J. & W. Strong	Wotton		6	Bell crank
1811	S. Plomer	Wotton	6, 10, 14		
	G.D. Harris	Uley	Not specified		
1813	H. & G. Austin	Wotton	10		
1814	G.W. & P. Playne	Longfords, Avening	8, 10, 12, 14	14	4-column beam crank
1815	J. Reedy	Wotton	6, 10, 14		
	S. Plomer	Wotton	6		
	Carrick & Maclean	King's Stanley	20		
1816	H. Hicks	Eastington	Second hand		
1817	S. Clissold	Ebley	10		
1818	Tattersall	Wotton	6		
	T. Sheppard	Chalford	8, 12		
	E. Sheppard	Uley	14, 16, 18		
	D. Lloyd	Uley	14		
	H. Hicks	Bond's Mill, Eastington	6, 14	10	6-column independent beam

Year	Customer	Location	Col1	Col2	Type
	Harris, Stephens, Maclean	King's Stanley	14		
1819	G.W. & P. Playne	Avening	14		
1820	G.W. & P. Playne	Avening	14		6-column independent beam
	G.H. Austin	Wotton		14	6-column independent beam
	J. Cripps	Cirencester		6	6-column independent beam
1821	G. Wallington	Dursley		14	6-column independent beam
	R.P. & G.A. Smith	Wallbridge, Stroud	20	24	Rotative beam crank
	T. Mercer	Kingswood		10	6-column independent beam
	D. Lloyd	Uley	24	20	Rotative beam crank
	E. Jackson	Uley	8		
	H. Hicks	Millend Mill, Eastington	14, 20	14	6-column independent beam
	Harris, Stephens & Co.	King's Stanley	24, 32, 40		
	Davis, Beard & Davis	Leonard Stanley		14	6-column independent beam
	H. Austin	Alderley		10	6-column independent beam
1822	Strange	Wotton	10	10	6-column independent beam
	Roger Smith	Southfield, Woodchester	10, 14	14	6-column independent beam
	W. Overbury	Avening	Second hand	14	Crank, ex Soho Foundry
	H. Hicks	Churchend Mill, Eastington	20, 34	24	Beam crank
	W. Helme	Stroud	12, 14, 16		
	Harris, Stephens	King's Stanley	32, 40	40	Beam crank
	H. & G. Austin	Wotton	14, 20, 24, 32		
	D. Lloyd	Uley		20	Beam crank
1823	W. & P. Playne	Avening	10	10	Independent
	E. & T. Neal	Wotton	10–30	20	Side lever

	E. Mason	Painswick	Not specified		
	W. Helme	Stroud		14	Independent
	H. & G. Austin	Wotton		32	Beam crank
	H. Wyatt	Vatch, Painswick		20	Side lever
1824	E. Davies	Stonehouse	30	30	Side lever
1825	Roach	Wortley, Wotton-under-Edge	20		
	E. & T. Neal	Wotton	40, 50	50	Side lever
	N.S. Marling	Lodgemore	30–50	40	Beam crank
	R. Ford	Wotton	20	20	Side lever
	S. Long	Charfield		20	Side lever
	N. Driver	Peghouse, Painswick		30	Side lever
	H. Wyatt	Vatch, Painswick		30	Crank
1826	H. Hicks	Meadow Mill, Eastington		30	Beam Crank
1827	Playne & Co.	Minchinhampton		6	A-frame
1827	W. Long	Charfield		20	Side lever
1839	Stanton & Sons	Stafford, Stroud		30	Side lever (?)

Source: Boulton & Watt Papers

BIBLIOGRAPHY AND SOURCES

MANUSCRIPT SOURCES

Public Records Office
Gloucestershire Archives, including a large MSS collection, formerly in Gloucester City Library (with the suffix 'gs' following the reference)
The Museum in the Park, Stroud
Stroudwater Textile Trust
Nailsworth Archives

NEWSPAPERS

Gloucester Journal (GJ)
Stroud Journal (SJ)
Citizen

PARLIAMENTARY PAPERS (PP)

BOOKS

Aldcroft, Derek H. and Fearon, Peter (eds), *British Economic Fluctuations 1790–1939*, London, Macmillan, 1972
Annual Register
Atkyns, Sir R., *Ancient & Present State of Gloucestershire*, 1712
Beacham, M.J.A., *Mills & Milling in Gloucestershire*, Brimscombe, Tempus, 2005
Bebbington, Alan (ed.), *A History of Uley*, The Uley Society, 2003
Bickley, F.B. (ed.), *The Little Red Book of Bristol*, 2 vols, 1900
Blunt, John Henry, *Dursley and its Neighbourhood*, 1877, Stroud, repr. Sutton, 1975
Bowden, Peter J., *The Wool Trade in Tudor and Stuart England*, London, Macmillan, 1962
Carrick, Pat, Rhodes, Kay, Shipman, Juliet, *Oakridge: A History*, Oakridge Historical Research Group, 2005
Carus-Wilson, E.M. (ed.), *Essays in Economic History*, Vol. 2, London, Edward Arnold, 1962
Carus-Wilson, E.M. (ed.), *Medieval Merchant Venturers*, London, Methuen, 1967
Finberg, H.P.R. (ed.), *Gloucestershire Studies*, Leicester University Press, 1957

Fisher, P. Hawkins, *Notes and Recollections of Stroud*, 1891, repr. Gloucester, Alan Sutton, 1986
Fosbroke, T.D., *Abstracts of Records and Manuscripts respecting the County of Gloucester*, 2 vols, Gloucester, 1807
Gloucestershire Inquisitiones Post Mortem, 6 vols, ed. G.S. Fry & E.A. Fry, Phillimore, 1893–1914
Gloucestershire Notes & Queries, 10 vols, 1881–1913
Harte, N.B. & Ponting, K.G., *Textile History and Economic History: Essays in Honour of Miss Julia de Lacy Mann*, Manchester University Press, 1973
Heaton, Herbert, *The Yorkshire Woollen and Worsted Industries from earliest times up to the Industrial Revolution*, Oxford, Clarendon Press, 1965
Hopf, Patricia M., *The Turbulent History of a Cotswold Valley: The Upper Slad Valley and The Scrubs*, Brimscombe, Nonsuch, 2006
Hyett, F.A., *Glimpses of the History of Painswick*, Gloucester, 1928
Hudson, Pat, *The Genesis of Industrial Capital: A Study of the West Riding Wool Textile Industry c.1750–1850*, Cambridge University Press, 1986
Hunter, D.M., *The West of England Woollen Industry under Protection and Free Trade*, 1910
Jenkins, D.T. (ed.), *The Textile Industries*, Oxford, Blackwell, Economic History Society, 1994
Jenkins, D.T. & Ponting, K.G., *The British Wool Textile Industry, 1770–1914*, Aldershot, Scolar Press, 1987
Kingsley, Nicholas, *The Country Houses of Gloucestershire*, Vol. 1, Cheltenham, Kingsley, 1989
Kingsley, Nicholas, *The Country Houses of Gloucestershire*, Vol. 2, Chichester, Phillimore, 1992
Landboc sive Registrum Monasterii de Winchelcumba (ed. D. Royce), 2 vols, Exeter, 1892–1903
Loosley, John, *The Stroudwater Riots of 1825*, Stroud Museum Association, Textiles Group, 1993
Mackie, John, *How to Make a Woollen Mill Pay*, London, Scott, Greenwood, 1904
Mahler, Oliver & Marshfield, Stephen, *Stroudwater Valley Mills*, Dursley, 1982
Mann, J. de L., *The Cloth Industry in the West of England from 1640 to 1880*, Oxford, Clarendon Press, 1971
Mills, S. & Riemer, P., *The Mills of Gloucestershire*, Barracuda Books, 1989
Morgan, John, *The Story of Erinoid*, Coleford, 2008
Palmer, Marilyn & Neaverson, Peter, *The Textile Industry of South-West England: A Social Archaeology*, Brimscombe, Tempus, 2005
Partridge, William, *A Practical Treatise on Dyeing*, 1823, repr. Edington, Pasold Research Fund, 1973
Perry, R., *The Woollen Industry in Gloucestershire to 1914*, Shrewsbury, Ivy House Books, 2003
Perry, R., *Wotton under Edge: Times Past – Time Present*, 1986
Phillimore, W.P., *Some Account of the Family of Holbrow*, 1901
Playne, Arthur Twisden & Long, Arthur Leslie, *A History of Playne of Longfords Mills*, Gloucester, John Bellows, 1959
Ponting, K.G., *The Woollen Industry of South-West England*, Bath, Adams & Dart, 1971
Ponting, K.G. (ed.), *Baines's Account of the Woollen Manufacture of England* (1875), repr. Newton Abbot, 1970
Power, Eileen & Tawney, R.H., *Tudor Economic Documents*, 3 vols, 1951
Randall, Adrian, *Before the Luddites: Custom, Community and Machinery in the English Woollen Industry 1776–1809*, Cambridge University Press, 1991
Records of the Corporation of Gloucester (ed. W.H. Stevenson), Gloucester, 1893
Ramsay, G.D.F., *The Wiltshire Woollen Industry in the Sixteenth and Seventeenth Centuries*, London, Frank Cass, 1965
Rogers, Kenneth, *Wiltshire and Somerset Woollen Mills*, Edington, Pasold Research Fund, 1976
Rogers, Kenneth, *Warp and Weft: The Story of the Somerset & Wilts Woollen Industry*, Buckingham, Barracuda Books, 1986
Rogers, K.H., *Woollen Industry Processes: I, The Domestic System*, Trowbridge, Friends of Trowbridge Museum, 2006

Rogers, K.H., *Woollen Industry Processes: II, The Factory Industry*, Trowbridge, Friends of Trowbridge Museum, 2008
Rose, Mary B., *International Competition and Strategic Response in the Textile Industries Since 1870*, London, Frank Cass, 1991
Rostow, W.W., *British Economy of the Nineteenth Century*, Oxford, Clarendon Press, 1961
Rudd, M.A., *Historical Records of Bisley with Lypiatt*, 1937, Stroud, repr. Sutton, 1977
Rudder, Samuel, *A New History of Gloucestershire*, Cirencester, 1779
Rudge, T., *The History of the County of Gloucester*, 1803
Skinner, Elizabeth, *Sheepscombe: One Thousand Years in this Gloucestershire Valley*, Sheepscombe Historical Society, 2005
Tann, Jennifer, *Gloucestershire Woollen Mills*, Newton Abbot, David & Charles, 1967
Tann, Jennifer, *The Development of the Factory*, Cornmarket Press, 1970
Valor Ecclesiasticus, temp. Henrici VIII auctoritate regia institutus, 6 vols, 1810–34
Veale, E.W.W. (ed.), *The Great Red Book of Bristol*, 4 vols, 1933–1953
Verey, David & Brooks, Alan, *The Buildings of England Gloucestershire I: The Cotswolds*, New Haven, Yale University Press, 2002
Victoria County History: Gloucestershire, vols: 2, 1907; 6, 1965; 10, 1972; II, 1976
Walmsley, Philip, *Stroud*, Stroud, Alan Sutton, 1994
Whitfield, Christopher, *A History of Chipping Campden*, Eton, Shakespeare Head Press, 1958
Witchell, M.E.N. & Hudlestone, C.R. (eds), *An Account of the Principal Branches of the Family of Clutterbuck from the sixteenth century to the present time*, Gloucester, privately printed, 1924
Woodcroft, Bennet, *Alphabetical Index of Patentees of Inventions 1617–1852*, London, EAM, 1969

ARTICLES; BOOK CHAPTERS

Beddow, Geoffrey, 'Dyehouse at Brimscombe Mill', *GSIA Journal*, 1976
Conway-Jones, Hugh, 'The Silk Industry in the Chalford Valley', *Glos. Hist. Studies*, XI, 1980
Cox, Christopher, 'A Distant Prospect, Being a Brief Look at Industrial Relations in the Stroudwater Area 200 Years Ago', *GSIA Journal*, 1989
Fairlie, A., 'Dyestuffs in the Eighteenth Century', *Economic History Review*, 17, no.3, 1965
Falconer, Keith A., 'Textile Mills and the RCHME', *Industrial Archaeology Review*, XVI, I, 1993
Falconer, Keith, 'Fireproof Mills – The Widening Perspective', Ibid.
Falconer, Keith, 'Mills of the Stroud Valley', Ibid.
Fuller, E.A., 'Medieval Cirencester', Bristol & Gloucestershire Archaeological Society Trans. BGAS, 54, 1932
Gill, G., 'Blackwell Hall Factors, 1797–1799', *Economic History Review*, 6, no.3, 1954
Griffin, Peter, 'The Mill at "The Barracks", King's Stanley', *GSIA Journal*, 2008
Haine, Colleen, 'Wool Drying Stoves along the Painswick Stream', *GSIA Journal*, 1981
Haine, Colleen, 'Cloth Mills along the Painswick Stream – Mills near the Centre of Painswick', *GSIA Journal*, 1982
Haine, Colleen, 'The Cloth Trade along the Painswick Stream. The Cloth Mills Pt III', *GSIA Journal*, 1983
Haine, Colleen, 'The Cloth Trade along the Painswick Stream: Pt IV', *GSIA Journal*, 1984
Haine, Colleen, 'The Cloth Trade along the Painswick Stream: Pt I', *GSIA Journal*, 1985
Haine, Colleen, 'The Cloth Trade along the Painswick Stream: Pt I', *Gloucestershire Historical Studies*, 1982

Haine, Colleen, 'Skinner's Mill', *GSIA Journal*, 1986
Hall, Michael, 'Rural Murals', *Country Life*, 30 August 1990
Hardwick, David, 'A Long Story – (of Mill and Mine Owners)', *GSIA Journal*, 2002
Hudson, Janet, 'The Early History of Two Stonehouse Mills: a re-interpretation', Trans. BGAS, 118, 2000
Jameson, Joan, 'The Eve of the Industrial Revolution: Access to land in the weaving parish of Bisley', *Gloucestershire History*, 24, 2010
Jenkins, D.T. & Malin, J.C., 'European Competition in Woollen Cloth', in Mary B. Rose (ed.), *International Competition and Strategic Response in the Textile Industries since 1870*, Cass, 1991
Kilburn Scott, A., 'Early Cloth Fulling and its Machinery', Transactions of the Newcomen Society, 12, 1931–32
Kingsley, Nicholas, 'Boulton & Watt Engines Supplied to Gloucestershire: A Preliminary List', *GSIA Journal*, 2005
Lindley, E.S., 'A Kingswood Abbey Rental', Trans. BGAS, 70, 1951
Lindley, E.S., 'Kingswood Abbey, its Lands and Mills', I, Trans. BGAS, 73, 1954
Lindley, E.S., 'Kingswood Abbey, its Lands and Mills', II, Trans. BGAS, 74, 1955
Mackintosh, Ian, 'Laying the Foundation: Stroud in the 16th Century', *GSIA Journal*, 1985
Mackintosh, Ian, 'The Metropolitan Town of the Clothing Trade', *GSIA Journal*, 1986
Mackintosh, Ian, 'History of Ham Mills 1600 to 1900', Typescript, Stroudwater Textile Trust, 1997
Major, J. Kenneth, 'Dunkirk Mills, Nailsworth, Gloucestershire', Typescript, 1979
Marling, S.S., 'The Woollen Industry of Gloucestershire: A Retrospect', Trans. BGAS, 36, 1913
Mayhew, Henry (ed.), 'A Visit to Hunt & Co's Lodgemore and Frome Hall Mills', 1862
Mills, Stephen, 'Beards Mill and the Mender's Shop (Leonard Stanley)', *GSIA Journal*, 1991
Mills, Stephen, 'An Unrecorded Mill near Slad?' *GSIA Journal*, 1993
Mills, Stephen, 'Oil Mill, Ebley (Ebley Corn Mill)', *GSIA Journal*, 1994
Mills, Stephen, 'The Mystery of Cloth Manufacture at Fromebridge Mill', *GSIA Journal*, 1999
Mills, Stephen, 'Millend Mill, Eastington – The Past and the Future', *GSIA Journal*, 2000
Mills, Stephen, 'The Rise and Fall of Henry Hicks, Clothier of Eastington', *GSIA Journal*, 2000
Mills, Stephen, 'Coal and Steam – the Arrival of Steam Power in Stroud's Woollen Mills', *GSIA Journal*, 2004
Mills, Stephen, 'Henry Hicks – A Man of Wide Horizons', *GSIA Journal*, 2008
Minchinton, W.E., 'The Beginnings of Trade Unionism in the Gloucestershire Woollen Industry', Trans. BGAS, 70, 1951
Minchinton, W.E., 'The Petitions of the Weavers and Clothiers of Gloucestershire, 1756', Trans. BGAS, 73, 1954
Moir, E.A.L., 'Marling & Evans, King's Stanley and Ebley Mills, Gloucestershire', *Textile History*, 2, 1, 1971
Moir, E.A.L., 'The Gentlemen Clothiers', in Finberg, H.P.R. (ed.), *Gloucestershire Studies*, Leicester University Press, 1955
Moir, E.A.L., 'Benedict Webb, Clothier', *Economic History Review*, 10, no.2, 1957
Palmer, M. & Neaverson, P., *The Textile Industry of South West England*, Tempus, 2005
Paterson, Nigel & Mills, Stephen, 'Cloth, Pins and Leather – An Examination of Frogmarsh Mill, Woodchester', *GSIA Journal*, 1997
Perry, R., 'The Gloucestershire Woollen Industry 1100–1690', Trans. BGAS, 1945
Randall, Adrian J., 'Work, Culture and Resistance to Machinery in the West of England Woollen Industry', in Hudson, Pat (ed.), *Regions and Industries*, Cambridge University Press, 1989
Rose, R.L., 'The Industrial History of Dudbridge', *GSIA Newsletter*, August 1966

Tann, Jennifer, 'Some Problems of Water Power: a Study of Mill Siting in Gloucestershire', Trans. BGAS, LXXXIV, 1965
Tann, Jennifer, 'Multiple Mills', *Medieval Archaeology*, 1969
Tann, Jennifer, 'Some Account Books of the Phelps Family of Dursley', Trans. BGAS, 86, 1967
Tann, Jennifer, 'The Bleaching of Woollen and Worsted Goods 1740–1860', *Textile History*, 1, no.2, 1969
Tann, Jennifer, 'Richard Arkwright and Technology', *History*, LVIII, 1973
Tann, Jennifer, 'Fuel Saving in the Process Industries during the Industrial Revolution: A Study in Technological Diffusion', *Business History*, 15, 1973
Tann, Jennifer, 'The Textile Millswright in the Early Industrial Revolution', *Textile History*, 5, 1974
Tann, Jennifer, 'The Employment of Power in the West-of-England Wool Textile Industry 1790–1840', in Harte, N.B. & Ponting K.G. (eds), *Textile History and Economic History: Essays in Honour of Miss Julia de Lacy Mann*, Manchester University Press, 1973
Walrond, L.F.J., 'Early Fulling Stocks in Gloucestershire', *Industrial Archaeology Review*, 1, no.1, 1964
Watson, C.E., 'The Minchinhampton Custumal', Trans. BGAS, 54, 1932
Wilson, Ray, 'Dunkirk Mills, Nailsworth: A New Chapter', *GSIA Journal*, 1988
Wilson, Ray, 'Dunkirk Mills, Nailsworth: A Progress Report', *GSIA Journal*, 1989
Wilson, Ray, 'Electricity Generation at Longfords Mill', *GSIA Journal*, 1992
Wilson, Ray, 'Coaley Mill – Corn, Cloth & Ironwork', *GSIA Journal*, 1996
Wilson, Ray, 'Dunkirk Mills, Nailsworth: Chapter Three', *GSIA Journal*, 1990
Wilson, Ray, 'Howard & Powell, Wallbridge Mill, Rodborough, Stroud', *GSIA Journal*, 2004
Youles, Tony, 'The Nailsworth Engineers pt 1', *GSIA Journal*, 1989

LEAFLETS, PAMPHLETS

Brooks, Richard, Observations on Milling Broad and Narrow Cloth etc., London, 1743
Mackintosh, Ian, 'Dunkirk Mills: a Brief History', Stroudwater Textile Trust, 2002
Mackintosh, Ian, 'Power and the Playnes', Typescript, Stroudwater Textile Trust, 1993
Stroudwater Textile Trust, Stroud Town, Walk 1, Step into the Picture, 2009
Stroudwater Textile Trust, Nailsworth Mills, Walk 1, Streams of Cloth, 2007
Stroudwater Textile Trust, Nailsworth Mills, Walk 2, Weavers' Yarns, 2008

PART II

MILLS AND ASSOCIATED BUILDINGS BY RIVER VALLEY

Note:
All MSS references, unless otherwise stated (e.g. Public Record Office – PRO), are to documents in Gloucestershire Archives. References with the suffix 'gs' are to documents formerly in Gloucester City Library, now in Gloucestershire Archives.

Where no six-figure grid reference is given, following the mill or workshop name, the precise location is unknown.

RIVER BOYD

Bitton Mill, ST 680698
In 1634 Thomas Flower held a corn mill and three fulling mills – probably three pairs of stocks. In the eighteenth and nineteenth centuries this was a corn mill.
(Smyth MSS, *A Book of Rentals*, 16067, f. 44 gs; D185 V/33)

Doynton Mill, ST 719744
Leased in 1647 to Thomas Harrord Junior by Robert Bushe of Bath. A dyehouse and a corn mill adjoined the mill, the latter occupied by a different tenant.
(Box 65gs)

LITTLE AVON AND TRIBUTARIES

Oil Mill, Berkeley, ST 676991
Daniel Marklove, clothier, owned Oil Mill in the late eighteenth/early nineteenth centuries, and this was probably his fulling mill. As the name implies, this mill was later used for oil seed crushing.
(PP 1802–03, VII, p.312)

Stone Mill, ST 688956
A leat was cut for a new mill *c.*1500. It was first of all 'a blademill, after a corne grist mill, after a

paper mill, nowe at this day both, anno 1639'. In 1634 John Bowser of Tortworth held a grain mill and fulling mill with two stocks under one roof in Stone. It was later a corn mill, containing one small broad waterwheel and a narrow 24ft-diameter wheel made by T. Bond of Burford. The building dates from 1729.
(Berkeley MSS, III, p.362; Smyth MSS 16067)

Damery Mill, ST 706943

In 1564 John Hicks owned a fulling mill in Tortworth. In 1632 William Hicks conveyed the fulling and grist mill to Arthur Hicks. By 1745 only the corn mill remained. It has been demolished.
(D340a/T172, T175, T137)

Avening Mill, ST 709942

In 1726 this was a fulling mill containing two pairs of stocks with 'the dyehouse and gardens thereto belonging'. It has been demolished.
(D340a/T142/17)

Unnamed mill in Tortworth (probable site), ST 684930

In 1637 a grist and fulling mill, occupied by Thomas Prowte, was leased to William Ames of Frampton Cotterell. It has been demolished.
(G.R.O. D340a/T127)

Huntingford Mill, ST 715935

A seventeenth-century rental records 'Elleanor Messlenger ... for a fullinge mill'. In 1824 the mill 'now used as a clothing mill' was occupied by Edward, Anthony, and John Austin and John Stubbs. Samuel Long owned the mill in the 1830s and it was occupied by John Holbrow and Rowland Nathaniel Long. It became a corn mill. The early nineteenth-century mill is now a hotel/restaurant.
(Berkeley MSS, III, p.236; RZ 41.1 (1-3)gs; D654/T11/4/T1; PP 1840, XXIV, p.399)

Charfield/New Mills, ST 723930

Called New Mills in the seventeenth century, its function was not specified. In the early nineteenth century it was in the hands of Samuel Long and was used for cloth manufacture. The water power was about 50hp in winter. By 1833 steam power had been installed. In 1840 there were sixteen handlooms. It was worked by members of the Long family until cloth manufacture ceased at the end of the nineteenth century. In 1891 there were 1,650 spindles and forty looms in the mill. There are three mill buildings, all built between 1815 and 1829. The earliest range has been well restored (see p.161). The large five-storey mill, minus three storeys, is a residence, the 1829 mill houses an engineering works.
(Smyth MSS, 16533 f.106; PP 1834, XX, p.252; PP 1833, XX, p.957; PP 1840, XXIV, p.396; Worrall's Directory 1891)

Ithell's/Bytheford Mill, ST 732932

Recorded in 1616, it was a forge house, corn mill and fulling mill under one roof and was sold in 1627 to Thomas Perry. In Perry's survey the site is called a 'grindstone mill and a blademill under one roofe'. By 1640, the mill comprised a fulling mill and grist mill. By 1795 the fulling and flock mill were conveyed to members of the Dyer family. It was a flock mill in 1867.
('Merryford Deeds', BGAS, LXXV, p.102; E.S. Lindley, Kingswood Abbey, BGAS, LXXV, p.92; Smyth MSS, 16533, f. 106; D654/11/23/T8; D654/11/ 21/T14; Directory 1867)

New Mills/Sury/Shewry Mill, ST 737930

Thomas Perry gave the mill to his son Samuel in the seventeenth century. Perry's IPM identifies one grain and two fulling mills in Kingswood called Shewry Mills. In 1741 the corn and fulling mill with the Rack Close were conveyed to the executors of Thomas Brown. In 1800 the site was described as 'stockmill in possession of Thos. England'. New mills were rebuilt in 1802 and bought by Humphrey Austin in 1806, being leased to George Austin in 1811. A detailed inventory of the interior of the mills was made.

In 1833 water was the only motive power and around 40hp could be generated. In 1844 the property was in the possession of Lewis & Dutton, clothiers of Wotton-under-Edge. Then for a short while, silk was manufactured here by Le Gros Thompson & Co., but from the 1860s to about 1870 it was owned by Samuel Long of Charfield Mills. It was then bought by Tubbs & Lewis and was no longer used for cloth manufacture. The well-restored, five-storey brick mill has an attractive central clock tower. Tie beams between each floor provide interesting detail. It is the headquarters of Renishaw plc (see p.159).
(Smyth MSS, 16533, f. 106; IPM (Index Library) II, p.101; Box 35, E.S. Lindley, 'Kingswood Abbey ...' BGAS, 74, p.98; D654/T11/4/T1; PP 1833, XX, p.954; Directories 1844, 1849, 1863, 1865–6, 1867, 1870)

Langford Mill, ST 745924

In the late eighteenth century Joseph Jones of Newnham leased 'all that fulling mill with two stocks' to Thomas Mercer of Kingswood, clothier. By 1801, Humphrey Austin was the owner, although Mercer was still the lessee. At this date Mercer entered into an agreement with Austin not to dam the water and prevent it from flowing down to Austin's New Mills below. Humphrey Austin settled the mill on his son in 1812. By 1839, the mill was occupied by J. Dancer and in 1851 by Llewellin Perrin. In that year, Perrin went into partnership with a Kingswood accountant, the firm being called Perrin & Chapman. The mill later became a silk mill and was bought by Tubbs & Lewis. The five-storey mill had a date stone of 1822 in the centre of the east wall (see p.161).
(D654/11/23/T14; D654/T11/4/T1; D654/T11/4/53; Directory 1839–40; D654/T11/20/B2)

Abbey/New Mill, ST 746922

A seventeenth-century rental notes: 'Next above that (Sury Mill) wch is now decayed was a mill called the abby mill where Wm Tanners dyehouse now is.' The mill was probably rebuilt shortly afterwards, becoming known as the New Mill, and, in 1640, belonged to John Smyth of Nibley by which date it was in poor condition and had more than one tenant.

In 1693 the mills were sold to Robert Daw of Bradley. Sometime after this they reverted to corn milling and remained thus until 1801, when 'All that water corn mill commonly called ... Abbey Mill' was let to Thomas Mercer at £21 p.a. It was then converted into a cloth mill. The power was said in 1834 to be 'principally water, [but] a small steam engine has been erected about 7 years ago'. By 1851 it was in the hands of Llewellin Perrin. From about 1865 to the 1870s it was occupied by George Knight & Stanton. The mills were burnt in 1898.
(Smyth MSS, 16533 f. 106; Smyth MSS, 16533 f. 107; P.C.C. 135 Evelyn; IPM (Index Library) II, p.167; D6T11/23/T14; D654/T11/20/B21; PP 1834, XX, p.337; Directories 1865–66, 1874)

Walk Mill, ST 750919

The mill was known as Walk Mill in 1537 which suggests that it was then a fulling mill. A 'recently erected fulling and gig mill', formerly owned by Richard Spicer, is referred to in 1687. The mill containing three stocks was advertised to be let or sold in 1730. In 1754 Thomas Tanner levied a fine on half 'the fulling mill with formerly three now two stocks called the Walk Mills'. By 1764 a flock mill had been added and Tanner was the sole owner of the mills. In 1839 the property was up for sale and comprised two adjoining mills, one five storeys and 76ft by 24ft and the other 75ft by 24ft and four storeys high. There were three waterwheels and three pairs of stocks. Ten ends of fine cloth could be made per week.

There was also a new reservoir. John Millman worked the mills in 1844 and, by the 1860s, Long & Wellings occupied the property. They were succeeded by Porter & Long and finally by Robert B. Cooper. The mills have suffered from fire.
(E.S. Lindley, 'Kingswood Abbey its Lands and Mills', BGAS, 74, 1955, p.53; Smyth MSS, 16533 f. 106; D1086, Kingswood; GJ, 17 November 1730; D654/TI 1/23/T12, T13; RX180.3(1); Directories 1844, 1863, 1865–66, 1870, 1879)

Nind Mill, ST 754915

In 1665 the mill, containing three stocks, was conveyed to Elizabeth Blagden and John Hicks. In 1687 Nind Mill was granted to Huntley, Bridges & Neal. In 1738 the 'fulling mill and other mills' at Nind were conveyed to John Nelmes and Thomas Hopton. The tenant was Edward Tanner. In 1808 the property was described as a fulling mill, gig mill and other mills, occupied by Vanes and Samuel Counsell. In 1817 the property was divided, John Blagden Hale retaining part of the mills. Hale confirmed that he agreed 'to erect for the lessees in the said premises which they are to have two waterwheels each six feet wide with three pair of stocks attached to one wheel and two gig mills to the other'. By 1824 another part of the mills was being let to John Partridge and George Hancock. This part comprised 'all those six lofts and one stock and gig mill being the old part of the mill called Nind Mill ... with a power to drive six engines and twelve shearing frames or cutters, one brusher and two tuckers'. Joseph Oldland, 'some time of Nind Mill ... clothier', was bankrupt in 1831. In 1834 Counsell & Millman, then probably sole occupiers, replied to the factory inspector's queries. The property comprised 'An old mill for milling or fulling etc. beyond our recollection' and a new part which was added in 1816 or 1817. They used 'water power only ... the power about 20 horse'. They carried out all the operations of cloth manufacture there. By 1839 there were twenty handlooms in the factory but only ten were in use. By 1849 Counsell had retired and the firm was known as Millman & Overbury. Overbury filed for bankruptcy in 1854. In 1865 it was Millman & Foxwell and by 1894, Millman, Hunt & Co. By 1897 cloth manufacture had ceased here (see p.91). In 1912 Nind Mill, by then used for flour and flock manufacture, caught fire. There were at least five wheels, three of which were approximately 8ft wide. The mills were fed by one of the largest ponds in the county.
(DI 1086; D1128; D654/111/73; LG, 7 October 1831; PP 1834, XX, p.336; PP 1840, XXIV, pp.396–97; LG, 20 February 1855; Directories 1849, 1865–66, 1894, 1897; *Miller*, 5 August 1912)

Hawkesbury Mill, ST 768869

This was a fulling mill in the reign of Edward II.
(PRO Court Rolls 175/41)

Wickwar Mill

This was a fulling mill in 1552 but ten years later had reverted to corn milling.
(D340a/T146)

DYEHOUSES, WORKSHOPS

Nind Dyehouse

This is first recorded in 1669 when it was settled on the marriage of William Ireland and Margery Tanner.
(D1128)

Picking and Burling Shops

These were let to William and Samuel Long in 1812 for £35 per year.
(Box 35gs)

Press Shops, near Walkmill Lane, approx. ST 749919

These had been occupied by John and Thomas Carpenter for many years prior to 1844 when the tenants got into financial difficulties.
(D654/T11/20/T3)

KILCOTT VALLEY, A TRIBUTARY OF THE LITTLE AVON

Hillsley/Byrettes/Austin's Mill, ST 770905

In the seventeenth century Robert Symonds died holding one messuage and a fulling mill in Hillsley called Byrettes. In 1825 the mill was occupied by Austin & Son of New Mills who were making cloth here. By the tithe survey it was probably a corn mill. The mill has been converted into a house.
(IPM II, p.107; RF156.18(8)gs; P320aVE 1/3)

Alderley New Mills, ST 775903

In 1636 William Crewe owned a fulling mill and the ford leading towards Kilcott. In 1666 the mill was occupied by Matthew Crewe. In 1736 William Larton is described as 'of New Mills, clothier'. He extended the premises buying 'several rooms of housing ... the sharrshopps which then was, and the chamber over them and the room wherein the cloth press then stood'. In 1822 they were occupied by Timothy Larton who had erected a considerable part of the buildings. The premises were described as: the 'mill-houses, fulling mills, gig mills, stocks, wheels, engines, machines, houses, outhouses, shops, lofts, chambers [and] rooms'. Timothy Larton was one of the first clothiers in the county to install looms in his factory. By 1806 he had twenty looms, sixteen of which were then at work. Shortly afterwards the mills were bought by Humphrey Austin. In 1840 Austin had fifty-nine handlooms at New Mills, thirty-two of which were not used. At the tithe map survey Austin owned and occupied part of the mills, the other part being a grist mill.
(IPM (Index Library) II, p.28; D1186 boxes 17, 18; D654/T11/14/T2; PP 1806, III, p.940; PP 1840, XXIV, pp.396–7)

Doges Mill, ST 785894

There is a tradition that this was once a cloth mill, perhaps confirmed by the field opposite being known as Rack Close.

OZLEWORTH BOTTOM, A TRIBUTARY OF THE LITTLE AVON

Grindstone Mill, ST 764913

In 1600 Matthew Pointz leased 'one smiths forge one grindstone mill and one fulling mill together in Alderley'. By 1640 Mary Barker was in possession of the site and she leased a dyehouse there to John Ireland in 1651. In 1692 the dyehouse was leased to Mathew Eastmead of Alderley, fuller, and by 1695 the mill was also in his possession. By 1740 the mill was solely a fulling mill and contained two stocks. At this date, the dyehouse was leased to a different tenant. It was possibly still a cloth mill in the early nineteenth

century and, at the tithe survey, it was occupied by Humphrey Austin. The mill is a small two-storey building of stone, lying parallel to the stream.
(E.S. Lindley, 'Kingswood Abbey its Lands and Mills', II, Trans. BGAS, 74, 1955, pp.51, 57; RF7.1; D1086 boxes 14 and 17; Alderley box gs; D1086, box 18; E.S. Lindley, *Wotton-under-Edge*, p.289)

Wortley/Broadbridge Mill, ST 767913

This was a grist mill and fulling mill, 'now or late in the tenure of Richard Taylor' in 1641. In 1649 it was occupied by Bartholomew Weeks, clothier. In 1798 the property contained two stocks and a gig mill, and was leased to Robert Roach of Wortley, fuller. When leased in 1827, it was described as 'late a fulling mill but now a water grist mill'. It was a small three-storey stone building.
(28899/A gs; SR187gs; 354(235)gs; Box 60 bdl 2gs)

Penley's Mill, ST 769914

The mill dates from at least the mid-sixteenth century and in 1608 Robert Plomer, also known as Seaborne, leased a messuage called The Milne and dyehouse to Robert Fowler, clothier, of Ozleworth. In 1674 it was still in the hands of the Fowler family and was leased to Daniel Adey of Wotton, clothier. In 1799 the fulling mill was leased to John Penley, of Uley, dyer. By the tithe survey, there was only a dyehouse at the site, which was occupied by Ann Penley.
(E.S. Lindley, *Wotton-under-Edge*, p.288; D1086, box 18; RF7.1; D1086 box 17)

Monks Mill, ST 774914

Formerly belonging to Kingswood Abbey, this was a fulling mill in 1537 and was let in 1557. In 1604 it was held by Huntley, Bridges & Neal. In 1612 the fulling and corn mills 'known as Munck milles', were re-leased to Christopher Neal. In 1613 Christopher Purnell, clothier, leased land to 'erect there a newe ffullinge mill'. This was an extension to Monks Mill. Later deeds refer to two mills here. Purnell bequeathed his part of the mill to his son Christopher in his will dated 1615. The whole mill passed into the hands of Richard Poole, who purchased the manor of Alderley in 1631. It appears that Poole bought out Neal and then allowed him to become a tenant there.

The mill remained in the hands of the Poole family until 1716, when Richard Osborne acquired it from William Parry. The property then comprised 'that fullinge mille or stocke and one corn or grist mill under one roof and one other fulling mill under one roof adjoining … with the storr lately built'. The mill was again extended in c.1768 and much rebuilt by John Osborne; Samuel Yeats, his son-in-law, succeeded to the mill. In 1810 Yeates and his son were in partnership. Yeates sold out in the 1820s and Penry Williams became the next owner. John Metivier and his brother became tenants in 1830. They were followed in 1835 by Thomas White and his son who mended and renewed the machinery. They occupied the mill until 1844. In 1839 they had fourteen handlooms in the mill, six of which were unemployed. The Whites were followed by Samuel Smith & Sons who were also working Nowell's Mill. The mill was held by Josiah Overbury in 1850 – he was bankrupt in 1854. In 1869 Rowland Smith moved to Stonehouse, taking machinery from Monks Mill with him. Monks Mill comprised a long east to west range of three distinct four-storey sections, part of the mills being driven by spring water and the rest by water from the main stream (see p.131).
(E.S. Lindley, 'Kingswood Abbey its Lands and Mills', Trans. BGAS, 74, p.54; D48/T33, Boxwell Deeds; IPM (Index Library) II, p.15; 28917 gs; 28899A gs; E.S. Lindley, *Wotton-under-Edge*, p.266; Smyth MSS, 16534; E.S. Lindley, 'History of Wortley', BGAS, LXVIII, p.50; Lindley, *Wotton-under-Edge*, pp.272, 277; PP 1840, XXIV, pp.396–7; LG, 20 February 1855; Box 60 bdl.gs)

Nowell's Mill, ST 778915

Edward Nowell, fuller, surrendered two fulling mills under one roof to the lord of the manor of Alderley, in 1669. In *c.*1725 the mill was held by Daniel Nowell. Between 1835 and 1850, it was owned by two Miss Burtons. In 1839, and probably until its closure, the mill was worked by the Smith family.
(D 1086, box 7; GJ, 15 November 1739; E.S. Lindley, *Wotton-under-Edge*, p.288; Directory 1839–40)

Hill Mill/Hell Mill, ST 783919

A fulling mill in Ozleworth was recorded in a pre-Dissolution survey of Kingswood Abbey lands *c.*1537. In 1699 Gabrielle Howe mortgaged 'all that messuage or tenement and a fulling mill and one grist mill being under one roof'. In 1706 the mill was described as 'late occupied by Nicholas Harris'. In 1710 it was sold to Daniel Hicks of Bristol. It remained a double mill until after 1721, after which it is unrecorded for several generations. William Hooper, miller of Hill Mill, went bankrupt in 1791. Partridge & Hancock were making cloth here in 1830 and William Jones was in occupation in 1842. Cloth manufacture probably ceased shortly afterwards. The leat was one of the longest in the county.
(E.S. Lindley, 'Kingswood Abbey its Lands and Mills', TBGAS, 73, 1954, p.188; D471/T46; D473/T1; D474/T5; LG 8–12 Feb 1791; Directories 1830, 1842)

Wortley Dyehouse

This is recorded in a survey of Hale lands made in 1674.
(28917 gs)

TYLEY BOTTOM, A TRIBUTARY VALLEY OF THE LITTLE AVON

Park Mill, ST 752922

This was probably 'Berkemill' mentioned in the Kingswood Abbey rental. In 1839 it was occupied by John Smith. At the tithe survey, William Long owned and occupied the premises. Long and Smith were joined by Wellings by 1856. In 1857 John Smith retired from his partnership with Long and Wellings. In 1870 Porter & Rogers were the occupiers, but in 1879 William Long was again recorded as occupier.
(Directories 1839–40, 1856, 1870, 1879, & PP 1840, XXIV, p.399; R. Perry, *Wotton under Edge: Times Past – Time Present*, 1986; LG, 21 April 1857)

Dudley Mill, ST 758925

Built in the early nineteenth century by the Revd Rowland Hill to provide an income for his almshouses it was first occupied by Mark Fowler. In 1836 it was leased to Samuel and William Alexander Long. At the tithe survey, the premises were empty. In 1844 cloth was being made here by Eli May and Samuel Smart and cloth manufacture probably ceased here soon afterwards. The mill has been demolished.
(D654/T11/16/T7; PP 1840, XXIV, p.399; Directory 1844)

Hack Mill, ST 755925

A mill existed here in 1537 and it may have been a fulling mill. However, between 1774 and 1850 it was a paper mill. It has been demolished; the pond, for a time, being the Wotton swimming pool.
(R. Perry, *Wotton under Edge: Times Past – Time Present*, 1986; ex. inf. Revd M. Chappell)

Loccumb Well Mill, ST 759932

Driven by a small spring; shown on the 1763 map of Wotton and, later, between 1798 and 1813, operated

by Austin's as a dyehouse. In 1830 Joseph Oldland was a woollen manufacturer here. It has been converted to residences.
(Directory 1830; ex. inf. M. Chappell; R. Perry, *Wotton under Edge: Times Past – Time Present*, 1986)

Cloud/Potters Pond Mill, ST 761933

In 1763 Daniel Adey was lessee of Cloud Mill. John Potter was granted a lease in 1784. It was probably a corn mill, for in 1812 Potter was said to be a miller. In 1812 William Potter bought the workshops which he had built prior to 1809 on the west side of Cloud Mill. He also connected a shaft to Cloud Mill waterwheel to provide power. In 1844 George Holloway and James Riddiver were manufacturing cloth here for a short while. The mill has been demolished.
(Box 60, bdl. 2 gs; D654/ TI1/8/T39; Directory 1844)

Moore's Mill, ST 761933

This was shown on a 1763 map, William Moore being the lessee. It may have been a fulling mill. Nothing further is known of the mill. It has been demolished.
(R. Perry, *Wotton under Edge: Times Past – Time Present*, 1986)

Neal's/Pounds Ground Mill, ST 762935

In 1823 John Tattersall, a dyer, leased a piece of ground and the lately erected mill. In 1824 he was bankrupt and his lands were conveyed to Edward and Thomas Neal, clothiers of Wotton. A clause in the conveyance mentions that the water used by the Neals for washing wool and cloth should not interfere with water pumped up for use in Tattersall's dyehouse. The Neal brothers had already built a mill and teazle house on part of Tattersall's land. During the 1825 weavers' riots, Neal's mill was attacked and in the same year the brothers got into financial difficulties. Later that year they became insolvent. By 1837 the property was sold. In 1838 clothing machinery, 'the property of a manufacturer declining in business', including carders, billies, jennies and cloth cutters, was offered for sale at Pounds Ground Mills. This site has been much altered.
(D654/111/167; RF354.13 gs; ex. inf. Revd M. Chappell)

Dyehouse/Holywell Mill, ST 763936

This may have been the fulling and grist mill known to have been near Coombe during Elizabeth's reign. In 1830 Lewis & Dutton were making cloth at Dyehouse Mill; they stated that the mill had been built in the nineteenth century, and that they had a 14hp steam engine. Dyehouse Mill became solely a dyeworks and was occupied by the White family until the late 1870s. The mill is a single-storey building.
(E.S. Lindley, *Wotton-under-Edge*, p.298; R. Perry, *Wotton under Edge: Times Past – Time Present*, 1986; Directories 1830, 1839–40, 1879; PP 1834, XX, p.289)

Strange's/Venn's Mill, ST 765937

In 1749 Daniel Adey leased a 'tenement and water corn grist mill called ... Venns Mill' to John Wallington and Philip Dauncey, clothiers. The lessees were permitted to convert part of the tenement into a dyehouse, provided that, at the termination of the lease, they restored it to its former condition. In 1812 the Strange family leased 'part of the mill house of William Strange ... now used as a fulling mill and factory and which room or loft is situate immediately over the steam engine contained in this building' to the firm of Austin & Co. The lessees also occupied shear shops and were permitted to have the use of the steam engine. It seems probable that the property continued to be divided. Robert Darrett was a cloth manufacturer of Holywell in 1822 and employed about 200 people. He had a 'works' at Holywell with a

16hp steam engine and water power equalling 30hp. By 1839 Overbury & Harris were cloth manufacturers of Holywell and Darrett disappears from the records. During this period, William Strange was a dyer in Holywell. When the cloth industry at Wotton collapsed, William Strange continued in sole possession of the mill which became a dyeworks. This was one of the last dyeworks in Wotton.
(E.S. Lindley, *Wotton-under-Edge*, p.152; D654/T1/T6; D1759; Directories 1822–23, 1839–40, 1870; PP 1833, XX, p.955)

STEAM MILLS

Britannia Mill, ST 762935
An early nineteenth-century brick mill, adjoining Pounds Ground Mill, but not driven by water power, owned by Neal and run in conjunction with Neal's Mill. It was probably used for preparatory processes or handlooms. Attacked by the weavers, in 1825, together with Pounds Ground Mill, it later became a printing works (see illustration). It has since been demolished.
(ex. inf. Revd M. Chappell)

Waterloo Mill, ST 759932
This mill was probably built by the Austin family in 1815. It was used by them until 1827 when they leased it to Richard Lewis and James Dutton. In 1827 the mill contained a 'steam engine, stocks, gigs, drums and shafts'. In 1830 Lewis & Dutton still occupied this mill as well as Dyehouse Mill but, in 1835, they were bankrupt and the mill was for sale. The premises were described as containing a high-pressure steam engine, stocks, gigs, a double washer, and an oven for heating plates. By 1840 the mill was still owned by John Austin and no occupier is recorded. The mill is a small, three-storey stone building with an engine house on the east side. It has been converted into residences. There is a date stone of 1815 on the west wall.
(D654/T11/4/T1; Directories 1830, 1840; IX 354(2)gs)

The Steep/Church Mill, ST 758933
The site was in the hands of the Austin family in 1763, but the main mill building was not erected until later. This was one of the first mills in the county to be built wholly as a steam mill and away from the valley bottom. In 1830 the four-storey mill was advertised to be let with the steam engine, air stove and other buildings. By 1837 the building was empty.
(E.S. Lindley, *Wotton-under-Edge*, p.304; G.C.L. RV354.2(2); PP 1840, XXIV, p.399)

Old Town Mill, ST 757935
This mill was erected in 1817 by the Austins, an unwise speculation. In 1830 Llewellin Ford occupied the premises, employing 150–200 people. A 20hp steam engine supplied power. There was a shortage of water for the engine as the only supply was well water. Austin manufactured broadcloth and kerseymere and the buildings were used for those purposes in 1825. By 1839 H. Beavan was occupying the mill. In 1840 he was joined by a partner, Stuart. The mill is a stone building of three storeys with a Cotswold slate roof.
(Directory 1830; PP 1833, XX, p.955; PP 1834, XX, p.290; Directory 1839–40; PP 1840, XXIV, p.399)

Sym Lane Mill
John Darrett was a clothier of Sym Lane in 1830. By 1837 the building was empty.
(Directory 1830; PP 1840, XXIV, p.399)

SEPARATE DYEHOUSES, SCOURING HOUSES AND WORKSHOPS IN WOTTON-UNDER-EDGE

Dyeworks below Strange's Mill, ST 765937
In the early 1870s Robert Vincent Perkins was a dyer at Holywell. The dyehouse has been demolished.
(Directory 1870)

Haw Street Workshops
In 1796 a workshop built by John Tipping and Abraham Owen was conveyed to George Owen. By 1821, Owen's workshops had been converted into tenements.
(D654/T11/31/T4; Directory 1830)

Bear Lane Workshops, ST 755934
In 1763 this site was occupied by Henry Winchcombe. It was probably the contents of this building which were offered for sale with Pounds Ground Mill in 1838. The workshops contained sleys, shuttles, a press shop, twenty-nine broadlooms, cloth presses, beams and scales.
(E.S. Lindley, *Wotton-under-Edge*, p.131; RF354.13)

Eight houses and weaving shops, ST 759928
These are shown upstream of Dudley Mill on the tithe map and were occupied by James Fowler and others.

High Street Workshops
These workshops were converted for cloth manufacture by Anthony and Humphrey Austin. A Rack Close is marked on the tithe map as lying between Long Street and Old Town.
(354(52) gs)

Workshops in Sinwell on part of a close called Long Leaze
In 1831 Charles King, a butcher of Dursley, came into the possession of part of Long Leaze, workshops and other buildings 'sometime since erected' by Isaac Palser and formerly occupied by Linnsell, John Hallis, William Durn, Sarah Shipway, Charles Cousins, William Moore and John Payne.
(Box 53, bdl. 7gs)

Workshops in Sinwell, erected on Rack Close
In 1763 a tenement with outhouses, buildings, workshops, a stove and summer house, which had been purchased by Edward Wallington from John Arrowsmith, were leased to Charles Adey, clothier. In 1834 'a Clothing Mill situate in the tithing of Sinwell' was for sale following a decree in Chancery in the case of Parrott v. Rickards.
(D654/T11/1/T8; LG 31 Jan, 1834)

Shear Grinding Mill in Sinwell
This was situated near a place called Rommakin in Sinwell and Bradley and was formerly occupied by James Walter, although by 1816 it was void.
(Box 60, bdl. 2gs)

Austin's Dyehouse, ST 759933
This was in the hands of the Austin family in 1798. By 1813 it was ruinous and was let to a carpenter.
(D654/111/175)

Pounds Ground Dyehouses, ST 763936
This site was occupied by Richard Lewis at the tithe survey; the map shows one large building and four small ones with a circular building, probably a wool stove, lying between the stream and the back brook from Dyehouse Mill. John Lewis was dyer here in 1822.
(Directory 1822)

Coombe Dyehouse, ST 770939
In 1808 Daniel Goodson Adey and George Austin conveyed land, on part of which stood a stove or dyehouse belonging to John Wheeler, to Samuel Vines. This had formerly been in the tenure of Ann Watts. In 1836 a grist mill at Coombe 'sometime since erected' by William Watts was leased to Joseph Wheeler.
(D654/T11/28/T1; D654/T11/21/T21)

Scouring House belonging to Richard Lewis, ST 767938
This small building is shown on the tithe map. Nothing remains.

Scouring House belonging to William Strange, ST 766939
This is shown on the tithe map. The three small ponds, fed by a spring, remain. The field below the scouring house was known as Dyer's Orchard.

THE DOVERTE BROOK, A TRIBUTARY OF THE LITTLE AVON

Wick Mill, ST 710968
In the early seventeenth century, Thomas Bowser rented the mill from the Berkeley family. In 1689 George Smyth appeared to own the property. The functions are not recorded. Eli Gazard, miller, insured the mill 'used as a grist mill and for the manufacture of flocks' in 1799. In 1816 Gazard is described as a paper maker. At the tithe survey Adonijah Harris owned the mill and it was occupied by Daniel Cooke. It had probably become a corn mill by the late 1830s. Two rack closes and a teazle ground are shown on the tithe map near the mill which provides further evidence of cloth making here.
(Box 42, bdl. 3; Smyth MSS, 16062, 16068; Royal Exchange policy 18804, 17 Jan 1799; Excise Letter no.98)

North Nibley
John Smyth described his home village of North Nibley in the following words:

> In the easterne part of the village of Nybley arise divers springs of excellent sweete water, which united and brought into one streame make a pretty river; whereon are seated seavern tuck mills and grist mills, most of them double mills before the said streame be passed through this village; the like whereto I knowe not within this county.

(Berkeley MSS, III, p.267)

Munday's/Pick's Mill, ST 726969

A seventeenth-century rental records that a fulling and grist mill 'under one roof of old thereto belonging commonly called Mundaies land' was conveyed to John Smyth. In 1687 the property, still a fulling and grist mill, was conveyed to Henry Adey of Dursley. The deed mentions that 'the roofe of the said mills is late demolished by fire'.

In 1794, the fulling and grist mills were sold to Giles Pick, miller. By 1806 the property was described as 'a mill containing one waterwheel, 2 pr of stones and Bolter'. The stone mill is three storeys high plus an attic. There was latterly one waterwheel and three pairs of stones. An office and miller's rooms adjoin the mill.
(Berkeley MSS, III, p.298; RV216.16; D654/ T11/8/T39; D654/111/33; P230/VEI)

Purnell's Mill, ST 729968

This was a corn and fulling mill owned by William Purnell in the early seventeenth century. Purnell also owned a rack ridge in Winley Field. The mill was bought by Anthony Hungerford in 1622–23 and the rack ridge and other lands were sold to John Curnocke. By 1698 the premises had been bought by Richard Tyler and were occupied by John Curnocke. In 1716 the mills were sold to Christian Harris of Wortley. In 1798 the mill was a fulling and flock mill and was occupied by Charles Organ. At the tithe survey, Llewellin Perrin occupied the mill and rack ground. He continued making cloth there until after 1856, but by 1863 he was described as a miller. The mill consisted of a single range, four storeys in height, with two short wings on the west side. There was also a range of workshops in the Rack Close.
(Berkeley MSS, III, p.277; IPM (Index Library) III, p.34; Charfield deeds gs; 19919/156; D654/T11/8/T41; Nibley deeds gs; Directories 1856, 1863)

Snitend Mill, ST 736967

In 1545 Robert Coldwell, master of Wotton-under-Edge Grammar School, leased the estate of Warren's Court 'and a fullynge mylle ... with reck lands' to William Thomas of Nibley. By 1599 the estate had been transferred to John Smyth, for in that year he sublet 'one tenement and one tucke mill under one roofe' to George James, clothier. In 1606 Smyth asked for the lease to be renewed. A survey of the grammar school estates, made in 1622, noted that Smyth held Warren's Court and the mill was sublet to Thomas Trotman. In 1677 Edward Smyth leased 'all that his fulling mill and grist mill under one roofe' to John Curnocke of Nibley, clothier. In 1718 an exchange of lands between the grammar school and Smyth's descendants took place and the fulling mill, held by Richard Martin, was conveyed to Nicholas Smyth Owen. A survey of the school lands made in the same year recorded that Trotman was still tenant.

In 1752 George Smyth leased the Rack Leaze and other land adjacent to the mill to John Hadley on condition that Smyth had access to his mill and the racks then standing in the Rack Leaze. In 1774 the mill, described as a corn mill in the occupation of David Organ, was leased to John Gazard of Wick Mill, Berkeley. It seems likely that the property had become a double mill by this date and that the corn and fulling parts were separately leased. In 1798 the property was for sale and was described as a 'duffell mill ... capable of being used for an extensive manufactory'. In 1801 Charles Organ leased 'all that cloth or fulling mill with the gig mill' to John Brown of Cam, clothier. In 1816 the mill, 'a flock and fulling mill ... in the possession of William Organ and others', was again up for sale. The sale bill claimed that 'This lot is very desirable for a fulling mill, or grist mill and capable of great improvement'. The mill has been demolished.
(216(24)gs; Smyth MSS, 16530 f. 22; RQ216.4 and 216.6; RX354.19; Box 101gs; RV216.I2; 1988(5)gs; Box 42, bdl. 3 gs; RV216.3; RV 216.1(2))

Gazard's/The Granaries Mill, ST 741968

In 1798 'a duffell mill and grist mill' occupied by 'Mrs. Gazard' were for sale. The property was 'particularly recommended to the attention of manufacturers there being a constant supply of water to the mills with which an extensive manufactory may be carried on, and having a good road leading to it'. In May 1800 Cornelius Gazard, flockmaker, insured his water-powered flock mill and grist mill. This may have been the mill which was worked by George Cumley in the 1830s. There were eight handlooms in the mill. There was a steam engine of 12hp at a mill in Nibley and it was probably at this site which appears to have been the largest mill in the parish. The mill has been demolished.
(RV216.3 gs; Royal Exchange policy 174796; PP 1840, XXIV, pp.396–7; PP 1839, XLII, pp.144–7)

Crowlbrook/Nibley Mill, ST 745967

In 1593 James Dupont leased a messuage and fulling mill (in the tenure of Robert Purnell) to John Purnell. In 1606 John Venn leased 'one ffulling mill lieinge in Horrend in Nibley' from James Dupont. This mill later came into the Smyth family and in 1712 George Smyth leased 'all that fulling mill … the racks thereunto belonging with the usual ways for bringing of clay and all other carriages to the said mill', to John Cox, fuller. The rent was £1 13s 4d p.a. In 1718 the mill, together with Snitend Mill, was owned by Nicholas Carpenter. In 1798 it was for sale, being advertised as a 'duffell mill having a constant supply of water and may be easily made commodious for an extensive manufactory'. In 1799 the mill was sold to William Purnell, and by 1816 it had been rebuilt and was again for sale. There were two pairs of stocks, one gig mill, with lofts over for machinery, then in the possession of Messrs Vizard and others. Vizard got into financial difficulties and mortgaged the property in 1818, and in 1823 the mill was for sale. In 1830 Samuel Plomer, clothier, was at Nibley Mill. At the tithe survey, Henry Gazard owned and occupied the mill. He continued manufacturing cloth there until after 1863, but he was described as a miller in 1879. The stone mill had a 20ft by 3ft waterwheel.
(216(25)gs; RX354.19; Box 41gs, bd1.1; RV216.12; Box 42, bdl.3gs; RV216.3; D1347, acc. 1603, Purnell; 216.1 gs; RV216.2; Directories 1830, 1856, 1863, 1879)

WORKSHOPS AND DYEHOUSES ON THE DOVERTE BROOK

Wick Dyehouse, ST 724971

In 1704 two clothiers, one from Wick and the other from Stinchcombe, agreed to erect a dyehouse in Wick alongside the stream running from Nibley. The agreement mentions 'the conveniency of the water for dyeing and washing their wooll'. The clothiers agreed to provide a copper or furnace there and to use the dyehouse on different days of the week. Samuel Webb of Wick was to use it on Monday, Tuesday and Wednesday and Francis Maninge of Stinchcombe for the rest of the week.
(Box 7gs)

Possible dyehouse in Nibley, ST 747967

This field is called Dyehouse Piece in the tithe terrier.

Press shop and shear shop in Nibley

In 1734 Anne Green, spinster and clothier of Nibley, sold her 'goods now being in the shear shop wool loft and roomes' (in Nibley). In the shear shop she had fourteen pairs of shears, two shear boards, two dubbing boards, one stage of handles and two cloth racks as well as other goods. In the press shop she had a cloth press with twenty-five dozen press papers.
(D654/11/25/F1)

THE EWELME OR CAM AND ITS TRIBUTARIES

Cambridge/Le Dustome Mill, SO 750037

This mill was part of the manor of Slimbridge. A survey of the manor made in Elizabethan times stated that William Bower held Le Dustome Mill in Cambridge. In 1634 the mill was described as a water grain mill. A mutilated portion of the presentments at the court in 1577 records that John Bower held a water mill and fulling mill in Cambridge. No other reference to a fulling mill here has been found. Later mills on this site were corn mills and, by 1821, edge tools. Finally there was a saw mill here.
(PRO E164/29/41; Smyth MSS, 16067; RF274.14(3); *Birmingham Gazette*, 19 February 1821)

Coaley Mill, SO 760024

A corn mill on this site was given by Robert de Berkeley to Stanley Priory. Nothing more is known of this site until 1674 when John Essington leased two fulling mills under one roof occupied by Samuel Whorton and a water corn mill occupied by Thomas Wilkins to George Dodson. In 1696 the property was conveyed to Essington's son. By 1791 the mills were occupied by Nathaniel Underwood and had been converted into iron mills. In 1821 the iron mills were still occupied by the Underwoods, who were described as edge-tool makers. Underwood was bankrupt in 1828. At the tithe survey the mills were owned by Joseph Longmore and occupied by Henry Savage. Five waterwheels and three large hammers were mentioned in 1841. 'Coaley shovels' were a well-known product of these mills.
(19919/68gs; 19919/75; LG Nov 1 1831; *Midland Counties Herald*, 23 September 1841; Box 18 gs)

Small unnamed mill in Coaley on a tributary of the Ewelme joining the Ewelme at Coaley Mill, SO 772018

This was owned by James Dauncey in the early 1820s when the property was described as a 'close of pasture in part excavated for a mill pond … together with the clothing mill or manufactory [sometime since a corn grist mill] then lately erected … and the dyehouse, stove, and other buildings'. The mill has been demolished.
(D1229)

Halmer Mill, SO 755020

A grist and fulling mill was estimated by Smyth to have been erected here in the 1560s. In 1652 the mill formed part of a marriage settlement when Daniel Tindall married Hester Adey. From 1678, and possibly earlier, the fulling and corn mills were both owned by the Tindall family but only the fulling portion was occupied by them, the corn mill being leased to local millers. In 1748 William Tindall was described as a miller, so possibly the whole mill was a corn mill at this date. In 1818 the property was insured with the Globe Insurance Company for £2,700 by Edward Sheppard and by this date it was a cloth mill again. There was a pair of stocks, gig mills, four shearing frames and a house and weaving house nearby. In 1837 the property was put up for sale on the bankruptcy of Edward Sheppard the elder and Edward Sheppard the younger. It was described as:

> A substantial, compact and desirable cloth fulling mill … The mill contains 3 waterwheels, 4 pairs of stocks and 3 gig mills, … and is capable of milling and roughing 12 ends of cloth weekly. The mill pond and reservoir [are] on a powerful stream, [covering] nearly 4 acres.

It was advertised in 1842 as 'hitherto used as a cloth fulling mill but may be easily converted to any other purpose'. The mill has been demolished.

(Berkeley MSS, III, p.158; Boxes 14, 15,16gs; RF65.30; 19919/51gs; LG July 14 1837; *Midland Counties Herald*, 16 June 1842)

Draycot Mills, SO 749012

This comprised two fulling mills in 1627 and belonged to John Browning. By 1696 there were two buildings: the 'fulling mill ... commonly called Draycots Mill and also all those [adjacent] two fulling mills and one grist'. A Rack Close was near the mills. At this date, the mills were occupied by Robert Smith of Nailsworth, clothier. In 1730 the mills, containing two three-quarter stocks 'the stocks newly put in' and a gig mill, were advertised to be let. They became the property of John Phillimore who bequeathed the mills to his son John. By 1787 they belonged to Mary Hodges who bequeathed the mills to her daughter Ann. In 1801 Draycot Mills were conveyed by lease and re-lease to Daniel Knight by Samuel White. At this date they were 'corn mills ... which said mills were then many years since turned into tenements and since re-converted into mills'. In the mid-nineteenth century the property was occupied by Samuel Pearce, a miller. When they were advertised to be let in 1844, the mills were described as 'newly built' with two waterwheels and five pairs of French stones. A four–five sack roller plant was installed by William Gardner of Gloucester in 1889 for James Workman. In 1892 the motive power was a steam engine, a waterwheel and a water turbine.
(IPM (Index Library), I, p.78; GJ 28 Dec 1730; 19919/37gs; D18/69/458 and 462; 19919/53gs)

Middle Mill, SO 749007

In 1816 Nathaniel Underwood of Cambridge Mill and Robert Davies Underwood of Coaley Mills conveyed 'all those iron mills (lately used as fulling mills)', which were formerly held by John Phillimore, Henry Hicks and now Nathaniel and Robert Davies Underwood, to Henry Vizard and Thomas Clark. In 1823 a 'water grist mill with 2 pairs of stones but capable of driving 4 ... lately used as a fulling mill ... [and] the Rack Close', rented by Dennis Potter and George Packer, was for sale. At the tithe survey, the mill was occupied by Thomas and William Richards. In 1879 Samuel Ford was in occupation and in 1881, when the mills were for sale, they were said to be worked by a 'backshot' iron waterwheel 16ft by 12ft and a new 80hp horizontal compound steam engine. There were seven pairs of stones and two pairs of patent porcelain rollers; also a small bakery.
(Box 16gs; RV65.3; LG, 24 January 1851; *Miller*, 21 February 1881)

Cam/Corrietts Mill, ST 754999

Thomas Fynymore was granted, sometime prior to 1533–38, 'oon water grist myll and oon mill called a giggemill being to gether Under on roffe in Came ... and a fulling mille being Under an other roffe With divers parcells of land'. In 1556 Henry Berkeley leased to Richard Tindall (Tyndale) his 'two mylles under one roofe that is to say a corn myll and a gygg myll' with a fulling mill in Cam. For a time the estate was in the hands of Edward Tyndale, brother of William, translator of the Bible. In 1600 Anne Countess of Warwick demised a water grain mill and two fulling mills called Corriettes to Arnold Oldisworth who leased them to William Harding. Harding's son William bequeathed the mills to William Purnell of Dursley. In 1673 the mills were in the hands of Benjamin and Benedict Biddle. In 1687 they were conveyed to Nicholas and Robert Webb of Wotton-under-Edge, clothiers. Benedict Biddle willed the grist mill and two fulling mills to the Webbs. In 1725 the fulling mill and gig mill were leased to Daniel Tyndale. In 1734 they were described as 'all that millhouse with a grist mill and stock mill therein'. In 1772 Samuel Phillimore conveyed to Richard and George Jones 'all that dyehouse formerly a grist mill with its appurtenances and also all those two water grist mills together with one fulling mill or stock under one roof'. The fulling and grist mill were separately tenanted. Cam Flour Mills were occupied by

John Bennett for a number of years and he remodelled them to the Gardner roller system c.1892 when the motive power was a 25hp compound condensing engine and a breast shot waterwheel. The firm went into voluntary liquidation (also owning Huntingford Mill) in 1903.

In 1830 Joseph Tippetts manufactured cloth at Cam Mill, and was still in occupation at the tithe survey. In 1851 the mills were sold to the Hunt family. W.T. Turner & Bros, woollen cloth manufacturers of Cam Mill partnership, was dissolved in 1853. Thomas Hunt was joined by A.B. Winterbotham in 1859 and in 1867 the machinery was valued at £3,906. There were 4,600 spindles, 110 power looms and four handlooms on the premises. The main mill building, five but formerly four storeys high, was probably built in 1815. It was of brick with iron-framed windows. A clock tower and other buildings were added later. The site had become extensive by 1904 (see p.87). This is one of the two remaining woollen mills in Gloucestershire, now owned by WSP Textiles, and specialises in non-apparel cloth for billiard and snooker tables.
(PRO C1/793/25 & I.H. Jeayes, *A Descriptive Catalogue of the Charters and Muniments at Berkeley Castle*, p.215; IPM (Index Library) III, p.76; Stroudwater Textile Trust MSS, Notes on Cam Mill; D654/11/8/T12, T6, T11; Glos N. and Q, X, 1904, pp.10–11; LG, 1 March 1853; Stroud Museum, Folder of materials on Winterbotham, Strachan & Playne; *Worrall's Directory 1891*; *Miller*, 6 June 1892, 6 April 1903; *Industrial Gloucestershire*, 1904)

Upper Cam Mill, ST 757992

A mill here was recorded in 1651 when Nathaniel Hicks leased 'all that water corn mill with the appurtenances and the said two fulling mills or stockes under one roofe in Cam', to Thomas Hicks. The mills were formerly occupied by Isaac Smyth. In 1711 Mary Andrews of Hereford leased a messuage in Upper Cam occupied by Henry Brown, clothier, with 'one press with an iron skrew and allso six plates with the papers and appurtenances to ye same belonging sufficient for ye pressing of two pieces of cloth att one time'. The deed mentions John Long, tenant of the fulling mill.

In 1762 Ann Barnes held 'a mill house with two fulling stocks and a gig mill therein' in the tenure of Henry and George Adey. In 1773, Samuel Adey went into partnership with John Wallington of Dursley. Their capital was £4,500. The trade was to be managed at this mill and in lofts, a stove and warehouses in Dursley. The mills were sold to John Purnell in 1774. The mills comprised a grist and two fulling mills in 1781. Shortly afterwards, they were bought by Nathaniel Lloyd of Uley, clothier. Daniel Lloyd succeeded to the business and, in 1814, insured the premises with Norwich Union. The industrial buildings – mill at Cam, stove; and a stove, shear shop, picking shop at Angerstone, Uley, with cloth room & warehouse and press shop/warping loft – were together insured for £6,800, a house and cottages for £700. There was additional insurance cover of premises with the Globe for £1,500 and 'the rest of the £198,000 is for stock in said buildings', demonstrating the necessity for a large working capital.

Daniel Lloyd got into financial trouble in the early 1820s and this mill was put up for sale or to be let in 1824, Lloyd retaining his workshops at Uley. The mill measured 137ft by 31ft and had:

> three floors for driving machinery, having a fall of 18.5ft with a reservoir and a powerful stream of water well applied on three waterwheels; also one 20hp steam engine (by Boulton and Watt) erected within the last two years, driving 8 pairs of stocks and five gig mills with corresponding machinery, capable of making 60 ends of fine Saxony broadcloth per week.

The mill was not sold and, on Lloyd being declared bankrupt, all the cloth-making machinery and stock in trade was put up for sale in 1826. The inventory of machinery shows that the Lloyds had readily adopted new mechanical inventions including scribbling and carding machines, jennies (including 4 x 200-spindle ones), Harmer and Lewis cutters.

By 1830, Nathaniel Iles Butler was making cloth at Upper Cam Mill and in 1839 Henry Dartnall was living there. Dartnall had sixteen handlooms in the mill in 1839, three of which were not in use. It was described as a fulling mill in 1845, but shortly after it was converted into a flour mill and has been demolished.
(D1229; Box 14,15gs; D149/124/65; Box 23, bdl. 6gs; 19919/40gs; RV65.3(4); R65.7; Directories 1830, 1839; PP 1840, XXIV, pp.396–7; Lodgemore Mill Archives)

Lower Mill, Dursley, ST 756989

This site, although probably older, is recorded in 1795 when John Harris's property was described. There was a new building containing two stocks and a gig mill, a scribbling machine room, carding machines and billies. In 1806 George Harris had four broadlooms and two or three narrow looms in Lower Mill. In 1842 Samuel Harris mortgaged the mill to secure £1,200. In 1844 the contents of the buildings were up for sale. These included 'a wool beater on a new principle by Turner', a 2hp steam engine, a 200-spindle mule and six power looms, three gig mills, a gig brushing machine and cloth cutters by Davies, Lewis, Gardner and Herbert.

In 1855 the trustees of George Harris leased the mill to Thomas and James Hunt and William Turner of Cam Mill, 'subject nevertheless to the requirements of the Dursley and Midland Railway Company of any portion of the premises'. Shortly afterwards cloth production ceased here.
(VE 1/1; PP 1806, III, pp.40–1; Lodgemore Mill Archives; Box 28, bdl. 19gs; RX115.13 (52))

Churn Works/Howard's Mill, ST 758984

This mill was probably worked by Edward Wallington in 1795. The mill contained two stocks and a gig mill with one machinery wheel. In addition, there were billies, scribbling engines, carding machines and jennies. In the nineteenth century the mill was owned and worked by the Howard family. When cloth manufacture ceased here, the mill was converted into a saw mill and became the Churn Works of Listers. It has been demolished.
(VE 1/1)

Townsend/Phelps's Mill, ST 759983

In 1614 a dispute arose in Dursley concerning the fouling of a watercourse flowing to a mill belonging to Richard Browning. George Bowser, a dyer, had committed 'a nusans ... in buyldinge a pigstye and an house of office' over the stream and had washed coloured wool in the Broad Well Stream above Browning's mill. The Broad Well Stream enters the Ewelme just above Townsend Mill. In 1724 Maurice Phillips conveyed 'all that fulling mill and grist mill under one roofe' to Thomas Phelps, clothier. The mill was in the hands of the Phelps family throughout the eighteenth century. In 1795 there were two stocks and a gig mill with a burling room and garretts above. The 'New Workshops' (about 50ft by 17ft) contained a shear shop, scribbling and carding machines, billies and jennies. There was a three-storey wool loft and four small cottages had also been taken into the works. By the end of the eighteenth century, the Phelps had got into financial difficulties and at the tithe survey the mill was managed by John Williams and Joseph Edmonds. Later, it became a flour mill and afterwards was demolished. The workshops mentioned in the survey of 1795 remain as the Lamb Inn at the bottom of Long Street.
(Box 24, bdl. 9gs; RF115.132; Box 25 bdl. 12gs; VE 1/1)

Vizards Mill (in Silver Street), ST 757981

This was the uppermost of the three mills on the Broad Well Stream. An indenture of 1677 records that Thomas Estcourt leased to Edmund Estcourt a fulling mill with two stocks and a grist mill. In 1690 Walter Estcourt demised the property to Edward Morse. This may have been the mill offered for sale

in 1741 described as 'a good new-built Fulling Mill with one stock and a constant supply of water', for which enquiries were to be made of S. Wallington. In 1783 the mill, containing one stock, was in the possession of Samuel Adey, although sublet to William Jobbins and Sarah Shearman. In 1792 'all that fulling mill formerly a grist mill' was owned by Adey, Wallington & Co. From the survey of 1795, the mill seems to have been small containing 'a small shear shop a cloth loft and stock mill 48 feet long and 20 feet wide'. In 1833 George Adey leased the mill, 'with the workshops and premises thereunto belonging … lately used as a clothing mill and factory' occupied by Charles and George Vizard, to Berkeley Wathen Bloxome. The premises were shortly afterwards converted into a brewery. All the buildings have been demolished.
(19919/80gs; Box 16; 23, bdl. 6, 21, 28 bdl 18 gs; GJ, 20 January 1741; VE 1/1)

Howard's Upper Mill in Back Lane, ST 757983

This mill may have been the shear grinding mill mentioned in the survey of Dursley. Apart from this, little is known of the site. It was worked by George and John Howard in the early nineteenth century and at the tithe survey was owned by John Howard. It later became a pin mill. The building was three storeys high.
(Dursley, with original information supplied by an old inhabitant and others, 1882)

Howard's Lower Mill in Back Lane, ST 758984

This was also worked by George and John Howard. At the tithe survey it was owned and occupied by John Howard. The mill has been demolished.

New Mills, ST 763980

In 1610 'All that fullinge myll and all that ther fullinge myll lately a grist myll with the appurtenances commonly called the Newe Mylls' was conveyed to Thomas Purnell for £310. Access was granted to the stove, pounds, frame and watercourses. In 1699 it comprised a corn mill, two fulling mills and a gig mill. By 1681 a dyehouse had been added to the site and the premises were occupied by William Tippetts. Tippetts went bankrupt in 1789. In 1795 John Tippetts was in occupation and the premises had become very extensive. The ground floor of the mill contained two stocks and a gig mill and 'a double gigg mill on a new construction 26 feet by 20'. Machinery included scribbling machines, billies, a carding machine and jennies. There were two coppers in the dyehouse and the press shop contained two presses and a handle stage. There was a stove for drying wool and an air stove for drying cloth. There were cloth picking and twilling lofts, lumber rooms, shear shops, cloth drawing rooms, another building 45ft by 18ft containing burling rooms and 'a cloth drying stove abt 138 feet long'. When Tippetts rebuilt the mills, presumably before 1795, he raised the weir by between 12ft and 14ft because the old mill had suffered from flooding by Phelps's Mill. This weir caused the water to rise in the wheel pits at Dursley Mill. By 1828 J.B. Foxwell owned New Mills and made cloth there for a short while, but by 1834 George Vizard occupied the premises. At this date, William Perrin agreed to sell New Mills to Henry Vizard for £2,700, including most of the fixtures. At the tithe survey, the mills were still owned by Henry Vizard but they were unoccupied. They have been demolished.
(Box 24gs; D146/124/47; D1681/T2; Box 23, bdl. 6gs; VE 1/1; D947/M2)

Rivers Mill, ST 765978

In 1783 this was described as 'formerly a fulling mill but since made use of as a wire mill'. In 1791 it was said to have been 'recently established' by Joseph Smith, paper maker (d. 1812). A deed of 1804 describes it as having been a fulling and gig mill. In 1826, on Cornelius Deasley becoming bankrupt, the mill was

for sale and 'well calculated for carrying on the clothing, dyeing or any other business'. Later it passed into the Lister family and wire cards for the woollen industry were made. The overshot waterwheel was supposed to have been one of the largest in Gloucestershire. The mill was destroyed by fire.
(Box 23, bdl. 6gs; D1347, acc. 1603; 1791 Directory; *Bristol Journal*, 8 February 1812; LG, 27 June 1826; Dursley, with original information … op. cit.)

Dursley Mill, ST 767978

In 1699 Daniel Knight leased a fulling mill, grist mill and dyehouse to Edward Dorney. In 1732 it was settled on John Andrews and Elizabeth Perrett, Andrews being described in 1769 as a maltster, although the property still comprised 'fulling mills and grist mills adjoining'. By 1809 the fulling mills had gone and the grist mill and dyehouse remained. The small stone mill, with some Tudor-style windows, is an interesting example of an early fulling mill. It is two storeys high with an attic and a loading door on the first floor (see p.40). It was fed by a long leat. A large, attractive, gabled clothier's house is nearby.
(D1681/T9; D897/1 and 2; Box 26, bdl. 15gs)

WORKSHOPS AND DYEHOUSES IN DURSLEY

Workshops near Woodmancote Court
Samuel Adey and John Wallington had workshops, lofts, a stove and a warehouse in Dursley, in 1773. There were offices, a dyehouse, wool loft, press room, counting house, cider mill house, stove room, burling room, scribbling room, three rooms for examining cloth, wool drying stove, store room and picking loft. There was a 100ft-long cloth stove with a shear shop and picking loft above.

There were at least two dyehouses in Dursley besides those on mill premises, both of which are recorded for the first time in the 1680s:

Broadwell Dyehouse
This was occupied by John Ireland in the 1680s as undertenant to Richard Smith.
(Box 22, bdl. 5gs; box 27, bdl. 16gs)

Dyehouse next to Dursley Poorhouse
This came into the Wallington family in the eighteenth century.
(Box 22, bdl. 5 gs; box 23, bdl. 7gs)

Shear shop, coal house and stove
This was owned by William Purnell in 1748 and was situated near his house.
(D149/124/56)

UPPER EWELME VALLEY

Eyles Mill, Wresden, Uley, ST 773980
Fed by a spring, this was a corn mill in 1566. Later, Thomas Browing was a weaver here. By 1655 Wresden was in the hands of John Eyles, who added the porch adjoining the mill house which bears his woolmark, the initials J.E.E. (John & Elizabeth Eyles) and the date 1687. Eyles was said to have been the 'first that ever made Spanish cloth in this parish'.

After Eyles's death, business at Wresden was carried on for a short while by John Bayley but soon the mill ceased to full cloth. John Bayley was succeeded by his grandson John Phillimore, the Phillimore family continuing to own Wresden until 1928. The small, stone, two-storey mill is one of the oldest in Gloucestershire. Notable features of the building are the circular openings in the north wall (see p.39). (Alan Bebbington (ed.), *A History of Uley, Gloucestershire*, The Uley Society, 2003)

Penley's Mill, ST 775977

This may have been the dyehouse occupied by John Penley who was described as 'of Uley, dyer' when he took a lease of Penley's Mill in Alderley (p.178). In about 1800 it was occupied by Sarah Gainey, dyer. John Penley (Jnr) operated the dyehouse, but his creditors closed the business down in 1822. John and family immigrated to Australia in 1838. At the tithe survey, the mill was a saw mill occupied by John Martin. Rockstowes House and the remaining building from Penley's Mill have been converted into a house.
(M. Lloyd Baker, *The Story of Uley*, p.52; Alan Bebbington (ed.), *A History of Uley, Gloucestershire*, The Uley Society, 2003)

Rockstowes Mill, ST 777976

Between 1790 and 1800 Rockstowes was operated by William Phelps of Dursley. It is likely that Edward Jackson either bought the mill or inherited it. In 1805 he began to build a house, Rockstowes Hill. In 1807 a newly erected mansion house, 'offices, dyehouses', at Rockstowes was offered for sale, the 'capacious' arched cellar under the house being divided into a dyehouse, counting house and stable. The mill remained in Jackson's hands and he increased the industrial buildings, including dyehouses, adding a steam engine in 1816. In 1819 he was bankrupt. In 1823 the mill was put up for sale and was described as a 'capital clothing factory in perfect repair with a powerful stream of water and two engines and all requisite buildings or workshops necessary for and capable of making about 50 ends of cloth every week'. The premises included a four-storey 'Blue Mill' containing spinning and shearing shops, a mill with two waterwheels which also contained spinning lofts and shearing shops, an engine house, air stove, workshops containing brushing and drawing shops, jenny shops, a 'capital fire-proof stove house warehouses and a watchman's house'.

In 1824 the mill was again for sale and the sale particulars mention 10hp and 8hp steam engines, six pairs of stocks and four gigs. In 1825 it was advertised again. This time the property was said to have been lately in the possession of Smith, King & Hanbury. There were, in addition to stocks and gigs, twenty-four shearing frames, thirteen scribbling and carding engines, eleven billies and nineteen jennies; the dyehouse contained seven blue vats. By 1830 Comley & Jones were making cloth at the mill and Henry Smith appears to have occupied the dyehouse.

After the demise of the cloth industry the mill was used as a water-driven corn and saw mill and, later, by the Rockstowes Dairy Factory Supply Co. producing dairy and agricultural equipment. Later still it was used for the manufacture of hats and protective industrial clothing. The main mill and the Blue Mill have been demolished but the wool warehouse and the building which comprised two houses with jenny lofts above remain. Both are now houses.
(M. Lloyd Baker, *The Story of Uley*, p.56; GJ, 26 January 1807; RV319.2gs; RX115.13(27)gs; RX319.1(15)gs; D1229; Directory 1830; Mills & Riemer, *The Mills of Gloucestershire*, 1989)

Marsh/Adey's/Bottom Mill, ST 784977

In 1814 and until after 1830, this mill was occupied by George Adey. By 1838 only part of the mill was working and this was occupied by Samuel Smith who had four handlooms, only two of which were in use. This was one of the last mills to work in Uley but cloth production had ceased by 1840. The tithe

map shows a 'factory' at the upper end of the mill pond. Marsh Mill became a saw mill for a time and in 1911 burned down. Some machinery (waterwheel and pitwheel) remained until removed in 1975. The 'factory' shown on the tithe map is now a house.
(Directories 1814, 1830; PP 1840, XXIV, pp.396–7, 400; Alan Bebbington (ed.), *A History of Uley*, The Uley Society, 2003)

Jackson's Mill, ST 786979
This was described in 1828 as 'all that newly erected millhouse or manufactory erected and built by one Edward Jackson ... together with the steam engine, three pair of stocks and one gig mill'. At this time, the property was occupied by Joseph Jeens. By the tithe survey the mill was occupied by Daniel Powell; the map shows a single building near the stream. The mill has been demolished.

Dauncey's/Blagden's/Top/Stouts Hill Mill, ST 787980
This was probably the earliest fulling mill in Uley 'late builded ... by Timothy Gyde' in 1689. Gyde lived at Stouts Hill, behind the mill. In 1784 the mill was occupied by Richard Blagden, clothier. A 'new' stone-built water-powered woollen mill called Blagdens, adjoining the 'old mill', was insured by Atlas Insurance for £400, the millwright's work being insured for £500 in 1808. The old mill, also water-driven, was insured for £200; the press shop and scouring house for £250. There was a steam engine and the room above was used for wool drying from the heat of the boiler. This was insured for £250, the fulling mill adjacent for £350. The mills were 'heated by upright pipes in brick flues and lighted by lamps or candles'. All the property was occupied by a member of the Dauncey family. In 1811 it was leased to Nathaniel and Daniel. At that date there was a fulling mill, gig mill, factory, dyehouses, stoves and other buildings. In 1828 the property was leased to George and James Dauncey at £270 p.a. The deed mentions that the Daunceys had a steam engine. The mills were insured by Sun Insurance in 1828 for £2,000. By 1838 fifteen handlooms had been installed in the factory. It was still operating in 1838 but by 1840 cloth production had ceased here. The mill later became a saw mill. A round wool drying stove, in ruins, is about halfway along the mill pond.
(D654/111/33; Directory 1784; Box 54, bdl. 10gs; D3549/39/5/8; Box 53, bdl. 7gs; PP 1840, XXIV, pp.396–7)

Sheppard's/Uley Mill, ST 794984
During the latter part of the eighteenth century this mill was owned and worked by John Holbrow. By 1789 he had moved to King's Stanley and although he still owned the mill, it was occupied by Thomas Maule. In that year, Edward Sheppard and Henry Hicks, both of Eastington, took a ten-year lease on 'all that cloth or fulling mill with the shear shops, dyehouse and other buildings'. This site grew to be one of the largest cloth factories in the county and Edward Sheppard became well known as chair of the Clothiers' Committee. A Boulton & Watt steam engine was installed in 1805 for working the fulling stocks. A detailed map of 1833 shows the layout of the property and the functions of each building. This was a vertically integrated manufacturing unit, all process of cloth manufacture being on site. Altogether, £50,000 was reputed to have been spent on buildings and machinery. In 1837 Edward Sheppard, father and son, were bankrupt and the property, 'together with 3 ... steam engines of 40, 28 and 20 horse power' with stoves, shear shop, dyehouses, workmen's cottages and outbuildings, was put up for sale. Most of the buildings were empty but part of the site was an iron foundry. By 1847 the much-contracted site was up for sale; only three of the former thirty buildings being shown on a sketch map of the site. The sale bill stated that Richard Clauburn was making agricultural implements there. In 1852 the mill was leased to Thomas and Henry Lusty, wood turners and timber dealers. The only industrial buildings which

have survived are the press shop and scouring house which, with the manager's house (Mill Farm), are converted dwellings.

(Box 52, bdl. 6gs; R.A. Pelham, 'Fulling Mills', S.P.A.B., p.11; D1011; D1388 SL3/50; LG, 21 April 1837; PP 1840, XXIV, p.453; RF115.31gs; PP 1840, XXIV, p.400; RX 319.6 gs; Box 52, bdl.6 gs)

WORKSHOPS, NON-WATER-POWERED FACTORIES AND SCOURING HOUSES IN ULEY

Angeston Workshops, ST 783981

This was a dwelling house called Angeston (Grange), part of which had been converted into workshops by 1795. In that year, the Bristol Fire Office insured the property for £700: the house for £150; wool lofts, press shop, spinning shop and lofts adjacent to the house for £130; another house, wool lofts and stable for £100; and 'a stove on the lower floor, and a shear shop, weaving shop and picking loft over the same … situated in the yard behind … the dwelling house' for £400. The following year Nathaniel and Daniel Lloyd, the owners, mortgaged the property. When they got into financial difficulties, the modest contents of the workshops were put up for sale, with those of Upper Cam Mill, in 1826. Included in the sale was white cassimere, purple, scarlet, drab, saxony blue, green ladies, saxony black, fawn stripe and medley cloth.

(Box 52 bdl. 6gs; R65.7)

Jeens's Factory, ST 785979

This is shown on the tithe map as 'Factory owned by Henry Moreland Jeens'. It was working in 1838 but by 1840 had ceased production.

(Alan Bebbington (ed.), *A History of Uley, Gloucestershire*, The Uley Society, 2003)

Dauncey's Factory, ST 789984

In 1814 a 'newly erected' five-storey building 67.5ft by 20.5ft, 'used in the clothing manufactory, with a shed for a horse wheel', was for sale. The sale bill announced that 'by the erection of a steam engine for which there is ample space, a sufficient and constant supply of water, a considerable trade might be carried on'. At the tithe survey the 'clothing factory' was owned by George Dauncey and occupied by Michael Norris. The premises were later used for wood turning. In 1991 the site was redeveloped for residential use and is now known as South Street Mill, an inscription on the end of the building reading 'South Street Mill 1814–1991'.

(RX319.1(10)gs; Alan Bebbington (ed.), *A History of Uley, Gloucestershire*, The Uley Society, 2003)

Scouring House at Cross Roads, ST 785978

This was used for wool washing, being just below the sheep dip, and was shared by Nathaniel Lloyd, Joseph Jeens and Edward Jackson Blackwell.

(Box 53, bdl. 7gs)

THE RIVER FROME AND ITS TRIBUTARIES

Framilode Mills, SO 751103

This former Gloucester Abbey corn mill had, by 1556, become a double mill – two corn and one fulling mills – when licence was given to Thomas Wynston of Herefordshire to grant his three water mills in Framilode to William Bulloke.

In 1729 Walter Yate, then owner of Framilode Mills, petitioned the House of Commons against the Stroudwater Navigation Bill. Apart from mentioning that his mills were 'ancient', Yate did not describe their function. The mills, containing three pairs of stones and a boulting mill, together with large granaries, were advertised to be let in 1736. In 1760 George Wilding became the lessee of 'certain grist mills and iron-slitting mills called Framilode'. In 1767 the mills were sublet to John Purnell, ironmaster of Frampton-on-Severn, and William Purnell, gentleman. The tenants were permitted to add forges and mills to the premises, but they were to leave intact the grist mill. Thereafter, the mills were used for rolling, slitting and tin-plate manufacture. In 1830 the mills were advertised to let. The mills have been demolished.
(D149/149/172, 173; Hockaday Abstracts, Frampton; Cal. Pat. Rolls, p. and M, III, p.133; J. H. of C. 1727–1732, p.509; GJ, 24 August 1736; R135.1 and 2gs; RR135.1(1-2)gs; Sun Fire Insurance policy 251943, 17 December 1767; LG, 10 February 1829; *Birmingham Gazette*, 23 August 1830, 14 September 1835)

Whitminster Mill, SO 760090
This was a fulling mill in the fourteenth and fifteenth centuries and, from 1764, a paper mill. The mill has been demolished.
(Alfred Shorter, *Paper Mills*, p.165)

Wheatenhurst Mill
There was a fulling mill here in 1336.
(IPM, V, p.258)

Fromebridge Mill, SO 769073
Godstow Abbey owned this mill which comprised two stocks, with corn and malt mills, at the Dissolution. In 1632 three grist mills called Fromebridge Mills were sold by John Arundell to Urian Wise. By 1713 there were four corn mills here in the tenure of Stephen Jenner. In 1735 the mills were bought by Dr T. Jenner. In 1760 the mills comprised 'a rolling mill, tilting mill and Block mill all under one roof' with a 'wire mill and offices adjoining' and a 'brass nealing house'. The mill reverted to corn milling and was still used as such in the 1960s. It has been suggested that the mill was used, for a while, by Charles Hooper as an extension of his Eastington business empire.
(*VCH Gloucestershire*, X, pp.148–52; D149/175; Box 33, bdls. 3 and 4gs; box 27, bdl. 17gs; Stephen Mills, 'The Mystery of Cloth Manufacture at Fromebridge Mill', *GSIA Journal*, 1999, pp.61–5)

Meadow/Eastington Mill, SO 779062
This is a nineteenth-century mill site, the first part of the mill probably having been built between 1815 and 1820. It was occupied by the Hicks family who were clothiers in Eastington for several generations. By 1839 Henry Fletcher & Son manufactured cloth here. Shortly afterwards, Charles Hooper leased the mill, Meadow Mill being used mainly for weaving. Steam power was installed and there was a 'coal pen' by the Stroudwater Canal. As trade contracted, more work was concentrated at Meadow Mill. In 1904 it was said that 'scarcely a year has elapsed without witnessing additions to the mechanical equipment'. By 1906 the mill had been closed.
(Directory 1839–40; tithe terrier, Eastington; A.E. Keys, *History of Eastington*, p.91; *Industrial Gloucestershire*, p.24)

Churchend Mill, SO 784057
This was a fulling mill with two stocks under one roof in the tenure of Thomas Hornchurche in 1575.

In 1597 it was in the tenure of Michael Clutterbuck. On the Stroudwater Navigation map of 1776, this is called 'Mr. James Mill'. In the early nineteenth century, the mill was worked by Henry Hicks & Son. By c.1833 the mill was owned by Charles Hooper. During Hooper's occupation of the mill it was used chiefly for spinning (see p.75). The mill was probably sold before 1879 and has been demolished.
(Box 29 (Eastington Church) gs; D1229; Directories 1820, 1830, 1879; PP 1834, XX, p.271; A.E. Keys, *History of Eastington*, p.74)

Millend Mill, SO 782054

Millend was, for several centuries, part of Leonard Stanley and the mill, by 1456, was in the hands of Leonard Stanley Priory. In 1552 this fulling mill, in the tenure of Agnes Clutterbuck, was granted by John Sandford to Richard Clutterbuck and his wife. In 1591 Richard Stephens confirmed the estate of Richard Clutterbuck 'in all those myls' with a horse house adjacent and 'liberty to dig clay for milling cloth'. The horse house may have been a shed for a horse wheel. In 1785 William Fryer leased the fulling mill to Edward Sheppard and Henry Hicks for twenty-one years. The mill was owned and worked by Henry Hicks and his sons from about 1806 until the early 1830s. In 1818 Hicks built a substantial five-storey stone mill on the site of the earlier fulling mill, the builder being Blackwell of Brimscombe. Within a few years Hicks had added a series of other buildings. A 14hp Boulton & Watt engine was purchased in 1821. By the 1830s Millend Mill had passed to Charles Hooper, formerly Hicks's manager. It was used mainly for fulling, bleaching and drying. Following Charles Hooper's death in 1869, Millend Mill was put up for sale by his son, Charles Henry Hooper. Three waterwheels were mentioned and the mill was said to be capable of employing 200 people. In 1871 the mill was again on the market and yet again in 1872 and 1887. The main building was described as 70ft by 37ft and five storeys high. There were four waterwheels and a 14hp Boulton & Watt engine.

The mill was converted to a flour mill after the 1872 sale, but in 1883 it was 'disused'. Milling recommenced and in the early twentieth century malting was added. There was a major fire in 1922 and the 1818 building was gutted but eventually repaired. In 2000 proposals were made for conversion into apartments but the site has become increasingly unsafe.
(Cal. Pat. Rolls Ed VI, IV, p.413; D540/T59; D1229; box 52, bdl.6gs; D2500; Directories 1820, 1830; A.E. Keys, *History of Eastington*, p.74; O. Mahler & S. Marshfield, *Stroudwater Valley Mills*, p.5; RX118.4, 6, 7; Stephen Mills, 'Millend Mill, Eastington – The Past and the Future', *GSIA Journal*, 2000, pp.6–18)

Bond's/New/Eycott's/Kinns Mill, SO 793053

New Mill was rebuilt in 1496 as a fulling and gig mill and had become a fulling and corn mill before 1542. When an exchange of lands in Stonehouse and King's Stanley took place between William Fowler and William Sandford in 1586, New Mill, comprising a corn mill and fulling mill with three stocks under one roof, remained in the hands of Fowler to whom it had formerly belonged. The mill remained in the Fowler family during the early part of the seventeenth century. By 1685 the mill – grist and fulling mill with four stocks – was in Elizabeth Smith's hands until, in 1714, John Bond, clothier, became tenant at the then-named Kinns Mill. In 1724 the mill passed out of manorial ownership to Daniel Webb and John Peach. By 1750, ownership had passed to Richard Pitt and, in 1774, to Henry Eycott, when it contained four pairs of stocks. In 1830 William Wood was making cloth here but, by 1837, William Playne Wise occupied the mill. He installed one power loom in April 1837. In 1840 the mills were leased to Robert Stephens Davies, William Cole and William Davies of Stonehouse Mills. The old mill contained five stocks (two iron) and a half stock and four gig mills. Between the old and the new mills there were three large undershot waterwheels, 30ft in diameter and 8ft wide. The new mill was five storeys. In 1844 Charles Warner was said to be manufacturing cloth at Bond's Mill although a directory of 1856 records

William White as in occupation there. In 1863 Charles Warner was again recorded as making cloth here. Shortly afterwards the property was taken over by Charles Hooper, weaving being the main process carried out in the mill. Hooper built cottages to attract workpeople. Cloth manufacture ceased here in 1934. It was later occupied by Sperry Gyroscope Co. Ltd.
(Janet Hudson, 'The Early History of Two Stonehouse Mills', Trans. BGAS, 118, 2000; Stroudwater Navigation map; D1917; D1815 Eycott; Directories 1830, 1839–40; PP 1840, XXIV, pp.396–7; Directories 1844, 1856, 1863, 1879; A.E. Keys, *History of Eastington*, p.74)

Beard's Mill/Leonard Stanley/Merretts Mill, SO 795049

In 1667 Stephen Fowler of Stonehouse granted Thomas Beard, clothier, liberty to cross his pasture, the Wyneyard, to his mill Merretts Mill. This was almost certainly Beard's Mill since Fowler's mill was the next upstream. The mill was called 'Mr. Beard's Mill' on the Stroudwater Navigation map. Members of the Beard family occupied the mill for around a century before they became the owners, probably when Nathaniel Beard was operating the mill in 1751. By 1820 the Beard family had entered into partnership with members of the Davies family for Davies, Beard & Davies were at the mill. In 1821 they installed a 14hp Boulton & Watt engine. John Beard was at the mill in the 1830s but cloth manufacture seems to have ceased for a while soon afterwards. By the 1870s, it was occupied by Charles Hooper, picking and dyeing being carried out. Cloth manufacture ceased here *c*.1906. The main water-powered mill was demolished in *c*.1908.
(D445/E3, D2080/283,435; R8118.3 (1-3); D1347 acc. 1603 sale particulars; D2080/283,435; LG, 3–7 May 1808; Stephen Mills, 'Beards Mill and the Menders Shop Leonard Stanley', *GSIA Journal*, 1991, pp.28–33)

Stonehouse Lower Mill, SO 801046

This was the Stonehouse manor mill in the fifteenth century, leased by William Carver. William Fowler owned this manorial corn mill in 1579. The mill did not become a cloth mill until sometime after 1683 when William Lye inherited the mill from his father, a miller. By 1697 the mill, with three stocks and a gig mill, was mortgaged to William Clutterbuck. In 1701 it was sold to John Arundel. At the end of the eighteenth century the mill was worked by William Hill. By 1820 Davies, Beard & Davies were manufacturing cloth here. The firm was known as Robert and Edward Davies & Co. by 1830; Davies, Cole & Davies by 1842; and Davies, Son & Evans by 1856. In 1889 there were 150 looms and 5,500 spindles at Stonehouse Upper and Lower mills. In 1891 only 120 looms are mentioned. The mill, with both water and steam power, was put on the market in 1904. The firm of Robert S. Davies & Sons ceased to manufacture cloth in 1906. The mill as shown in *Industrial Gloucestershire* shows the premises to have been very extensive (p.142).
(D445/T12; Stroudwater Navigation map; Directories 1784, 1820, 1830, 1842, 1856; PP 1840, XXIV, pp.396–7; Chance & Bland, *Industrial Gloucestershire*, 1904)

Bridgend/Nashes Mill/Dyeworks, SO 804045

Humphrey Osborne leased the mill to William Nicholson, clothier, in 1579. In 1588 the mill was sold to Jasper Selwyn. In 1616 the owner of Lower Mill was ordered to clear his ditch at the upper end of Richard Halliday's rack close, belonging to this site, which strongly suggests this was a fulling mill by that date. In the seventeenth and eighteenth centuries the mill was worked by the Nash family; it was probably the mill of Giles Nash, mentioned in 1637. Six generations of Giles Nash were connected with this mill, the Giles who died in 1767 being said to have made a fortune in the trade. By 1775, much of the scarlet dyeing for the Clutterbucks of Stanley Mills was being carried out here. In 1804 John Dimock leased the mills from the Elliot family. During the nineteenth century, the dyeworks was occupied by Gainer & Bishop, and later by Joseph Gainer. The large and unusual brick house remains, but the dyeworks has been demolished.

(Janet Hudson, 'The Early History of Two Stonehouse Mills', Trans. BGAS, 118, 2000; 'The Clutterbuck Diary'; Rudder, *A New History of Gloucestershire*; *VCH Gloucestershire, 10*)

Stonehouse Upper/Reddall's/Corneham Mill, SO 806047

In 1507 this was a corn mill tenanted by Richard Mill. In 1525–26 the mill, comprising a water (corn) mill and fulling mill, was leased by the Abbot of St Peter's, Gloucester, to John Sandford. At his death, William Sandford held a fulling mill with wool loft above it in Stonehouse. The following two references are to mills in Stonehouse, probably Stonehouse Upper Mill. In 1685 a fulling mill was occupied by William Mill. In 1724 a fulling mill was conveyed to Daniel Webb and John Peach in trust for Samuel Peach. In 1776 the mill was called 'Mr. Reddalls Mills'. By the early nineteenth century the property was run together with Lower Mill by members of the Davies family. In 1847 the Upper Mill was destroyed by fire and was rebuilt in 1875 'when it was materially enlarged and improved'. The *Industrial Gloucestershire* engraving shows the new mill to have had waterwheels, although the steam age was well advanced by then (see p.144). Cloth manufacture ceased here in 1904. The new mill was a four-storey brick building with lofts and a centrally placed staircase on the west wall.
(C. Swynnerton, 'The Water Mill of the Abbots of Gloucester in Stonehouse', Trans. BGAS, XLVI p.50; IPM (Index Library) I, p.201; D445/T12 and T6; Stroudwater Navigation map; Chance & Bland, *Industrial Gloucestershire*, p.25)

Stanley/Giles/Clutterbuck's Mill, SO 813043

In 1563 the Earl of Arundel sold a grist mill and fulling mill with three stocks to William Fowler and William Sandford, both clothiers of Stonehouse. In 1629 Margaret Clutterbuck re-leased a tenement called Giles Meese and 'all those … fulling mills [and] corne mills' to John Clutterbuck. In 1660 there were 'three fulling mills one grist mill and a gigg mill', a warping room, shear shop and dyehouse. In 1688 the mills formed part of the marriage settlement of Jasper Clutterbuck and Katherine Nash. In 1720 the mills were occupied by William Payne, clothier. In 1731 Jasper Clutterbuck advertised 'A clothing mill' driven by an overshot waterwheel that was 'so completely finished by the famous Mr Chinn that in the driest summer there can be no want of water', containing a whole stock and a half stock, a gig mill and 'conveniences for dyeing, racking and pressing'. In 1761 his son Jasper was in possession of the property and a knapping mill had been added. The direct Clutterbuck connection with Stanley Mill ended on Jasper's death in 1782, his only living child, a daughter, having married John Hawker, the Dudbridge dyer. In 1783 the property was conveyed by John Hawker to John Holbrow of Uley for £4,340. In 1786 John Holbrow leased the grist mill containing two pairs of stones to John Cam, with liberty to draw one head of water from the pond for every two which Holbrow or his tenants could draw for the cloth mill.

In 1802 the mill came into the hands of Paul Wathen and in 1808 'all those three fulling mills, one grist mill, and one gigg mill … under one roof', together with dyehouses, wool stoves and other buildings, and 'new erected cloth rooms, counting house, workshops and buildings adjoining' were sold to Joseph Wathen. In 1813 the old mill had been demolished and a new mill begun which was sold by John Hawker and Joseph Wathen to George Daniel Harris and Donald Maclean. The mill was being rebuilt as the fine fireproof factory which stands today (p.84). A partnership was formed between Harris, Maclean, Charles Stephens and John Carrick. Carrick dropped out after five years and in 1827 Harris also left the firm. In 1833 Maclean, Stephens & Co. described their new mill to the factory inspectors: 'The mill is fireproof, the floors being composed of brick arches on iron beams, over which is laid stone paving. The rooms are 12 feet high from the paving to the crown of the arch above.' There were five waterwheels on a fall of 16ft, one wheel was of 30hp, four of 25hp each and a steam engine of 40hp. The premises were lit by gas which was generated there. In 1836 when Donald Maclean withdrew from the partnership the mill with millwork was valued at £40,532 2d and the manager's house, lodge, cottages, and old mill at £2,050. Three valuations were undertaken of the machinery, the sums ranging from £1,830 to £4,034.

In 1840 there were 100 handlooms, ninety of which were in use. In 1842 Charles Stephens conveyed Stanley Mills to Nathaniel Samuel Marling, who paid £27,000 for it and worked it on his own until 1853. In 1854 the property was leased to N.S. Marling's nephew, Samuel Stephens Marling, who had been at Ebley Mill and in 1851 had gone into partnership with Josiah Greathead Strachan, trading under the name Marling, Strachan & Co. Thus, Stanley and Ebley were worked as one concern. In 1854 Nathaniel Summers Marling, N.S. Marling's son, joined them. In 1858, when the agreement between S.S. Marling and Strachan was due to expire, they agreed to extend it, at the same time admitting William Henry Marling. When Strachan retired in 1863, the business continued under the name of 'Marling and Co.'

In 1865 there was a short attempt to run Stanley and Ebley as separate concerns. Samuel Stephens Marling, William Henry Marling and Samuel Stephens continued at Ebley as Marling & Co. and William Henry Marling remained at Stanley where he went into partnership with Alfred Selfe. The partnership was dissolved in 1869 and Stanley and Ebley Mills were reunited. When Samuel Stephens died in 1883, Stanley was run by Leonard, Woolbright and Co. The link between the two mills was close. The firms agreed to avoid competition by making different types of cloth. Stanley specialised in worsteds and combination goods. Sir William Henry Marling agreed to the sale of Stanley and Ebley Mills to Marling & Co. Ltd, and the firm continued under this name until 1920 when it was sold to P.C. Evans & Sons Ltd.

Between 1854 and 1871 Samuel Stephens Marling spent £7,000 on additions and improvements to Stanley Mill. In 1854 a large power loom shed, steaming and gig mill house and a new boiler shed with two high-pressure boilers were built. Ten years later a coal wharf was built at Ryeford and dyeing vats and furnace at the mill. In 1867 new high-pressure condensing engines and, in 1868, an 80hp turbine were installed. By 1889 Marling & Co. possessed 180 looms and 4,500 spindles in their mills at Stanley and Ebley. Marling & Evans ceased trading in 1989. Marling Industrial Felts continues at the site. The main mill, five storeys high and built of brick with stone quoins, is iron-framed. Traceried iron arches supported by cast-iron columns support the brick-arch floors. The windows are a most attractive feature of the mill, some being of Venetian pattern, while others are long with vertical glazing bars only.

(D445/T8; D873b; 1347 acc. 1603; D873 T51, 59; D1159; D4644 2/101 (acc 6408); Box 36 gs; GJ, 15 February 1731; 11 May 1736; E.A.L. Moir, 'Marling & Evans, King's Stanley and Ebley Mills, Gloucestershire', *Textile History*, 2, 1971, pp.28–56; PP 1834, XX, p.256; PP 1840, XXIV, pp.396–7; Kelly's Textile Directory, 1889)

Ryeford/Phillips Mill, SO 815045

This was a corn mill in the hands of the Gibbs family in 1507 but, by 1600, it was a fulling mill. It is referred to tangentially in the will of John Phillips, clothier, dated 1692. In 1717 'all that fulling mill sometime containing three stocks and one gig mill' formed part of the settlement on the marriage of Thomas Phillips, clothier. In 1804 'all that cloth mill' was sold to Saul Lusty of Neath. In 1819 the messuage and the mill, 'pt of which was then used as a grist mill and the other pt a cloth mill' with a scouring house on the site of the old mill, was conveyed to R. Hyde. Samuel Evans 'late of Ryeford Mill, clothier' was declared bankrupt in 1826. In 1830 Reuben Hyde was working the corn mill and Henry Miles the cloth mill. Cloth manufacture probably ceased soon afterwards. In 1853 it was entirely a corn mill with two waterwheels and six pairs of stones. It was a saw mill in the 1880s. Much of the main mill building has been long demolished, the chimney being demolished in 1964.

(19919/188gs; PCC Prob.11/400; LG, 9 May 1829; LG, 20 October 1848; Directory 1830; RX 289.1gs)

'Barrack Mill', SO 809038

This was probably the unnamed 'clothing' mill, containing two pairs of stocks and two gig mills with a powerful steam engine, together with a new-built grist mill, engine, and 18ft-diameter waterwheel, lately occupied by Sir Paul Baghot, which was offered for sale in Leonard Stanley in 1830. The brick- and stone-built mill, together with eight cottages and a mansion house, was the property of William Holbrow. It has been demolished.

(Q/RI 89; D1347 acc. 1603; Peter Griffin, 'The Mill at "The Barracks", King's Stanley', *GSIA Journal*, 2008)

Unnamed mill in King's Stanley

This small mill, reputedly built by James Hogg *c.*1804, has been attributed to the ownership of Jeptha Young who wrote *Lays for the cottage and Rural poems or Rhymes from the loom*. At the tithe survey, the mill was owned by Charles Beard. By 1861 it was a house. It has been demolished.

(Mahler & Marshfield, *Stroudwater Valley Mills*, p.7)

Hoggs Mill, SO 823045

It is not known if this was a cloth mill but it was owned by Stephen Clissold, a clothier, and partly occu- pied by George Howell at the tithe survey. The mill has been demolished.

Oil Mill, SO 825045

In 1723 William Adderley, mercer, mortgaged the newly built mill 'for the making of rape and linseed oyl' to John Bond. In 1731 'a very good and well-built oil mill' was advertised for sale or letting and soon afterwards had been converted to fulling and was owned by the Rimmington family. In 1764 the mill contained four stocks and two gigs. The Rimmingtons went into partnership with Richard Flight and were bankrupt in 1786. The mill then contained two stocks. In 1820 James Lewis, clothier, was here. He died in 1826 and was succeeded by his son J.F. Lewis. In 1833 Lewis told the factory inspectors that 'the bulk of the weaving is now carried on in the factories where the hours are limited'. In 1834 he said: 'Some part of these mills must have been erected 100 years or more, and no part of them later than 30 years.' The power was water only. Oil Mill ceased to be a cloth mill *c.*1840. In 1857 Joseph Eycott, farmer, owned Oil Mill, then described as a flour mill.

The mill consists of a single range lying approximately north to south, with wings at each end. It is three storeys high with a loft. The walls of the ground floor are of stone while those of the upper floors are brick (see p.160). After having remained empty for some time the mill, now called Snow Mill, has a new lease of life for the innovative manufacture of artificial snow.

(D149/T/888; D873/T20; D2219/9196/12/1; GJ, 15 July 1731; Directories 1820, 1830; PP 1833, XX, p.946; PP 1834, XX, p.286; Stephen Mills, 'Oil Mill, Ebley (Ebley Corn Mill)', *GSIA Journal*, 1994, pp.63–5)

Ebley/Deerhurst's/Maldon's Mill, SO 829046

This was a double mill site comprising a corn and fulling mill which were let to different tenants in the early sixteenth century. In 1537 Richard Barrow and Joanna, his wife, leased 'a water mill called a grystmyll in Stonehouse called Derehurst's Mill … the tuckmill excepted' to William Bennett of Ebley. The grist and fulling mills were joined under the same tenant in 1538, when Joanna Coke leased the fulling mill to William Bennett for ninety-nine years. In 1580 John Berrowe of Awre leased the water mill to Thomas Bennett, and built Ebley Court in 1587. By 1597 Leonard Bennett was in possession of both the corn and fulling mills. His daughter married William Selwyn and, through her, the mills went to the Selwyn family. By 1681 the cloth mill was occupied by John Turner. In 1685 Selwyn & Turner of Ebley Mill and Thomas Warner of Dudbridge Mill came to an agreement concerning the height of the water between the two mills. In 1689 Selwyn demised the mill, 'shierhouse' and dyehouse to Turner. A date plaque of 1597 and the initials W.S. incorporated into a later building at Ebley may indicate a date of some rebuilding of the mills.

In 1791 the cloth mill and the grist mill were conveyed to William Bray. In 1799 the mills were sold to Samuel and Stephen Clissold for £1,908 8s 7d. Stephen Clissold began the new mill between 1818 and 1820 but it was sold before completion. It was he who purchased 7 acres of land for a reservoir to provide water for the four or five waterwheels. The 'old mill', said to date from 1800, had a fine Venetian window, other windows being elliptically headed with a single mullion, later mirrored in the fenestration of the new 1818 mill. A detailed map of the site was made by Baker in about 1820. This shows both the old and new mills and describes the function of each workshop. There were stoves, a gig mill, teazle house, counting house, scouring house, dyehouses and warehouses. In 1827 the property was sold to William Evans, Ronald Robson and Thomas Severs. By this date, the grist mill had gone and there were stocks, a gig mill and vats, as well as a counting house. John Figgins Marling was the tenant in 1834. By 1840 there were seventy-one handlooms at the mill but twenty-nine of them were not in use. In the spring of 1840, Thomas and Samuel Stephens Marling moved to Ebley. S.S. Marling grew the business which became one of the largest in the district and steam power was added.

In 1851, Samuel Stephens Marling went into partnership with Strachan, Marling contributing value in kind and Strachan contributing cash. Strachan retired in 1862, and the business became known as Marling & Co. once more. In 1852 a huge hole was blown in the front wall of the pedimented block and in 1859 the block was destroyed by a massive fire which caused some £20,000 worth of damage. Clissold's mill was unaffected and in 1865 G.F. Bodley, who had been employed to build Selsley church, was commissioned to design a new building. The turret to the new building was aligned with Selsley church. In 1865 the company was dissolved and re-formed at Ebley by Samuel Stephens Marling and Samuel Stephens, without William Henry Marling who went to Stanley Mill. This firm was dissolved in 1869 and Stanley and Ebley were united. When Sir Samuel Stephens Marling died in 1883, his son William Henry Marling and Samuel Stephens continued at Ebley as Marling & Co. but William Henry made Stanley Mill the centre of his operations. In 1920 the partnership of Marling & Evans was formed, although the firm was wholly owned by P.C. Evans & Sons. Ebley Mill became the carding and spinning department for Stanley Mill in the 1930s. The mill was driven by electric power from 1939 and the last surviving waterwheel and steam engine were scrapped. All the processes for manufacturing military cloth were undertaken at Ebley in the Second World War, but the mill returned to its role as a carding and spinning mill for Marling & Evans in 1945. After the war there was heavy investment in machinery – the first high-speed mule in the world was installed and improved by the mill engineers. And, in the 1960s, the mules were replaced by spinning frames which were appropriate for the synthetics for which the company became known. Competition, depression and a shortage of capital investment led to the closure of Ebley Mill in 1981. The mill was purchased by Stroud District Council in 1986 and was sensitively restored and opened in 1990.

The Old Mill, together with some workshops, was demolished in 1965. The New Mill is an impressive building of five floors with a slate-roofed turret at the north end. The windows are of two lights, divided by wide stone mullions, each light being subdivided by two glazing bars. The large reservoir has been drained but the steep banks are easily discernible. The clothier's house, Ebley Court (formerly Skinner's Place), was destroyed by fire in 1968 (see pp.132, 162).

(289)(21), (24), (43), (45), (46)gs; D873b; D873 E9, E13; RF201.3(2); 289 (63) and (64)gs; E.A.L. Moir, 'Stanley and Ebley Mills' (typescript); PP 1834, XX, pp.254–5; PP 1840, XXIV, pp.396–7)

Cherynges/Norris's Mill, SO 835046
This was a fulling mill by the 1480s. In 1716 James Small of Ozleworth, clothier, leased a 'fulling mill and gyggmill and cloth press therein' to Stephen Dangerfield of Randwick, clothier. In 1718 Dangerfield transferred the lease to Daniel Fowler. It later became known as Norris's Mill; members of the Norris

family held the mill for around 100 years. It became incorporated into the Dudbridge Mill complex. (D149/190/407, 408; D67/Z46; Prob.6/15)

Dudbridge Upper & Lower/Chance's Mill, SO 835047 (for Dyehouse see p.233)

This is a large and complex site comprising two distinct mills, one of which, probably the older, was known as Lower Mill. A water corn mill was here from an early date, and a fulling mill was at the site by 1685 when the owners of Dudbridge and Ebley Mill came to an agreement over the height of the water between their properties. In 1708 Thomas Warner re-leased the Spillman's Court estate with a messuage called Dudbridge, occupied by Henry Halliday, and later by Daniel Chance, with three fulling mills, one corn mill, one gig mill and a dyehouse adjacent to the messuage with eight racks in the Rack Close. D. Chance was a woollen manufacturer at Dudbridge in 1784. In 1790 a 'roomy' cloth mill (with one and a half stocks, a gig mill and shear shops) was advertised to be let. It was said to be near to Hawker's works, 'an eminent and well-known dyer and also close to his knapping mill'. In 1814 George Adey & Co. were described as woollen manufacturers of Dudbridge. In 1827 'Desirable Clothing Mills' at Dudbridge were advertised, the mill containing four waterwheels, stocks, gig and machinery mill with an 'abundant' water supply. At the tithe survey the site was described as a grist mill occupied by Peter King and sundry tenants.

From 1856 onwards, David Apperly appears in directories as occupying Dudbridge Mills. It is probable that J. & D. Apperly had been here for some years before. The firm was apparently established in 1794 although no directory includes the name Apperly amongst the woollen manufacturers of Stroud and district until towards the middle of the nineteenth century. Replying to the factory inspectors of 1833, John Apperly said that the greater part of the mills had been erected within the previous three years but that water power only was used. During the 1850s William Clissold was associated with the firm. In 1868 they sold the southern part, Lower Mill, to Tubbs & Lewis, elastic braid manufacturers, who in turn sold the mills to Thomas Bond Worth of Stourbridge for carpet manufacture. At this date the Lower Mill contained self-acting mules, 60in carding engines and condensers with a capacity to make forty ends of fine cloth per day. There was a 50hp steam engine and two waterwheels each of 15hp. In 1872 Sir Alfred Apperly introduced the 'latest automatic equipment for spinning, carding and weaving'. There were ninety looms and 6,950 spindles at the mill in 1889. In 1902 Lower Mill was reacquired. The engravings in *Industrial Gloucestershire* show the site to have been a very large one with several long, three-storey buildings and many two-storey ones. By c.1914 the mills had electricity for power and lighting. The firm went into receivership in 1931 and the mill closed in 1933, Redler Conveyors of Sharpness occupying part of the site. Since then the site has been much altered, the remaining buildings having been sensitively restored for housing.

(Revd C.E. Watson, 'The Minchinhampton Custumal', Trans. BGAS, LIV, p.265; 289(63)gs. See Ebley Mill above; GJ, 22 February 1790; D1159; P272a VE 1/1; LG, 11–15 May 1756; GJ, 22 December 1827; GJ, 14 March 1868; Directories 1784, 1814, 1820, 1839–40, 1856, 1863; *Industrial Gloucestershire*, p.26; PP 1840, XXIV, pp.396–7; PP 1834, XX, pp.265–6; *Stroud Official Guide*, c.1914; Kelly's Textile Directory 1889)

Hawker's Mill (Dyeworks), see p.233, Nailsworth Valley

Fromehall Mill, SO 843048

This is one of the oldest recorded sites, a fulling mill being mentioned in 1218. It belonged to the Halyday family and in 1687 comprised two stocks and a gig mill.

In 1545 William Halliday, clothier, was assessed at £42. In 1678 Michael Halliday, clothier, levied a fine on 'one fulling with two stocks and one gig mill'. By 1687 the number of stocks had been increased to five. In 1693 the mill was leased to Nathaniel Beard and the stocks had been reduced to two. In this year,

Halliday's daughter, Ann, married William Essington of Bowbridge and, on Michael Halliday's death, Essington controlled Fromehall Mill. The mill then passed to Thomas Phillips. In 1736 Phillips sold the property to John Shipway of Rodborough for £11,085.

Fromhall Mill, containing three pairs of stocks, gig, 'lofts for machinery', shear shop, wool lofts, press-house and dyehouse, was advertised to be let in 1808. In 1828 the mill was leased to Nathaniel Samuel Marling and was occupied by Thomas Workman. There was a grist mill at the site too, which Marling let to Thomas Creed of Painswick in 1828.

In 1846 William Hunt became owner of the property. The mill was run by Hunt, together with Lodgemoor Mill, until 1865 when the mills were sold to Josiah Greathead Strachan for £30,000. In the year of the 1862 Exhibition Henry Mayhew published a detailed account of Fromehall and Lodgemore mills. Fromehall specialised in the manufacture of delicate-coloured cloths. In 1889 Strachan & Co. of Lodgemore and Fromehall mills had 157 looms and 8,300 spindles in twenty-seven frames between the two mills. In 1891 there were 2,640 spindles and fifty-seven looms at Fromehall. 'The Old Mill' is now occupied by workshops. It is stone-built and appears to be of mid-eighteenth-century date. It is four storeys high and is lit by small horizontally headed two-light windows divided by wide, stone mullions on the ground, second and third storeys and by three-light windows on the first floor. A large, stone, nineteenth-century building adjoins the mill at right angles on the north side, replacing the old dwelling house.
(D149/21; Revd C.E. Watson, 'The Minchinhampton Custumal', Trans. BGAS, LIV, p.370; D892/T68; D67/Z46; GJ, 9 May 1808; E.A.L. Moir, 'Lodgemore and Fromehall Mills' (typescript); D149/272/497; Kelly's Textile Directory 1889; Worrall's Textile Directory 1891)

Lodgemoor/Merretts/Upper and Lower Latemore Mill, SO 845050

In the sixteenth century a track crossing the Frome at Lodgemore was the boundary between the lands of the Spillman and de Rodborough families. Each family owned fulling mills on either side of the track and, in time, the mills together became known as Lodgemore.

Lower Latemore was reputed to be older than Upper Latemore/Merretts Mill. In 1608 the owners of the two mills were involved in a dispute over water supply. There was a mill containing stocks and full-ing mills, a mill house called the grist mill, formerly known as Loft or Meese, and a place called Welkins containing a gig mill and a mill for grinding corn. In 1776 the mills were in the Halliday family. In 1794 Richard, Henry and John Cooke were in possession of the mill which was occupied by John Jones. The Cookes got into financial difficulties and William Whitehead became the owner of the mill.

The contents of workshops at Lodgemore were put up for auction in 1808 by the assignees of George Hawker. These included billies, jennies and 100 pairs of shears.

Lodgemore Mill suffered a series of fires in the nineteenth century, the first of which was in 1801. Other fires occurred in 1807, 1811 and 1871. The 1811 fire was reported in a number of provincial newspapers, being discovered between 3 and 4 a.m. Within half an hour 'the whole building presented one complete body of flame, and in a short time the roof and side-walls fell in with a tremendous crash'. A grist mill adjoining was partly saved by the arrival of the fire engine. The cloth mill was said to have been 'recently erected at a cost of nearly £8,000, exclusive of the machinery'. In 1814 a new mill was built, having both steam and water power. Whitehead leased the mill, including the new one, to Nathaniel Samuel Marling. Whitehead got into financial difficulties and was bankrupt in 1827. William Playne and Robert Jeffries Brown bought the premises consisting of a 'capital mill', presshouse, gig mill and a new erected mill with steam engine at a rent of £1,800 p.a., Marling continuing as tenant. In 1839 William Hunt of Dyehouse Mill became tenant of Lodgemore Mill; the mill was sold to him seven years later for £11,750. Lodgemore Mill was described in detail in 1862, when Henry Mayhew's reporter(s) undertook

a three-day visit, describing the manufacturing processes in detail (see Part I, Chapter 8). In 1865 the mill was sold, together with Fromehall Mill, to Josiah Greathead Strachan who introduced Hebdon's Patent Cloth Testing Machine from Yorkshire in 1870. In 1871 the mill was completely destroyed by fire. James Ferrabee submitted plans for a new mill which was begun in 1875. It is built of red and blue brick and had two 35hp engines. In 1889 there were 157 looms and 8,300 spindles at the two mills. In 1891 Lodgemore had 5,660 spindles and 100 looms, more than half the total. In 1890, Josiah Greathead Strachan sold out to Strachan & Co. In 1920 Strachan & Co. merged with Hunt & Winterbotham of Cam and Playnes of Longfords. The mill, now owned by WSP Textiles Ltd, continues to manufacture non-apparel cloth including cloth for tennis ball covering. Several buildings escaped the last fire, including the very fine stone mill house and adjoining industrial buildings which back on to the Fromehall mill pond.
(E.A.L. Moir, 'Lodgemore and Fromehall Mills' (typescript); D149/21; Jennifer Tann, 'Some Problems of Water Power …', Trans. BGAS, LXXIV, pp. 53–77; D67/Z 46; D873 E4; GJ, 23 May 1808; Kelly's Textile Directory 1889; Worrall's Directory 1891; Mahler & Marshfield, *Stroudwater Valley Mills*)

Cuttles Mill, see Slad Valley, p.228

Wallbridge Mill, SO 848050 (for Dyehouse see p.219)

In 1470 Thomas Bigge, who had acquired the mill from John Bigge, conveyed a fulling mill to his wife Alice. In 1608 the mill was in the hands of William Trotman and was rebuilt soon afterwards. The property was in the possession of Thomas Webb who, in 1633, was described as a clothier of Wallbridge. By 1718 the mill had passed from Thomas Webb to his daughter Anna, wife of Samuel Sheppard of Minchinhampton. Sheppard died in 1724 and the mill passed to the younger Sheppard son, Thomas, who died in 1757. In 1761 three fulling mills, one gig mill adjacent and rack land were leased to the clothier Samuel Watts. The premises had formerly been occupied by Thomas Theyer, then Samuel Hawker, then William Hill. In his will, Samuel Watts bequeathed to his nephew Richard Wallbridge fulling mills; and to his nephew Nathaniel a dwelling house, dyeing furnace, brewing furnace and cloth rack. Nathaniel Watts was the first to introduce the fly shuttle to the region in 1793. The magnificent painting of Wallbridge in the Museum in the Park, Stroud, thought to have been painted between 1789 and 1797, depicts Wallbridge as a complete (and picturesque) industrial estate. The multi-gabled mill, from the visible roofline, resembles a domestic building of the period, while the canal bends around the single-storey dyehouse (to avoid cutting through the mill estate). Richard Watts, dyer, sold the mill in 1820 to the brothers Peter, Richard and Gustavus Adophus Smith. The site comprised dyehouses which were lately occupied by Richard Watts as a dwelling house and were then converted into a woollen manufactory, with a rack stove and wool stove. Both Peter Smith and G.A. Smith were woollen manufacturers of Wallbridge in 1822–23. The Smiths continued there until the early 1840s when John Howard occupied the premises. Four power looms were installed in 1837. The partnership between John Howard and George Howard was dissolved in 1851, John Howard continuing at the mill. In 1871 the five-storey mill with 24hp engine was put up for sale but withdrawn from the market. In 1891 there were twenty-two looms there. Howard & Powell, clothiers and dyers, ran Wallbridge Mill from the late nineteenth century until 1960 when the mill closed. When the railway was being built along the Stroud Valley, the company bought Wallbridge Mill, demolished part of it and then sold it back to the manufacturers. The mill was demolished in 1963–64.
(IPM (Index Library) III, p.112; S.P.D.CCXLI, p.36; D1347, acc. 1603, Stroud; D67/Z/55-7; PP 1840, XXIV, pp.396–7; LG, 2 Jan 1852; *Sussex Advertiser*, 30 April 1872; Directories 1822–23, 1830, 1839–40, 1842, 1844; Stroud Museum 1962, 148; Worrall's Directory 1891; Ian Mackintosh, 'Into the Picture', STT; Ray Wilson, 'Howard & Powell, Wallbridge Mill, Rodborough, Stroud', *GSIA Journal*, 2004, pp.31–41)

Capel's/Orpin's/Arundel Lower Mill, SO 854046

This was a fulling mill in 1513. Richard Arundel rented the mill from Over Lypiatt Manor in 1581 and, in 1608, purchased it. In 1654 Daniel Capel occupied the mill and in 1710 Daniel Capel, 'phisitian', leased a newly erected 'stove and workhouse for the clothing trade' to Richard Capel of Stroud, clothier. William Capel was here in 1784. In 1820 John Capel occupied the mill. In 1839 M. Bowerbank made cloth and kersey at Capel's Mill but apparently he only occupied part of the premises as Thomas Vick, miller, was also here. By 1842 Matthew Grist worked the mill at first, apparently for making cloth, although shortly afterwards he changed to shoddy manufacture. When the railway was built along the Stroud Valley, several buildings were demolished as the viaduct crossed the mill site. The mill has been demolished.
(*VCH Gloucestershire, 10, Stroud*; D24/T6; Directories 1784, 1820, 1839–40, 1842, 1856; Mahler & Marshfield)

Arundel Mill, SO 855045

In 1381 William Huckvale, tucker of Over Lypiatt, worked a fulling mill at this site. In 1585 John Huckvale retired to Huckvale's Place and sold the mill, comprising two fulling and one corn mill, to Richard Arundel. Richard Arundel owned the mill in the early seventeenth century. In 1749 the fulling mill, containing three stocks and a gig mill with a dyehouse, was occupied by Samuel Wallington. In 1785 Thomas Crozier Arundel leased the mill to John Woods. In 1795 the mill, occupied by Benjamin Cooke, was conveyed to Josiah Paul Paul and Thomas Gregory. In 1807 James Arundel leased a messuage with wool lofts above, a fulling mill, dyehouse and tenters on the Rack Hill to Rowles Scudamore and Walter Watkins. The premises were divided by 1820, for the dyehouse was occupied by D.W. Smith & Son, blue dyers, and the mill by John Gordon, woollen manufacturer. By 1830 Ann Gordon, probably John's wife, was making cloth at the mill. Four power looms were installed in 1837. Smith & Gyde, 'black and medley dyers', were in sole occupation of the premises by 1842. The directory of 1863 records two Gydes in occupation, one a dyer, the other a manure manufacturer, but by 1879 only the dyeing business remained, the firm then being known as Gyde & Clough, and in 1894 as Gyde & Co. The mill has been demolished.
(PRO E315/394/ff 117, 118. D1347, acc. 1603, Stroud; D892/T80/18; Directories 1820, 1822–23, 1830, 1839–40, 1842, 1856, 1863, 1879, 1894)

LIME BROOK (A SMALL TRIBUTARY OF THE FROME)

Weyhouse Mill, SO 861044

This was a grist mill in 1690 and in 1734 was leased by James Clutterbuck to Stephen Power, a shear grinder, then in 1756 to his son Stephen who was in the same trade. By 1842 the mill had been demolished.
(D1157; D1756/6; D1157)

Newcombe's/Sandys' Mill, SO 859044

In 1804 Thomas Newcombe erected a water mill on the site of a leather mill. In 1822 the mill contained two gigs, machinery workshops and 'Mustard Warehouse'. It was still owned by Newcombe and the annual value was £77. Some time later the mill came into the hands of John and James Sandys and the mill thereafter was known as Sandys' Mill. In 1838 Charles Stanton was said to have been a former occupier but the mill was then empty. In 1859 the Sandys leased the mill with the steam engine and reservoir to Walter and John Stanton, by whom it was then said to be occupied. It has been demolished.
(D38A/12; P.320 A; VE1/7; D1159; PP 1840, XXIV, p.398)

RIVER FROME (MIDDLE TO UPPER SECTIONS)

Bowbridge Mills and Dyehouses, SO 857043

This was a complex site and large industrial unit for several hundred years, being in multiple ownership for part of the time, sometimes between members of the same family. A fulling mill was recorded here in 1513. The seventeenth-century watercourse leases record that Nathaniel Workman and Roger Fletcher were at Bowbridge Mill.

In 1680 Thomas Weary leased 'all that corn or grist mill two fulling mills one gyg mill, one furnace and all the vates therein or thereunto belonging' to William Essington. Also conveyed were 'all those other two fulling mills and one gyg mill' formerly held by Nathaniel Workman and then by William Fletcher.

In 1735 Richard Essington and others demised a messuage at Bowbridge, and a 'cottage or tenement formerly a grist mill and all that fulling mill containing two stocks one half stock and one gig mill' occupied by John Gainey. In 1775 the premises were conveyed to John Partridge. A new mill was built in 1780 and new dyehouses in 1795 and 1802. In 1806 Nathaniel Partridge conveyed to Thomas Partridge the cottage, formerly a grist mill, and the two fulling and one gig mills, or as much as was then standing and 'also all that new erected fulling mill' with lofts, counting houses, dyehouses, workshops and other buildings erected by John Partridge, lately occupied by him and then by Nathaniel Partridge. In 1808 John Partridge of Painswick, dyer, and Joseph Partridge of Bowbridge, dyer, drew up articles of partnership.

In 1810 part of the mills were let to Samuel Clutterbuck, the main part of the premises remaining in the hands of various members of the Partridge family. The premises were essentially divided into three and a clause was inserted in the deed whereby if there was a shortage of water, the parties were to stop work until there was sufficient for them to carry on again. By 1833 John Partridge was fulling solely by steam power. John Holbrow became the sole owner of the property sometime before 1832 and he let it to William Henry and Charles Stanton. In 1832 they sublet 'all that clothing mill' containing five pairs of stocks and two gig mills to John William Partridge, clothier and dyer. When J. Partridge replied to the factory inspector's questions in 1833, he said that he manufactured broad superfine and other cloths for an extensive dyeing establishment. He also carried on the business of a public dyer. He had four factories and extensive dyehouses. One factory was built in c.1795, and the power was wholly provided by water. Factory 2 was 'a very old building but altered in 1824 to its present purpose' and the motive power was wholly steam. Factory 3 was erected in 1802, and the power was partly water and partly steam. Factory 4 was erected in c.1780 and was driven by water power. Overall the water power was around 30hp and steam 18–20hp. In 1840 J. Partridge continued as a cloth manufacturer and dyer. In 1856 Joseph Partridge was recorded as a dyer of Bowbridge and Nathaniel Partridge as a dyer of scarlets and manufacturer of dyers' spirits.

In 1863 Merrett & Chambers, blue, black and medley dyers, and Partridge & Co. were in occupation of the site. Partridge & Co.'s stove and dyehouse were worth £130 per annum and the mill and dyehouse occupied by Merrett & Chambers were worth £305 per annum. Clutterbuck Chambers and Nathaniel Partridge were at Bowbridge Mills in 1874 (see p.72). In 1894 Strachan & Co. purchased part of the site (probably the Partridge portion) to undertake dyeing for Lodgemore and Fromehall mills, and William Clutterbuck Chambers remained in possession of the other part. They remained in occupation until the early twentieth century. The two older mill buildings were demolished in the 1950s; another in the 1960s and the main dyehouse and block facing Butter Row was demolished in 1971. On the upstream side of the Rodborough road, one eighteenth-century stone mill remains, converted into housing. It is two storeys high and roofed with slate with horizontal-headed, three-light windows which are divided by wide stone mullions.

(PRO E315/394/1; *VCH Gloucestershire*, X; Boxes 78; 69gs; D1801/2; D1815, Stanton; D1241, bdl. 28;

D1512; P.32 on VE1/8; PP 1833, XX, p.931; PP 1834, XX, p.281; PP 1840, XXIV, pp.400–1; Directories 1830, 1839–40, 1842, 1856, 1863, 1874, 1879, 1894, 1910).

Stafford Mill, SO 859038

Richard Stafforde was fined for making defective cloth in 1561. Between 1650 and 1680 the mill was owned by Richard Stafford and in 1708 and 1717 a messuage called Stafford Mill, sometime held by Richard Stafford, was leased by John and Thomas Griffin to Charles Coxe. In 1720 'one fulling mill' was leased to Lawrence Clutterbuck and Samuel Capel, together with Griffin's Mill. In 1731 William Griffin leased the mill to John Prinn for a year. In 1793 the mill containing two and a half stocks, a gig mill and dyehouse with fixed copper, lately the property of Thomas Griffin, was sold to William Stanton. The mill remained in the Stanton family until it closed. A steam engine was installed in 1831 and a large power loom shed built. In 1833 William Stanton told the factory inspectors that 'a portion of Stafford Mills was erected more than 40 years since, other parts have been erected 18, 15, 10, 7, 5, and 2 years since'. Stanton was one of the first Gloucestershire clothiers to adopt power looms. He installed the first in 1836, and by 1838 he had twenty-four power looms in use and four out of use, with sixty-seven handlooms in use and twenty-one out of use. The mill was still working as a cloth mill in 1879, but it ceased in 1886. Bailey's paints were made on part of the site from the late 1880s and by 1889 they occupied the entire site (the firm moved to Griffin's Mill in 1960). The mill has been largely demolished, although the chimney remains; the site is now occupied by a number of small businesses.
(D181/III/T59; D1815, Stanton; PP 1833, XX, pp.283-4; PP 1840, XXIV, pp.396–7, p.398)

Griffin's Mill, SO 859035

Richard Fowler sold the mill to John Griffin in 1599 and in 1627 the property comprised a grain mill, fulling mill, gig mill and a house. In 1638 Thomas Griffin was seised of the property comprising a grain mill, two fulling mills and a gig mill. In 1720 it was leased, together with Stafford Mill, to Lawrence Clutterbuck and Samuel Capel, and in 1731 it was leased to John Prinn. In 1764 Thomas Griffin conveyed the property then consisting of two stocks, a gig mill and dyehouse to Joseph Godfrey and Alexander Hamilton. In 1794 John Partridge conveyed the property containing two and a half stocks, one gig mill and a dyehouse to Nathaniel Cooke of Woodchester. In 1814 Thomas Caruthers was making cloth at the mill, and Thomas Howell was making cloth there by 1820. In 1828 some machinery was offered for sale. Edward Shipway, clothier, of Griffin's Mill, was bankrupt in 1836.

By 1838 the mill was occupied by Bayliss who was using it as a saw mill. A couple of years later, Holmes was occupying the mill. In 1846 the mill was up for sale and was probably bought by Nathaniel Samuel Marling, who owned it in 1853. In 1856 William Barnard occupied the mill. In 1858 Nathaniel Samuel Marling leased the mill to Henry Hooper, timber dealer. Hooper is described as a stick manufacturer in 1863. The mill remained a timber and stick mill for many years. During the early twentieth century it was an aircraft factory. It is now home to Bailey's Paints and several other businesses. The main mill is a late nineteenth-century, three-storey brick building.
(P.H. Fisher, *Notes and Recollections of Stroud*, p.211; IPM (Index Library) III, p.61; D181/III/T59; T61; D1815, Stanton; D1241 No. 113; D445/E8; D445/T35; GJ, 9 February 1828; LG, 12 February 1836; Directories 1814, 1820, 1839–40, 1863; PP 1840, XXIV, p.398)

Ham Mill, SO 861032

In 1599 the 30-acre Doleman's Ham estate was purchased by William Webb of Nailsworth from the Tayloe family. William Tayloe held 'a mill, a tuck mill and a gyggemill and a close of land' on the Lypiatt manor estate in the 1550s, but this may not have been Ham Mill. In 1601 William Webb re-leased

Doleman's Ham Mill to Nicholas and Thomas Webb. In 1685 there was a court case concerning Samuel Webb's right to the title of Ham Mill comprising two fulling mills, one gig mill, one grist mill, five racks, a dyehouse and presshouse. The mill remained in the hands of the Webb family until the middle of the eighteenth century. In the Civil War Samuel Webb obtained letters of safe conduct from Prince Rupert and Prince Maurice. In 1685 the mill contained 'two stocks for fulling one gig mill and one grist mill with a dyehouse presshouse and five racks'.

The Winchcombe family purchased part of the estate in 1742 and operated the mill into the 1780s. The family had made its money by good marriages, good trading and careful negotiation with the last of the Webb family. James Winchcombe, father and son, grew the cloth business at Ham Mill. Thomas Gyde was a customer of the dyehouse, the dyehouse being one of those in Stroudwater which undertook scarlet dyeing. After Winchcombe's death his heir sold Ham Mill.

In 1813 the 'capital mill and factory of great power and extent' was advertised for sale and bought by the Wathen family, although William Marling occupied the mill. William Marling took his sons Samuel Stephens and Thomas Marling into partnership and Ham Mill, it would seem, became a training ground for a remarkable clothing family, the mill being run in conjunction with Fromehall Mill. In 1833 the firm told the factory inspector that they carried out all branches of the manufacture of superfine cloth except dyeing. 'The premises have been employed in the manufacture of woollen cloth more than 100 years, a second mill erected in 1814, a third in 1825, a fourth in 1832.' There was a steam engine of 30hp and three wheels of 10hp each.

Ham Mill was a focus of the weavers' riots in 1825. The mill caught fire in 1841; fifteen of the forty-five power looms installed in 1836 were destroyed. In 1840 Thomas and Samuel Stephens Marling moved to Ebley Mill, but in the following year Thomas returned to Ham Mill.

By 1846 Ham Mill was in the hands of Nathaniel Samuel Marling, who leased 'All that mill lately used as a clothing mill', together with the three waterwheels, to a wood turner. He did not remain in occupation for long, for Thomas Sampson, shawl manufacturer, was named as tenant by 1853. By the 1860s, Alfred Ritchie was making cloth at Ham Mill. In 1866 there was another disastrous fire. The newly erected power loom sheds were saved as there was little wind but the main mill roof collapsed in 35 minutes. By 10 p.m. the mill was a smouldering ruin, 'the tenter and drying machine and shed … alone of the mill buildings escaped destruction'. Richie purchased the ruin, commissioning James Ferrabee to build a new mill on the site of the old. Spinning mules were located in the long range, powered by a large ground-floor waterwheel. A new detached mill house was also built on the south side of the stream. Brick, single-storey loom sheds, together with a stable and coach house, were erected and a pair of fine iron gates placed at the entrance to the site.

The deed of partnership for the firm of Alfred Ritchie & Co., dated 1870, mentions a loom shed, wool sorting and burling shops, wool stoves and dyehouses, carpenters' and blacksmiths' shops, willy house, boiler house, yarn and stock warehouses and a 40hp steam engine. In 1880 Alfred Ritchie and others leased the mill to Elliott and Ritchie for seven years. In 1889 there were eighty-five looms and 4,560 spindles at Ham Mill. In 1891 the number of looms was put at seventy while the spindles had apparently been increased to 4,590. Cloth manufacture ceased in 1899. Thomas Bond Worth & Son, carpet manufacturers, bought the mill in 1900, ceasing manufacture at the site by the end of the 1990s. The mill is empty.

(P.H. Fisher, *Notes and Recollections of Stroud*, p. 211; PRO E315/394; D1842/43; D873b; D873 E10D45/E8; D1159; Hawker MSS; E.A.L. Moir, 'Lodgemore and Fromehall Mills' (typescript); PP 1834, XX, p.276; PP 1840, XXIV, pp.396–8; Directory 1856; Kelly's Textile Directory 1889; Worrall's Directory 1891; Ian Mackintosh Typescript notes)

Thrupp/Sewell's/Huckville/Wathen's Mill/Ferrabee (Phoenix) Ironworks, SO 863029

There was a medieval fulling mill at this site, associated with Huckville's Court. In his will of 1540, Thomas Sewell, clothier, bequeathed to his son Thomas 'my dwelling house, fullyng mylles and dyeing house' together with his 'vattes and furneys with sheares and other shoppe stuffe'. In 1608 Thrupp Mill was owned by a clothier named Richard Sewell. On his death in 1635 he owned two fulling mills, one gig mill and one grain mill called Huckvales Court. In 1708 Huckvale's Court and Sewell's Mill were conveyed to Jeremiah Davis and Richard Baker, and in 1752 let to Samuel Wathen, a member of the well-known family of clothiers, who subsequently purchased the estate. After Samuel Wathen died his widow, Elizabeth, let 'all that clothing mill' which contained five pairs of stocks, with the gig mill house, presshouse, dyehouse, stove, picking shops and rack meadow, to John Ferrabee in 1828. The two waterwheels at the east end of the mill building and the stocks were to be removed. A foundry was to be erected at the east end of the mill. Ferrabee became well known for his waterwheels, textile machinery and the manufacture of Budding's lawn mowers. In 1856 the Ferrabee brothers moved to Port Mill and George Wailes & Co. took over the Ironworks, selling them to Bruton & Sons in 1872. The mill has been demolished.
(Thames & Severn Canal map; GJ, 5 January 1828; D873/T86; D846/III/29; Mahler & Marshfield)

Hope/Gough Mills, SO 864026

In 1705 members of the Sewell family were associated with the site, although Joseph Gough, clothier, then worked the mills. Gough had married the widow of Richard Sewell. The site at this date comprised three fulling mills and one gig mill. Daniel Gough was here in 1748. The mill was rented to John Cambridge in 1768. In 1792 the site comprised three fulling mills and one gig mill, later four fulling mills and one gig mill. Between 1805 and 1812 Samuel and Nathaniel Wathen built a new mill, retaining two waterwheels and renamed the mill Hope Mill. In 1814, Nathaniel Wathen came to an agreement with John Lewis of Brimscombe Mill concerning water supplies.

In 1826 the movable machinery was offered for sale in 1828 and in 1829 the assignees of Nathaniel Wathen conveyed the 'ancient mill ... containing two fulling mills and one gig mill ... called ... Gough's Mill ... [and] All that newly erected fulling mill ... called ... Hope Mill to Robert Bamford'. The deeds of 1814, 1825 and 1826 describe Gough Mill but call the property Hope Mill. Robert Bamford, when replying to the factory inspector's queries, said that the mill was 'built at once. The rooms being large, lofty and airy. Began to be employed for the present purpose in July, 1829.' The power was 'about 17 horse water power'. Bamford was described as a woollen yarn manufacturer, list manufacturer and wool stapler in the 1830s. He probably left the mill in the 1840s.

In 1863 Charles Barton was a silk throwster at Hope Mill. A deed of 1870 shows that, after the brief spell of being used only for silk throwing, Hope Mill was used once more for cloth manufacture. The partnership of Edward Barnard Sampson and P.J. Evans, both described as fancy woollen cloth manufacturers of Hope Mills, was dissolved in 1870. It seems probable that the mill was divided and occupied by two firms in the late 1860s. Part of the site was occupied by Edwin Clark, steam launch manufacturer, in 1878 and, on his death in 1897, the business was acquired by L.J. Abdela of Manchester who built canal and river boats here. The two upper storeys of the surviving eighteenth-century mill were taken down in the 1960s. The site is now partly a fixed caravan site and partly an industrial estate.
(D67/Z/32; & IPM (Index Library) III, p.144; D892/T10/3; D1159 acc. 1753; D1712/1; D829/T53/10; D873b; D1159; GJ, 2 February 1828; PRO E315/394; PP 1834, XX, pp.266–7; Directory 1863)

Brimscombe Upper and Lower Mills, SO 866024

These two mills were run together under a single owner or tenant. In 1539 John Bigg leased two fulling mills at Brimscombe and in 1594 Ursula Bigge of Stroud owned the mills and Bigges Place. When Roger Fowler died in 1626 his will mentioned the house called Bigges Place, together with two fulling mills, a gig mill and corn mill. In 1648 Henry Fowler of Minchinhampton conveyed 'all those ffullinge mille gig mille and grist mille and barn', all of which were occupied by Henry Mayo, to William Webb of Stroud, clothier. In 1705 the property, occupied by Thomas Pointin, was conveyed to Brice Seed. In 1733 Brice Seed and Walter Davis conveyed 'all that fulling mill consisting of three stocks and a gig mill ... and the new erected building standing near to the mill pond designed for a grist mill but now converted into a messuage or tenement', to John Dallaway who occupied the premises at the time.

In 1760, the site comprised fulling, gig mills and 'the knapping mill house', a scouring dyehouse and blue dyehouse. In 1790 William Dallaway conveyed Bigges Place, the range of buildings lately erected by William and James Dallaway, the fulling and gig mills, the knapping mill, the scarlet, blue and scouring dyehouses, the shear shops and buildings called the wood house, warehouse and coal ash house to Joseph Lewis, miller. In 1838 Joseph Lewis had one power loom and thirty-one handlooms employed and twenty-nine unemployed handlooms at the mills. After Joseph Lewis's death the mills were worked by Archer Blackwell and John Remington, later by William, John and George Lewis, and by 1840 by William Lewis alone. Lower Mill was rebuilt in 1843 and in that year Lewis went bankrupt and the mills were offered for sale. The property was bought by John Ferrabee, who also bought all the machinery in the mills except the stock, gigs and gearing. A detailed inventory of the machinery was made and this is the earliest date for which there is any information about the uses to which each of the two mills were put. The 'Large Mill', presumably Upper Mill, was devoted to wool and yarn preparation and dyeing, while weaving was located at Lower Mill. In 1845 the mills were again for sale. Two steam engines are mentioned in the particulars and John Webb was said to be occupying the mills, while Christopher Smith occupied the dyehouses. The property was bought by Samuel Stephens Marling who leased it in the same year to Thomas William and Thomas Coke White of Monks Mill, Alderley. Upper Mill had an iron waterwheel and a 40hp steam engine. There were eight pairs of iron fulling stocks and four broad gig mills. Lower Mill contained three waterwheels, a 20hp steam engine, and six pairs of iron stocks, six gig mills. There was no mention of looms.

In 1858 Samuel Stephens Marling leased Brimscombe Upper and Lower Mills to P.C. Evans and John William Bishop, clothiers. P.C. Evans was the son of Aaron Evans who worked one of the mills in the Toadsmoor Valley. The schedule which follows on the deed shows that the uses of the various floors in the Upper Mill probably remained much the same. The fulling stocks had been removed and two fullers by Ferrabee and two by Buck had been installed in their place. One waterwheel and the stocks had been removed from the Lower Mill. An indenture attached to the deed records alterations made by Marling at the cost of £3,865. He erected a new loom shed and a building around 64ft by 27ft which was to be a new willy and oiling house and mechanic's shop. The open air stove was converted into a closed warehouse 'by building up the former chequered brick work walls'. He agreed to erect a building at the Lower Mill where P.C. Evans could install his dyeing machinery. In 1889, P.C. Evans & Sons had 124 looms and 6,400 spindles at the two mills. More details are given in Worrall's Directory two years later: there were 5,670 mule and 818 doubling spindles for woollen yarn and 1,540 spinning and 770 doubling spindles for worsted yarn. The looms had been increased to 144. In 1920 the firm amalgamated with Marling & Co. Ltd at Stanley and Ebley. Little work was done at Brimscombe after that and the mills were eventually sold. The two mills are illustrated in *Industrial Gloucestershire*. The Upper Mill consisted of a large, wide range with a tall chimney adjacent and a number of smaller buildings. The main mill building has been demolished. The former dyehouses, adjoining the pond and what was probably the

counting house, remain. Lower Mill consisted of one main range of three distinct sections. This mill was burnt down between 1918 and 1930 when it was used as a teazle store for P.C. Evans & Sons. After the Second World War Upper Mill was Lewis & Hole's iron foundry, Lower Mill being an electro-plating works. In 1967 Perolin Chemicals acquired the site. Little remains of the original textile factory buildings. Bigges Place was demolished for road widening.
(P.H. Fisher, *Notes and Recollections of Stroud*, p.8; PRO E315/394; D873b; D873 E12; D181/III/T9; PP 1840, XXIV, pp.396–7; D1159; G.R.O. D1241, Sale Particulars; Kelly's Textile Directory 1889)

Brimscombe Port/Field's/Hopton's Mill, SO 868023

The seventeenth-century watercourse leases record that John Field held a watercourse from Brimscombe as far as Brimscombe Bridge. In 1744 Samuel Peach held a fulling mill here and in 1762 it was in the hands of John Peach but when, in 1786, the mill was sold to the Thames & Severn Canal, it was a corn mill. A 'Mansion House', together with a corn mill at Brimscombe Port, was offered for sale in 1807 with the suggestion that it could be converted to a clothing mill. William Dallaway does not seem to have worked this mill with the Upper and Lower Mills but his successor, Joseph Lewis, owned it (then comprising a fulling and corn mill) in 1815. William Lewis probably occupied Port Mill and ran it as a cloth mill, together with Upper and Lower Mills. In 1843 the mill, together with the Upper and Lower Mills, was put up for sale by the creditors of William Lewis. Port Mill was said to contain an 18hp steam engine. The mill was sold to N.S. Marling who leased it to P.C. Evans, then to the Ferrabees. A mansion house stood near the mill and had been absorbed into the mill buildings. Ferrabee got into financial difficulties and engaged a Mr Fox as partner. James Ferrabee went bankrupt in 1866. There was a serious fire at the mill, following which it was rebuilt as the handsome structure of today. The mill was put up for auction in May 1872. One of the buildings was three storeys high, 70ft long, 35ft wide and built around seven years previously. Another was three storeys high, 100ft long and 24ft wide. There was a 20hp steam engine and two waterwheels.

The property was bought by P.C. Evans for £5,300 as an extension of his business at Brimscombe Mills. Evans retired in 1880 and his three sons continued the business. After the amalgamation with Marlings, after which the firm became known as Marling & Evans, Brimscombe Port Mill was virtually closed. In 1940 they sold the mill to G.A. Hensher who sold it to Bullock, Parson & Co. in 1944. In 1948 the mill was sold to Benson's Tool Works. There are now several non-manufacturing businesses in Port Mill, including The History Press. The buildings are attractively grouped facing Brimscombe. Large stone quoins are a decorative feature of the mill.
(PRO E315/394; GJ, 7 September 1807; G.R.O. D873b; D1159; D1241; F.T. Hammond, Brimscombe Port Mills, typescript in Stroud Museum; 1872 Sale Bill in Stroud Museum)

Bourne/Grime's Mill, SO 873022

In 1597 Henry Poole of Sapperton sold the Bourne estate, including two fulling mills and four grist mills in Nether Lypiatt Manor. In 1619 William Elkington of Minchinhampton conveyed to Robert Gyde of Bisley one grist and one fulling mill. Two fulling mills and 'Bourne land' were leased in 1654 to Thomas Freame and William Gryme. In 1820 Thomas Hill was making cloth at Bourne Mill, possibly as a tenant of Aaron Evans who lived at Bourne House at the time. At this date the corn and woollen mills were separately tenanted. By 1839 John Webb was making cloth at the mill but a Mr Holmes was also using the mill as a saw mill. By 1842 the mill was in the possession of N.S. Marling who, in 1854, leased 'all that newly erected mill called the Bourne Mill' to John Webb. The stock mill contained a waterwheel with iron arms, iron wings and segments and three pairs of iron stocks. The gig mill house contained an iron waterwheel, fed by an iron pipe which took water from the pond, and two gig mills. There was a mosing gig house, a shearing shop, a wool stove and five racks in the Rack Field. In 1856, however, Dangerfield & Foot, silk throwsters, were said to be at Bourne Mill. They were still there in 1867, but by 1865 Richard Grist was using part of the mill as a flock mill. The mill was later used for making walking sticks and is now occupied by several businesses, including a cycle/snowboarding firm. The waterwheels were removed and the pond filled in during the early 1960s. The main mill is a well-restored, four-storey narrow stone building. A small three-storey stone building lies at right angles to the main mill, probably the gig mill house mentioned in 1854.

(Box 78gs; PRO E315/394; D1801/12; D1347 acc. 1603 Stroud; Directories 1820, 1839–40, 1856, 1867, 1879; PP 1840, XXIV, p.398; D1159; D873b)

Dark/Leversage's/Winn's Mill, SO 875020

There was a fulling mill here by 1579. Roger Fowler, clothier, held the mill in 1622 and, in 1671, there were two fulling mills, two grist mills and a gig at the site. In 1751 William Prinn conveyed 'two tuck mills or fulling mills one grist mill and one gigg mill thereunto near adjoining', all occupied by Timothy Ratten, to Daniel Lysons, Nicholas Hyett and Charles Barrow in trust for partition. In 1756 the property, to which another grist mill had been added, was conveyed to Peter Leversage and James Canter. In 1756 the cloth and the corn mills were separately let and in 1784 a shear grinding shop and a dyehouse were added. The several occupiers were John Jones, John Wynne, John Hooper and John Alans. A knapping mill and scouring house had been added when the premises were sold to Michael Hodgson in 1786. In 1801 the mill and shear grinder's mill, occupied by William Winn, were leased to Joseph Cambridge, clothier. In 1805 the property was conveyed to Thomas Howell, mealman. In 1820 the property was conveyed to William Read King and Charles Newman. In 1844 Thomas Jones was making cloth at Dark Mill. In 1860 'all that clothing mill commonly called the Dark Mill', with what was formerly a grist mill but now a saw mill, a blacksmith's shop and the Rack Hill, were occupied by John Webb. In 1870 Reeves & Co. were making gun felts at Dark Mill. By 1874 William Farrar, a dyer and coal merchant, occupied the mill. Later it became a saw mill and was finally absorbed into Critchley's pin-making business. The mill was demolished in 1964.

(D1347, acc. 1603, Stroud; *VCH Gloucestershire, X*; D892/T80/18; PP 1840, XXIV, p.398; Directories 1844, 1870, 1874)

Wimberley/Delamere's/Wetmore's Mill, SO 877021

In 1200 a fulling mill at this site was let to Robert Delamere. By 1625 the mill was in the ownership of Robert Ridler. In 1638 John Ridler and Thomas Tayloe of Bisley, clothiers, let 'Wymberley' Mills comprising a grist mill and a fulling mill with three stocks, together with three racks, to Nathaniel Ridler, clothier, of Over Lypiatt. The fulling mill was owned by Samuel Whitmore of Hyde Court, Minchinhampton, in 1765. The 1820 directory describes Thomas Jones and John Webb (Senior) as of Wimberley Mill. In 1830 John Webb Junior was making cloth there. Wimberley Mill was advertised for sale in 1842 as 'Those desirable Clothing Mills … working 3 pairs of stocks, with the Gig Mill etc'. By 1845 David Farrar, woollen manufacturer and wool dyer, occupied the cloth-making premises and he continued there until after 1867, Critchleys using part for pin manufacture. By 1879 the mill had become a stick and saw mills worked by Liddiatt & Co. By 1894 Critchley Bros had returned and taken over the whole site for pin manufacture.

(D4289 T4; Juliet Shipman, *Chalford Place: A History of the House and its Clothier Families*, 1979; Directories 1830, 1867, 1879, 1894; tithe terrier, Minchinhampton; *Midland Counties Herald*, 8 September 1842; Letter in Stroud Museum 1845)

St Mary's/Clark's Mill, SO 885023

This is one of the best-documented mill sites in Gloucestershire. In 1388 the mill, together with houses and land, was granted to the Chantry of St Mary in Minchinhampton church. At the Dissolution of the chantries in 1547 a fulling mill is mentioned and in 1548 Francis Halliday rented a fulling mill called St Mary's Mill. The original mill was somewhere near the bridge and was directly fed by a leat. The mill comprising two fulling mills, a grist mill and a gig mill were sold by Edward Halliday to Henry Winchcombe of Upton St Leonard, clothier, in 1594. In his will dated 1640 Henry Winchcombe bequeathed the mills to Henry Ockold of Gloucester; Edward Arundel was the tenant. In 1589 the mills were mortgaged by Ockold and his son to M. Bulstrode. Bulstrode transferred the mortgage to Nathaniel Ridler in 1691. In 1704 the mill was a double fulling and corn mill. By 1714 the mills, solely devoted to woollen manufacture, comprised three fulling mills and a gig mill. In 1721 Richard and Nathaniel Cambridge sold the mills to Samuel Peach, John Small of Rodborough being the tenant. In 1741 Samuel Peach the elder conveyed the mills to his son, Samuel. An additional pair of fulling stocks had been added by this date, making four in all. In 1763 Peach came to an agreement with John Iles of Chalford concerning the height of the water between St Mary's Mills and Iles Mill.

In 1769 Edmund Clutterbuck and Nathaniel Peach, trustees of Samuel Peach, conveyed the property to William Innel and William Sevill. Thomas Fry Clark became the owner but was bankrupt in 1780 and in 1782 the mills, 'together with the new erected tenements or workshops and buildings', were conveyed to Bryant & Haythorne. In 1795 Lucas, Hopton & Bryant conveyed St Mary's Mills, consisting of 'a grist mill, a fulling mill, two stocks and a gig Mill' with workshops, dyehouses and stove houses, to Monkhouse Tate on Bryant's bankruptcy.

Tate was bankrupt in 1812 and the mills were conveyed to London merchants who leased the premises to Nathaniel Partridge in 1816. In 1818 the mills were for sale and were acquired by Richardson Borradaile, acting for William Cotton. In 1819 the mills were leased to Samuel Clutterbuck of Bowbridge who began a major rebuilding programme by pulling down the 'very old works' and building the main mill, more or less as it appears today (with the addition of a clock and cupola) to the west of the original mill. One range, dating probably from the seventeenth century, was retained from the earlier industrial buildings and still survives. By 1833 the stocks had been increased to four pairs and the grist mill had vanished. The power was 'water 30 horse, steam thirty horse'. In 1834 Samuel Clutterbuck mortgaged the mill and workshops and by 1840 he was no longer making cloth here. It was probably in the 1830s that St Mary's Mills were offered for sale or to be let. The undated bill gives a detailed description of the site as well as showing engravings of the mill complex (see p.82). The site was described as comprising 'a Gentleman's residence … Powerful Mills and Factory building … Manager's House … 12 cottages for workpeople'. The main mill was 80ft by 32ft, five storeys high, 'with an excellent supply of water'. The engine house containing a 30hp engine adjoined the mill. At right angles to the water mill lay a four-storey building, 60ft long and 22ft wide. Opposite the mill lay another five-storey building, 70ft by 22ft, and adjoining this was a rack house and wool stove. In 1851 the mills were vacant and advertised to be let. The fall to the two iron 13ft-wide and 14ft-diameter waterwheels was said to be 7ft. There were two steam engines, of 3hp and 25hp.

In 1874, Mrs Clutterbuck and others leased the mills to W.G. Grist who was already occupying part of the premises for flock manufacture. Frederick Wiggin & Co. were tenants of a part. By 1903 the mill was occupied by the Chalford Stick Co. and 100 people were employed. Sticks continued to be made here until the mid-1960s when a small printing works was set up. The *c.*1820 mill is one of the most attractive Gloucestershire textile mills, the eliptically headed windows divided by stone mullions, reminiscent of those at Ebley Mill. The ground floor of the mill now houses one remaining waterwheel and a display of some textile machinery, besides a late nineteenth-century Tangye compound steam engine.

(D67/Z/32; D1815, Clutterbuck; LG, 19–23 December 1780; D1332/F1; Box 39gs; PP 1834, XX, p.258; *Midland Counties Herald*, 2 January 1851; RV69.1)

Clayfield Mill, SO 887024

This was a grist mill in 1704, powered by a spring which enters the Frome near the lane leading to Brookside and Iles Mill. In the mid-eighteenth century it was owned by members of the Ballinger family and it is reputed to have remained a corn mill until at least 1790. It was a cloth mill by 1820 and in 1830 Nathaniel and Joseph Jones were at Clayfield Mill. The waterwheel was an external one, the tailrace being taken under the canal and discharged into the River Frome. The early eighteenth-century building was gutted by fire and reconstructed c.1862. The building, now a house, was reduced in size when the A419 was widened (p.163).
(Hammond MSS Box 20; TS 182/5; Mahler & Marshfield, *Stroudwater Valley Mills*; ex. inf. Mr M. Mills)

Brookside SO 887024

This handsome seventeenth-century gabled house on the west side of the lane leading to Iles Mill was the clothier's house for Iles Mill. In 1782 Abraham Walbank, attorney, conveyed it to his daughter who, in 1811, sold it to Joseph Iles, clothier. Iles added a south wing, housing a cloth mill and engine house, and it remained a mill until 1882. In that year Handy and Jesse Davis demolished the wing and the property reverted solely to domestic use.
(Hammond Archives; Mahler & Marshfield, *Stroudwater Valley Mills*)

Iles Mill, SO 888024

In some ways resembling the St Mary's Mill estate, the Iles Mill estate, with its fine clothier's house, mill manager's house and associated industrial buildings, comprises a self-contained unit. In 1608 the mill was worked by Thomas Butt, clothier. John Iles, clothier, owned Iles Mill in the 1730s when he came to an agreement with Samuel Peach of St Mary's Mills concerning the water supplies of the two mills. In 1806 the mill came into the ownership of the Ballinger family. During the early to mid-nineteenth century the mill was worked by Thomas Jones. In 1840 Joseph Pitt, Devereux Bowley and Jacob Wood conveyed the cloth mill and engine house erected by Joseph Iles to Charles Lawrence. In 1848 John Ballinger bequeathed the mill, occupied by Thomas Jones, to his son Henry. Thomas Jones still occupied it in 1856 and by 1863 walking sticks were being made there. In 1879 Charles Grist was making flock and shoddy at Iles Mill. The mill burned down in 1913. A three-storey stone building with a tiled roof associated with the mill survives as residences. The mill manager's house at right angles to the former mill survives.
(D1815, Clutterbuck; PP 1840, XXIV, p.399; M.A. Rudd, *Historical Records of Bisley with Lypiatt*, p.328; Directories 1856, 1863, 1879)

Wool Warehouse/Silk Mill, SO 889024

Built c.1830 by Handy and Jesse Davis as a wool store it was converted in the 1840s to a silk throwing mill by Samuel Hook. It was subsequently converted to residences.
(Mahler & Marshfield, *Stroudwater Valley Mills*; Jennifer Tann, 'Worms to Riches', *Warp & Weft*, 2011, STT)

Tayloe's/Woolings/Ballinger's/The Great Mill/Belvedere Mill, SO 892025

In 1558 William Tayloe did service for a tuck and gig mill. The mill remained in the hands of the Tayloe family for over 200 years. In 1620 Woolings was granted to Robert Hone, from whom it passed to William Tayloe. In 1683 an indenture describes the site as a mill containing two stocks and a gig mill. In 1775 Hester Tayloe sold the fulling mills with stock and gig mill called The Woolings to Charles Ballinger. In 1794 Tayloe's Mill, with rack and stove, was occupied by Handy and Jesse Davis, together

with a grist mill with two pairs of stones occupied by Thomas Trotman and a dyehouse formerly occupied by David Farrar. An indenture of 1857 between Samuel Attwood of Cheltenham, mealman, and Edward Archer, Gent., conveyed Tayloe's or The Great Mill, formerly a cloth mill but lately rebuilt by Samuel Attwood as a corn mill, with four pairs of stones. Steam power was added in the 1880s, and the mill is now well restored and occupied by an electronics company. The well-conserved stone mill comprises a three-storey range astride the stream with a wing running west from the south end.
(M.A. Rudd, *Historical Records of Bisley with Lypiatt*, pp.309, 310, 328; PRO E315/394; F. Hammond MSS, Boxes 3, 20; Directories 1820, 1867)

Butts Mill, SO 892024
Located immediately south of Tayloe's Mill on a spring which entered the Frome by Tayloe's Mill pond; built in 1784 and demolished 1844–45, being in the path of the GW Railway.
(Hammond MSS)

MILLSWOOD SPRINGS

Toghill's/Sevillowes/Seville's Upper/Little/Lower (New)/Mr Hunt's Mill, SO 893026
By the 1820s there were four mills at this site, to the north of the main road, powered by two channelled streams emerging, respectively, from Chalford Hill and Millswood.

In 1807 Charles Innell, clothier, enfeoffed a recently erected mill to William Toghill and Thomas Park. In the following year a mill lately erected and called the New Mill, then occupied by John Webb and Thomas Cormeline, was conveyed to William Toghill. The mill was occupied by Toghill or his tenant, Joel Chew. In 1820 William Toghill transferred the mill to William Davis. In 1821 Thomas Clark and William Toghill mortgaged workshops and other buildings lately erected with a cloth mill, engine house, steam engine, wool lofts, drying house and dyehouse and workshops (formerly two cottages), the latter formerly occupied by Joel Chew. Toghill was bankrupt in 1826.

The two more recently built mills were Seville's Upper/Sevillowes and New Mill. One was described in 1826 as 'part newly erected, part ancient'. Little Mill, 'lately occupied by Joel Chew', was put up for auction with Toghill's Mill in 1830. The latter was described as 'A capital clothing mill … capable of making 40 pieces of cloth per week'. The mill contained four pairs of cast-iron stocks, gig mill and a 12hp high-pressure steam engine (by Wolff). Daniel Cox agreed to rent the mill.

The mills were purchased by Handy Davis. In 1837 William Davis conveyed the mill, occupied by Daniel Cox, to his daughter. By 1838 Handy and Jesse Davis were bankrupt and the mills, occupied by John Gardner, were, again, up for sale. New Mill contained two iron and one wooden waterwheels, six pairs of stocks, four gig mills; 'the little Mill' was purchased by Handy Davis. In 1841 the premises, formerly the property of William Ireland (deceased), were for sale, following a Chancery suit. The bill of sale included a map of the site. There were nineteen lots in all, including 'mill, dyehouse … all that commodious mill (called) … Seville's Upper Mill, [and a] newly erected mill 3 storeys high with the engine house attached'. A small part of a mill is now a cottage (see Map 6).
(D1388 No. 81; D1241 Box 4 bdl 15)

Brownshill Workshop
The assignees of Joseph Gill, bankrupt, advertised scribbling and carding machines and other tools, as well as a stock of cloth and wool. The machinery was driven by a horse wheel.
(GJ, 12 December 1808)

Unnamed Mill, SO 893025

The oldest surviving building on the current Chalford Industrial Estate (part of it appears to be Tudor) was driven by the Millswood springs. The assumption is made that this was a fulling mill. The name unknown, it has not been possible to match it with any deeds. A fine 1850s painting of the mill hangs in the Museum in the Park, Stroud (p.41), showing an iron low breast or undershot waterwheel on the east side.

RIVER FROME

Chalford/Bliss Mill, SO 893025

This mill site (Chalford Industrial Estate) now covers the sites of five independent mills. The original Bliss Mill was located at the west end of the estate as it abuts on to the lane to Hyde Hill. It was part of the estate of Thomas Mill who, in 1455, leased the mill to John Field, tucker. By 1439 it was a fulling mill being leased from Bisley manor by Alexander Robull. From 1516 onwards, the mill, comprising a fulling and grist mill, belonged to Corpus Christi College, Oxford. In 1524 the college granted the mill to Walter Compton of Bisley and the mill was rebuilt in 1539. In 1647 the mill lease was taken over by the Ridler family, members of which were tenants of the properties for over 150 years. The mill is shown on an Elizabethan map. In 1708 the property was in the tenure of Joseph Blisse and comprised a fulling mill, two dyehouses, presshouse and racks. In 1751 the Ridler estate was valued at £24,337 although by then the mill was 'much out of repair'. In 1794 Corpus Christi College let Chalford Place with the fulling and grist mill to William Hunt Prinn, Charles Ballinger becoming a subtenant. In c.1808 the estate was described as 'a mansion house, two cloth mills, warehouses' etc. In 1820 S.R. Pattison was making cloth at Chalford Mill, part of the property being let to a subtenant, Daniel Abraham Cox, clothier, millwright and victualler. Cox was still at Bliss Mill in 1838, but cloth manufacture probably ceased there shortly afterwards. Between 1840 and 1856, William Dangerfield was making bone umbrella handles here. In 1875 Dangerfield, stick and timber merchant, leased Bliss Mill, together with the adjoining mills. There was a disastrous fire on the estate in 1890, after which 'it became necessary to reconstruct a large part of the buildings'. Dangerfield died in 1894 and, after several changes, the firm was taken over by A.C. Harrison & Co. Ltd in 1903. The site is illustrated in *Industrial Gloucestershire* (p.98). In 1970 it became an industrial estate with several tenants. Little of Bliss Mill survives, save part of the clock tower (see Map 6).
(D181/III/T56; Juliet Shipman, *Chalford Place: A History of the House and its Clothier Families*, 1979; Directories 1820, 1856, 1879; D1347; Chance & Bland, *Industrial Gloucestershire*, pp.39–40)

New Mill (Chalford), SO 894025

In 1852 the mill came into the possession of Nathaniel Jones. In 1864 Jones sold New Mill to James Apperly, machine manufacturer. By 1875 the mill was owned by William Dangerfield and had been absorbed into the Bliss Mill complex. This was probably one of the buildings damaged by fire in the 1890s and subsequently rebuilt. The building still stands, looking much the same as it does in the engraving of the site made in 1904.

The mill is a long, three-storey stone building with one small, centrally placed dormer window and two small pediments on the north side (see Map 6).
(D1159; D181/III/T10 & T12; D1241; D1388; D1347)

Mugmore Mill, SO 894025

Mugmore Mill was built in 1803 by John David Webb. It was powered by water from the Black Gutter which flows to the Frome from Minchinhampton Common. In 1838 a 'clothing mill … now in the

occupation of Mr Daniel Webb, … driven by fine spring water' was offered for sale. From then onwards it probably continued in the same ownership as New Mill and, in 1864, was sold to James Apperly with New Mill. By 1875 it had been absorbed into the Bliss Mill complex and was worked by William Dangerfield as a timber works. A small building can be seen in the engraving of Bliss Mill in 1904 lying next to New Mill and this was probably Mugmore Mill (see Map 6).
(D1241; D1159; D1347; *Midland Counties Herald*, 30 August 1838)

Spring Mill, SO 895025

The mill was recorded in 1711 when it was leased together with a messuage and dyehouse. William Gardner was making cloth here in 1820. By 1830 Edward Lambert Trotman was using Spring Mill as a cloth mill. Cloth manufacture probably ceased here in about 1840, and by 1856 Joseph Jones was using it as a silk mill. Francis Jones was here in 1863 and by 1865 Charles de Bary and William Rudulph were making silk here. In 1864 Nathaniel Jones, by then the owner of the mill, sold it together with New and Mugmore mills to James Apperly. In 1867 the mill was sold to William Dangerfield and it became a timber mill. It appears to have consisted of one range of two distinct halves, each half being two storeys high, the roof level being higher in the eastern half (see Map 6).
(Directories 1820, 1830, 1856, 1863, 1865; D1159; D1347)

Randall's/Wood Mill, SO 895025

In 1860, the Great Western Railway Company conveyed 'All that fulling mill with the steam engine, and boiler, gig mill house, press shops, workshops, scouring house erected some time since' on the site of an old fulling mill called Wood Mill or Randall's Mill to Nathaniel Jones. The deed recites that the property had for many years belonged to John Trotman Lambert. In 1860 Nathaniel Jones was using the mill as a silk mill and another part of the premises was unoccupied. The site of the stove for dyeing cloth had been pulled down. The mill was sold by Jones to James Apperly in 1864. In 1875 it was in the possession of William Dangerfield. The large three-storey mill has been demolished (see Map 6).
(D1159; D1347)

Halliday's/Stoneford/Smart's/Bidmead's Mill, SO 896025

There was a fulling mill here in Tudor times. In 1608 two fulling mills were sold by Mr Smart to Mr Batt. In 1726 Samuel Halliday conveyed a fulling and gig mill called Stoneford Mill to William Bawden. In 1813 Richard Hawkeswell and John Driver Bidmead assigned the property to James Croome. John Driver (Bidmead) was at Halliday's Mill in 1820. In 1821 Bidmead, broadcloth manufacturer, was bankrupt and the fulling and gig mill with dyehouse and workshop were conveyed to John Restall, aka Baker. In 1823 the estate was vested back to Bidmead. Joseph Chapman acquired the mills leaving them to his younger son. In 1830 James Smart was making cloth at Halliday's Mill, by which time there was a steam engine there, and in 1844 William Smart occupied the mill. This mill survived the depressions of the 1830s and '40s and William Smart continued there until 1858 when he re-leased the property, consisting of a fulling and gig mill with stove house, to Obadiah Smart. There was an 8hp steam engine by 1845. It later became a silk mill and was in the hands of William Sydney Cox until 1902. In 1920 Peter Waal and colleagues set up their arts and crafts furniture-making business here; it closed in 1938. The elegant Georgian clothier's house lies adjacent to the present mill. An early photograph shows the present mill building adjoining the original small, two-storey mill astride the stream (now gone) (p.78). The nineteenth-century mill, housing several small business tenants, consists of one main four-storey range lying astride the valley, the first two storeys stone and the upper two brick.
(M.A. Rudd, *Historical Records of Bisley with Lypiatt*, p.309; D1159; box 78gs; Directories 1820, 1830, 1844, 1856, 1858–59; Hammond MSS, Box 20)

Workshop in Chalford, possibly on Chalford Hill

In 1808 machinery and utensils belonging to D. Gardiner, deceased, were put up for auction and included scribbling machines, carding engines, spindle jacks, jennies, willy, and twenty-five pairs of shears.
(GJ, 31 October 1808)

Seville's/Innel's Mill, SO 902025

In 1608 Edward Smart of Througham enfeoffed a messuage and two tucking mills, late in the tenure of Henry Custos and then of Edward Newarke, to Richard Batt. The mills were situated at Stephen's Bridge, Bisley. In 1649 Jane Watkins conveyed the mills (which had been given to her by her father Richard Batt) to Walter Sevill. In 1663 they were confirmed in trust for Walter Sevill's use during his lifetime but afterwards were to be the property of John Sevill. (During demolition a stone was found at the site with the inscription W. Seville 1708.) In 1712 John Sevill sold the two tucking mills and gig mill to Thomas Harmer of Minchinhampton, clothier, for £675. Between 1865 and 1867 Nathaniel Jones was making cloth there but by the 1870s the mill was used for silk throwing. In 1873 Nathaniel Jones leased the property to William Chapman, silk throwster. There were three sections to the premises, part of the old cloth mill remaining unaltered. The mill buildings comprised a triangle, the middle block lying astride the stream; the most modern block lying parallel to Chalford High Street. Part of the nineteenth-century mill buildings remained until 1964 when they were demolished.
(D181/III/1, 2, 4, 6; D1241 bdl. 51; Directories 1820, 1830, 1842, 1844, 1858–59, 1865)

Valley/Tyler's/Moreton's Mill, SO 905025

This was a fulling and corn mill in 1785, identified as Moreton's Mill in 1789 and was a cloth mill occupied by Nathaniel Bennett in 1820. Jeremiah Ockwell Restall, farmer and timber dealer, was here in 1856. The two-storey mill was demolished in *c.*1920.
(Hammond Papers Box 8; Directory 1820)

Ashmeads Mill, SO 909028

Charles Innel, clothier, occupied this mill in 1820. He sold the mill, together with the waterwheel and steam engine, in 1825. By 1870 it was a flock and shoddy mill occupied by Nathaniel Teakle. By 1879 it was a silk mill occupied by John Knight. The mill premises have been largely demolished and partly rebuilt as a residence. There is a large, overgrown pond and what was probably an air stove abutting the road.
(Directories 1820, 1870, 1879; Mahler & Marshfield, *Stroudwater Valley Mills*)

Twissell's/Frampton's/Blankett's/Baker's Mill, SO 915029

In 1478 Frampton's Mill was held by John Quarron. William Twissell held the mill, known as Blankett's, in 1571 and it was probably a fulling mill, the Twissell family having been clothiers for several generations. They lived at Frampton Place and the mill lies in the valley below the house. The mill was later a corn mill. The attractive gabled mill is a typical seventeenth-century fulling mill of domestic proportions. A later two-storey industrial building lies parallel to the main mill.
(PRO E 315/394 ff.119–20; *VCH Gloucestershire, Bisley*)

Puck Mill, SO 922029

A grist mill was built here *c.*1572 and by 1708 it comprised a fulling mill, dyehouses and presshouses and was settled by the clothier Joseph Bliss on the marriage of his son. In 1749 the mill was conveyed by William Bliss and others to James Bidmead, baker of Bisley, who sold it to the Thames & Severn Canal

in 1791. A grist mill by 1804, it was owned by successive millers but by 1856 it was a silk mill operated by Blower & Smart. The mill had fallen into disuse before 1865 and has been demolished. An old photograph shows the mill to have had a circular wool stove.
(D181/111/T56; TS 211/2; *VCH Gloucestershire*; Directory 1856)

Rookwoods Mill, SO 927052

At the end of the eighteenth century this was a cloth mill belonging to James Butler. He sold the mill to William Tyler in *c.*1793 who rebuilt it. From *c.*1800 onwards the mill was used for corn grinding and by 1844 part of it had been converted into a tenement. The mill has been demolished.
(D1801/1, 7; D185/VII/I)

Miserden Mill, SO 944088

In 1832 'a capital overshot corn and flock mill drawing 2 pair of stones' with a drying house and dwelling house with lofts above was advertised for sale as part of the Miserden Park estate. The flock mill probably replaced a fulling mill. A former woollen mill at Miserden was mentioned in the 1839 Parliamentary returns.
(D1388, bills of sale no. 38; PP 1839, XLII)

INDEPENDENT DYEWORKS IN THE FROME VALLEY

Bridgend Dyeworks, ST 804045

See p.197

Blue Row, Dudbridge, ST 834045

This was owned by the Hawker family *c.*1810.
(D1181, Account books Hawker and Richards)

Wallbridge Dyeworks, ST 848050

A lease of 1820 describes dyehouses which were lately occupied by Richard Watts as a dwelling house and then converted into a woollen manufactory. In 1820 Peter Smith & Co. were blue and black dyers of Wallbridge. Wallbridge Dyeworks does not appear in directories after the 1830s.
(D1347 acc. 1603)

Stroudwater Dyeworks, ST 853047

These were established near Capel's Mill. In 1879 William Bishop, wool dyer, occupied the works. The firm was known as William Bishop & Son in 1894.
(Directories 1879, 1894)

Arundel Dyeworks, ST 855045

See p.205

Bowbridge Dyeworks, ST 857043

See p.206

Farrar's Dyeworks, ST 890024

This dyehouse, dating from at least the mid-seventeenth century, was part of the Woolings estate, although separately leased. It was occupied by David Farrar at the end of the eighteenth century and, at the tithe

survey, a David Farrar occupied the premises. The long, single-storey stone building with circular stone-ringed openings under the eaves survived into the 1960s but has been demolished.
(Hammond MSS, Boxes 3, 20)

TRIBUTARIES OF THE RIVER FROME: OZLEBROOK/CUCKOLD'S BROOK

Ruscombe Mill, SO 861032

This may always have been a corn mill although leased to Daniel Gardner, clothier, in 1648 and to his son Giles, a clothier, in 1677. A Daniel Gardner sent his cloth to a fulling mill in Painswick in the 1720s. It had been demolished by 1819.
(E 179/113/35A rot.5; P320A/VE 1/1-3, 7)

Puckshole Mill, SO 834059

This was owned by Thomas Ellery in 1822 and contained one stock, a gig and other machinery. Harman & Adey were making cloth here in 1871 when they went bankrupt. By the late nineteenth century it was a corn mill (known as Vale Mill). It has been demolished.
(*VCH Gloucestershire, XI, History on Line*)

Paganhill/Little/Bournes Mill, SO 834055

This may have been the mill mentioned in a deed of 1679 when Robert Dring of Middlesex sold a former all grist mill, then being a fulling mill, to Thomas Warner of Paganhill, clothier. In 1822 it was owned by John Phelps and let to Thomas Steel. In 1840 it was owned by John Delafield Phelps and occupied by Stephen Clissold. It was worked by members of the King family between 1897 and 1906. The mill has been demolished.
(D1347 acc. 454, Stroud; P 320A/VE 1/1-3)

Ozlebrook Mill, SO 835053

This may have been the corn mill advertised to be let by Mrs Field of Paganhill in 1736. There is the tradition that this was a cloth mill at one time. William Clutterbuck, a shear grinder, occupied Ozlebrook Mill in 1830. By the tithe survey it was a corn mill. It has been demolished.
(GJ, 21 December 1736; Directory 1830)

THE PAINSWICK STREAM

Stratford Mill, SO 847054

There was a fulling mill here in 1597. In 1609 Edward Stratford of Paganhill died seised of a water and fulling mill occupied by Thomas Merryett. His heir was his grandson, John. In 1627 John Stratford and Alice, his wife, sold 'a water myll conteyninge a tucke myll and a griste myll' situated below Stratford bridge to Gyles Davyes. In 1688 Giles Gardner, clothier, was owner of the Stratford estate which had two fulling and one gig mills. In 1735 the mill was held by William Little, baker. The mill became a corn mill and in 1801 was described as 'eligible either for a miller or a clothier'. In 1807 Callowell was extending a grist mill 'partly converted to uses in the clothing manufactory' near Stratfords, then in the tenure of Giles Hutchings. Biddle & Bishop occupied the mill for corn grinding in 1842. In 1843 the mills were

advertised, having two 14ft-diameter overshot waterwheels, one 7ft, the other 5ft wide, and three steam engines of 50hp driving thirteen pairs of French stones and other machinery. The mills were 'judged to be the most powerful in the West of England'. The mill was destroyed by fire and the site is now covered by Stroud Tesco store. The 'Mansion House' is the Museum in the Park.
(D149/277/567, 568; LG, 5–9 August 1788; GJ, 7 September 1807; *Midland Counties Herald*, 3 August 1843; Miller, 20 February 1899)

Salmon's/Bliss's Mill, SO 847061

A mill existed here in 1439 in the tenure of William Bliss and in 1486 Thomas Blysse owned the mill but there is no record of its function. A fine sixteenth-century clothier's house adjoins the mill and contains two date stones of 1593 and 1607, together with a cloth mark embodying the initials E and F. In 1608 Edmund Fletcher and Thomas Fletcher were recorded as clothiers here. Thomas Fletcher was seised of a fulling mill in Painswick in 1621. In 1723 John Ellis, clothier, worked the mill and in 1735 it was occupied by John Pinfold. In 1749 John Pinfold was named clothier of Salmon's Mill, Pinfold being here between 1754 and 1760. In 1786 the estate was offered for sale and included three stocks, one gig mill, a grist mill, and workshops. The sale may mark the end of cloth production at this site for by 1799, William Drew had Salmon's Mill and was still named as owner in 1839. Biddle & Bishop were tenants from 1832 to 1844. In 1839 N.S. Marling owned the mill and in 1844 he leased it to William Hall. There were two waterwheels, one 14ft, the other 13ft. Godsell & Sons, corn millers of Salmon Springs, were here in 1884. The mill has been demolished.
(IPM (Index Library) I, p.6; Colleen Haine, 'The Cloth Trade along the Painswick Stream', *Gloucestershire Historical Studies*, XIII, 1982; D25/48, 95; D873 E3; Baker's map; LG, 22 August 1884)

Grove/Hermitage Mill, SO 848064

In 1763 Elizabeth Capel granted a lease of a cloth or fulling mill at the site of an oil mill. By 1820 it was a paper mill with John Ward as tenant, but by 1827 it was, and continued to be, a grist mill. By 1836 the mill was still in the hands of the Capel family and from then onwards a series of tenants held the mill. It was put up for sale in 1858, with waterwheel and a 12hp steam engine. In 1865 Stratford Flour Mill Co. was at Grove Mill. It continued as a flour mill until after the First World War. Nothing remains of the mill.
(D1388; GJ, 24 April 1858; Colleen Haine, 'The Cloth Trade along the Painswick Stream', *Gloucestershire Historical Studies*, XIII, 1982)

Rock Mill, SO 848068

The initials E(dward) G(ardner), the date 1681 and a clothier's mark are above the door of the mill house. Daniel Gardner was recorded as a clothier in 1699. In 1738 the fulling mill, 'now in possession of Mr William Packer', was advertised to let. By 1743 the mill was tenanted by Anthony Bidmead, clothier. Between 1754 and 1785 Zachariah Horlick owned Rock Mill. In 1766 it was occupied by Edward Gardner. Job Gardner of the Rock Mill died in 1768. John Adey is recorded as of Rock Mill, clothier, in 1791 and 1797. In 1810 Rock Mill, containing two stocks, a half stock and a drying house, was put up for sale. It was for sale in 1811 and again in 1812. The mill occupiers, trading under the name H. Miles & Son, clothiers, were bankrupt in 1826. Their stock for sale included scribbling and carding machines, shearing frames, billies, jennies and a press. Henry Hicks owned the mill until 1837 with John Wathen as tenant from 1833–37 and then owner until 1841. At the tithe survey, Edward Bloxome owned the mill but it was occupied by Joseph Wathen. In 1847 it was offered to be let by the Stroudwater Pin Company, stock and gig water power being 'capable of milling and rowing 6 pieces of cloth per week'. There was also a 12hp steam engine. It was a pin mill in the mid-1850s. In the early 1860s the mill began to be used

for dyewood-grinding, first by John Thomas & Company, and later by Thomas Gould. In 1889 Tabram & Co. were manufacturing flock here. In the late nineteenth century the mill is reputed to have been grinding corn. It was demolished but there is a fine gabled clothier's house.
(GJ, 24 October 1738; F.A. Hyett, *Glimpses of the History of Painswick*, p.76; Baker's map; RR229.23gs; LG, 18 February 1826; Colleen Haine, 'The Cloth Trade along the Painswick Stream', *Gloucestershire Historical Studies*, XIII, 1982; Directories 1856, 1863, 1865–66, 1867, 1879; Kelly's Textile Directory 1889)

Pitchcombe/Jenner's/Wade's Mill, SO 852075

In 1771 Pitchcombe grist mills comprised three pairs of stones, a boulting mill and two large rooms for granaries. In 1796 William Cook was tenant and it was probably then that the mill was converted to a cloth mill, for the partnership between John Cook and James Wood of Pitchcombe, cloth manufacturers, was dissolved. The partnership between Jeremiah Cother and Abraham Lloyd, clothiers of Pitchcombe Mill, was dissolved in 1801 and Cother's estate, goods and stock in trade were advertised for sale in 1806 following his bankruptcy. 'A large mill' and a 'Capital clothing mill … called Jenner's Mill' containing two stocks and a gig mill were itemised. The machinery included scribbling and carding engines, jennies and shears. The mill may not have been sold. Charles Davis occupied the mill between 1810 and 1814; Davis, Beard & Co. tenanted the mill to 1823 and in that year 'Jenner's Mill, commonly called Pitchcombe Mill', was advertised for sale containing three stocks and two gig mills. Benjamin Wood was tenant from 1825–26 and William Fluck from 1828. In 1833 Fluck told the factory inspector that the mill had been erected at different dates and that the power was '16 horse water power in winter varying to about 8 horse power during the … summer'.

Three power looms were installed in May 1838 and, in addition to these, there were thirty-five handlooms, two of which were not in use. By 1839 Fluck had thirty-eight looms at Pitchcombe Mill, three of which were power looms, two of which were unemployed. William Freeman was making cloth here in 1849. The 1842 Directory has Ebenezer Durdin, umbrella and parasol stick manufacturer and manufacturing chemist at Pitchcombe Mill, possibly on a divided site. By the end of the nineteenth century the mill had been fully converted to corn milling and in the 1930s was still in operation but was disused shortly afterwards and demolished.
(Baker's map; LG, 10–14 March 1801; LG, 6–10 May 1806; GJ, 7 July 1806; GJ, 3 March 1823; Rept Factory Comms. 1834, pp.21–2; Handloom Weavers Rept, 1839; Colleen Haine, 'The Cloth Trade along the Painswick Stream', *Gloucestershire Historical Studies*, XIII, 1982; LG, 29 December 1843; LG, 22 July 1845; LG, 26 May 1848; Directories 1830, 1849, 1879; PP 1834, XX, p.262; PP 1840, XXIV, pp.396–7)

Pitchcombe Upper/The Little/Little's Mill, SO 848079

The mill is located on the Pitchcombe Stream. Jeremiah Stanley (d.1814) was described as 'respectable clothier of Pitchcombe' and his property was advertised for sale including the 'valuable clothing mill' with a 20ft-diameter wheel. The mill was repeatedly put on the market, being termed a 'clothing mill' into the 1830s. But it had been converted to corn milling by 1838. In 1850 it was used for hook and eye manufacture but soon became a corn mill again. It was used by a firm of builders and carpenters between 1870 and 1927 and is now dwellings.
(GJ, 15 August 1814; GJ, 13 March 1815; GJ, 4 September 1820; Colleen Haine, 'The Cloth Trade along the Painswick Stream: Pt IV', *GSIA Journal*, 1984)

Small's/Ayer's/Seagrim's/Lenall's Mill, SO 855083

While the Pallings were well established as Painswick clothiers in the seventeenth century, the first evidence of a member of the family holding this mill was in 1717 with Thomas Palling. In 1735 Sarah Palling, widow, surrendered the property to her son and, at this date, it comprised two fulling mills, one gig mill, a lately built shear shop and a dyehouse. From 1781–90 William Carruthers was ratepayer for the mill. Carruthers died in 1790 and Small's Mill, comprising four stocks, gig mill and presshouse, was let to James Stanley until 1805. From 1814–20 John Palling was tenant of the mill – for many years he also had King's Mill.

In 1820 Nathaniel Jones owned the mill and Mr Mills occupied it. Matthew Rice of Small's Mill, clothier, was here in 1832 and was succeeded by John Papps, clothier. In 1837 the mill was for sale and contained, amongst other fittings, billies, scribbling and carding machines, a mule and broadlooms. Rice told the factory inspector in 1838 that his power was '14 horse water power in winter, varying to only 7hp during ... summer'. In 1839 the mill was for sale and it was stated that 'a large sum of money has been spent in the last 3 years enlarging the mill'. Philip Foxwell was bankrupt by November 1840. Aaron Shipton and Joseph Wise Jenkins, clothiers, of Small's Mill were bankrupt in 1847. By 1863 it was a saw mill until 1910. Soon afterwards the mill became derelict. By the 1960s only two storeys of the mill remained but, since then, the waterwheel has been restored and a dwelling built on the mill site.
(Baker's map; RR229.23gs; PP 1834, XX, pp.262–3; LG, 13 April 1847; Colleen Haine, 'The Cloth trade along the Painswick Stream', *Gloucestershire Historical Studies*, XIII, 1985; Directories 1863, 1879)

King's/Lower/Paulin's/Culvert/Packer's Mill, SO 859089

A mill existed here in 1495 but its function is unknown. Richard Packer bought the mill, then a corn mill, in 1625 and by 1671 it was a corn mill, tuck mill, gig mill and mosing mill, worked by Thomas Packer, who was succeeded by his son Daniel. By this time there was a dyehouse at the site. In the mid-eighteenth century the mill was worked by John Packer; and subsequently by Richard Packer (d.1774). In 1787 a corn mill was advertised, the fulling and corn mills having separate occupancies. Members of the Palling family were at King's Mill immediately prior to the nineteenth century, although William King, clothier, was a tenant. John Packer of Painswick, clothier, was bankrupt in 1816 and in 1817 the mill was offered for sale, with stocks, gig mills and waterwheels. By 1820 and for many years the mill was owned by John Palling and his son William. In 1855 Henry Fletcher, manufacturer of scarlet and billiard cloth, occupied the mill but on his bankruptcy in 1858 the mill was for sale. There were 'two excellent and powerful waterwheels'. Fletcher's machinery and stock was offered for sale separately and included machinery driven by steam power. In 1859–60 Edward P. Sampson made shawls at the mill. By 1863 the property was used as a pin mill and continued as such into the early 1900s, using water power only. Most of the manufacturing part of the premises has been demolished but part remains incorporated into King's Mill House. This is a most attractive, well-conserved, stone L-shaped building.
(D1449/F113; Colleen Haine, 'The Packers of Painswick', *Gloucestershire Historical Studies*, IX, 1978; F.A. Hyett, *Glimpses of the History of Painswick*, p.85; GJ, 20 January 1817; Sale Bill, Stroud Museum; Colleen Haine, 'Cloth Mills along the Painswick Stream – mills near the centre of Painswick', *GSIA Journal*, 1982; Directory 1863; Chance & Bland, *Industrial Gloucestershire*, p.18)

Spring's Mill, SO 863090

This is one of the numerous small mills which existed in tributary valleys, often driven by springs (although this mill name originates with the Spring family, rather than the source of power). The first evidence of a mill here is from 1802 when a house for machinery with a waterwheel newly erected and capable of driving two carding or scribbling machines was advertised to be let by Daniel Spring, mill-wright. It was advertised again to be let in 1805. In 1817 Charles Gyde was tenant for the dyehouse until 1836. Nothing remains of the mill.
(GJ, 15 February 1802; GJ, 30 December 1805; Colleen Haine, 'Cloth Mills along the Painswick Stream – mills near the centre of Painswick', *GSIA Journal*, 1982)

Mason's/Cook's/Reed's/Painswick Mill, SO 868093

A date stone of 1634 and the cloth mark of the Webb family, incorporating HW, over the doorway of the old mill house place Henry Webbe here, although at this date no mill is recorded. By 1700 John Harris was recorded for Cook's Mill and in 1725 Sarah Cook advertised the mill with two stocks, gig mill, dyehouse, furnace, 'and a good place for spinning'. She died in 1741, leaving the mill, a grist mill and cider mill attached to her eldest son Richard.

John Gardner was tenant between 1784 and 1798 and in 1790 John Merrett was tenant. Mr Mason was a tenant between 1796–98 and, from 1798, the owner. The Mason family owned the mill for the next half a century or so. In 1804 the mill and 'utensils in the clothing business' was for sale and advertised to be let three years later. Jacob Chamberlain was recorded at the mill in 1810 and was bankrupt a year later. In 1812 the mill containing two pairs of stocks and gig mill was offered for sale. It did not sell. Joseph Wight, a broadcloth and cassimere manufacturer, was tenant between 1826 and 1840. No tenant was found and until 1843 the mill was vacant. In the 1860s it was worked as a silk mill by N. & J. Jones and by 1870 was a pin mill, continuing as such until 1920. A small industrial wing attached to the mill house is all that is left of the mill, being now included in the seventeenth-century mill house.

(GJ, 12 April 1725; Baker's Map; RR229.2gs; GJ, 23 March 1812; Directory 1839–40; tithe terrier; Painswick Annual Register, p.213; Colleen Haine, 'Cloth Mills along the Painswick Stream – mills near the centre of Painswick', *GSIA Journal*, 1982)

Cap Mill, SO 869094

The mill has been associated with the Webb family. Jasper Selwyn was at Cap Mill between 1690 and 1709 and it was worked as a cloth mill by the Packer family in 1729. Daniel Packer died in 1739 leaving the mill to his nephew John Packer. The mill containing two stocks, a gig mill, with dyehouse, was leased to William Knight in *c.*1772. John Packer was declared bankrupt in 1816, Cap Mill and King's Mill being offered for sale. In 1820 Samuel Wood owned the mill. In 1832 Nathaniel Iles Butler, clothier, owned the mill.

Butler was still working the mill in 1839, but was bankrupt in 1841. By the 1850s Watkins & Okey had moved here to manufacture pins. In 1867 Alfred Keene, wood turner, occupied the premises. The stone mill was one of the smaller fulling mills in Gloucestershire. The attractive gabled mill house adjoins the site.

(Colleen Haine, 'Packers of Painswick', *Gloucestershire Historical Studies*, IX, 1978; F.A. Hyett, *Glimpses of the History of Painswick*, p.85; Baker's map; RR229.23gs; (H) 93.3gs; Directories 1839–40, 1856, 1867, 1879; Chance & Bland, *Industrial Gloucestershire*, p.18)

Brookhouse/Savory's Mill, SO 871095

This was a cloth mill by 1751. The mill contained two pairs of stocks and a gig and was in the hands of the Knight family towards the end of the eighteenth century. In 1802 Robert Wight occupied the mill. He was bankrupt in 1832. By 1822 there was both steam and water power. In 1836 the mill was said to be 'dreadfully out of repair … it will take some weeks to put it in driving condition'. Offered to be let in 1839, the mill had two high-pressure steam engines, one of 6hp, the other of 24hp. In the early 1840s the mill was worked by John Driver. By 1849 William Savory owned the mill, for in this year he leased the ground floor which contained a waterwheel and two pairs of stocks to William Palling, clothier. The mill was used for corn grinding in 1867 but by 1879 was converted to pin making by Henry Savory and is depicted in an early photograph.

(Baker's map; Directories 1830, 1839–40, 1867, 1879; D361b/T8; RR229.23; tithe terrier, Painswick; D1159; Colleen Haine, 'Cloth Mills along the Painswick Stream – mills near the centre of Painswick', *GSIA Journal*, 1982)

Loveday's/Bayliss's Lower/New Mill, SO 874097

Thomas Loveday was a clothier here in 1715 and John Gardner, broadweaver, was probably a tenant. In 1792, until the early 1820s, William Loveday held the fulling mill, gig mill and grist mills. Following Loveday's death, the mill, containing two stocks, a gig mill, scouring house, malt house and cottages, was offered for sale by William Bayliss. Bayliss either bought the mill or it could not be sold, for he remained owner and worked this mill, together with Upper Mill, until his death in 1837. By 1839 Philip Foxwell was working the mill which was then called New Mill. Foxwell was bankrupt in 1840 and the machinery and stock were put up for sale. It remained empty for three years and, later, became a corn mill and was worked by John Fayers. The stone mill adjoins the mill house. A fire severely damaged the mill and house in the 1920s, parts of the old mill being later demolished. Another section was restored with the old mill house, converting it to a residence.
(RR229.23gs; Directories 1830, 1867, 1879; GJ, 28 November 1840; D1338 S.L. 39; Colleen Haine, 'Cloth Mills along the Painswick Stream – mills near the centre of Painswick', *GSIA Journal*, 1982)

Zachariah Powell's Mill, SO 874099

This was a small mill driven by the waters of a spring which enters the Painswick stream between Bayliss's Upper and Lower Mills. It was owned and occupied by Zachariah Powell in the early 1820s. The mill, containing two pairs of stocks, was put up for auction in 1837 and was reputed to have been demolished in the 1860s.
(Baker's map; GJ, 18 March 1837; Directory 1820; RR229.2gs; Colleen Haine, 'The Cloth Trade along the Painswick Stream: The Cloth Mills. Pt III', *GSIA Journal*, 1983)

Bayliss's Upper/Lodge Mill, SO 877103

William Bayliss lived at Castle Hale, Painswick, in 1795, and was making goods for the East India Company in 1812. In 1818 Edward Bayliss, clothier, mortgaged lands, a tenement, 'Clothing Mill' (formerly a water grist mill). This was worked by William Bayliss in 1820 – he went bankrupt in 1823. A fine coloured plan and elevation show the mill, containing stocks, gig and washer, to have been seven bays in length, three storeys high with a dormer-lit attic, and mullioned leaded windows (see p.75). There was a separate 'steaming house'. In 1838 'Two Cloth Manufactories' owned by the Bayliss family (Upper and Lower mills) were advertised to be let, being capable of making twelve to fourteen ends of fine cloth per week. Philip Foxwell owned and occupied the mill in the early 1840s. In 1855 Thomas Cook, flock and wadding manufacturer, had the mill. There was a wool stove at the upper end of the mill pond. The mill has been demolished.
(Baker's map; RR229.23; Museum in the Park, 1967, III; GJ, 10 March 1838; Colleen Haine, 'The Cloth Trade along the Painswick Stream: The Cloth Mills', *GSIA Journal*, 1983; tithe terrier)

Damsell's or Cox's Lower Mill, SO 877111

The Gardner family held the mill in the seventeenth century; Richard Gardner, clothier, dyeing in 1728. In 1723 William Lediard was at Damsell's. In 1740 the property was surrendered to William Packer. In 1741–45 William Damsell held the mill. William died in 1754. In 1818 Elizabeth Cox and William Cox leased the mill to John Cox and Weston Hicks of Oliver's Mill. By 1842 cloth manufacture here had ceased. By 1856 the mill was occupied by Charles Gardner, maltster. The small two-storey stone building is one of the most attractive small mills in Gloucestershire.
(D1629; Colleen Haine, 'The Packers of Painswick', *Gloucestershire Historical Studies*, IX, 1978; Colleen Haine, 'The Cloth Trade along the Painswick Stream: The Cloth Mills', *GSIA Journal*, 1983; RR229.23; LG, 8 July 1851)

Oliver's/Cox's Upper Mill, SO 879115

A water mill at this site is recorded from the fifteenth century. In 1785 it was sold to John Cox, clothier of Painswick. This was worked by John Cox and Weston Hicks prior to 1818 and continued to be worked in conjunction with the Lower Mill until the late 1830s. The mill, with two waterwheels and a 17ft fall of water, stocks and gig, scouring house and wool stove, was vacant by 1839. J. Cox, miller, was at the mill in the 1850s and '60s. Nothing of the mill survives.

(RR229.23gs; PP 1839, XLII; GJ, 16 March 1842; Directories 1856, 1863; Colleen Haine, 'The Cloth Trade along the Painswick Stream: The Cloth Mills', *GSIA Journal*, 1983)

Tocknell's Mill, SO 882118

James Tocknell, clothier, died in 1602. While he may have used this mill as a fulling mill, no documentary proof has been found. Adjoining is a handsome clothier's-type house, Tocknell's Court. In 1820 William Cox occupied the mill which was then a corn mill.

(Baker's map; RR229.23gs)

Eddels'/King's Mill, SO 884121

In the parish of Cranham, and formerly a corn mill, this is likely to have been a fulling mill by the end of the eighteenth century when copyhold was granted to Thomas Eddels of Minchinhampton, clothier. In 1801 his 'cloth mill' was for sale, together with utensils. Benjamin Wood was tenant of the mill in 1801. Thomas Eddels was bankrupt in 1805.

(D936 M 1/ 2; GJ, 3 August 1801; GJ, 7 January 1805; tithe terrier; Colleen Haine, 'The Cloth Trade along the Painswick Stream: The Cloth Mills', *GSIA Journal*, 1983)

THE WASH BROOK, A TRIBUTARY OF THE PAINSWICK STREAM

Washbrook/Gyde's/Merretts Mill, SO 857095

This was a grist mill in the seventeenth century; by the early eighteenth century two separate grist mills are recorded. In 1799 John and William Bayliss, clothiers of Painswick, had bought the estate except for where a mill formerly stood, implying that Upper Mill had been demolished. In 1802 John Bayliss is described as a woollen manufacturer here and from 1809–16 he held a mill and lands, plus 'New Mill', suggesting that another mill had been built on the site of the former Little Mill. In 1823 Edward Bayliss was bankrupt and in 1824 'Two excellent clothing mills' in different tenancies were put up for sale. By c.1828 Washbrook Mill, the larger of the two, had become a grist mill and was advertised to be let in 1844. Part of the mill with the house comprises a dwelling.

(RR229.23gs; D1214 box 6 bdl 51; GJ, 5 July 1824; *Midland Counties Herald*, 29 August 1844; Colleen Haine, 'The Cloth Trade along the Painswick Stream: Pt IV', *GSIA Journal*, 1984)

Little/Upper/New Mill, SO 858097

The mill was owned by Edward Bayliss in 1820. At the tithe survey it was occupied by Nathaniel Iles Butler, who also owned and worked Cap Mill. It was advertised to be let in 1844, 'worth the notice of a Flock, Paper or Cutling Manufacturer'. The mill was derelict for much of the twentieth century but there is now a residence on the site named 'Washbrook Mill'.

(Baker's map; D1241 bdls. 28, 51; *Midland Counties Herald*, 29 August 1844)

Lower Doreys/New Mill, SO 858098

The mill was run by Thomas Wood in conjunction with Upper Doreys Mill at one stage. In 1825 a 'capital cloth factory' was advertised for auction, comprising a fulling mill with five pairs of stocks, three gigs, a washer, a 40ft-diameter waterwheel, a 20hp steam engine, a rack stove, a wool stove and 'a new built machinery mill' for carding engines. The mill did not sell and Thomas Wood still owned it in 1826 but John Pegler occupied it. From the late 1820s the mill appears to have been let in several different tenancies. In 1839 N.S. Marling owned New Mill, continuing as owner until 1865. In 1846 Robert Hogg was described as clothworker of Dorey's Mill. By 1856 it was a corn mill and was demolished, possibly in the 1870s.
(F.A. Hyett, *Glimpses of the History of Painswick*, p.79; GJ, 31 January 1825; GJ, 3 & 17 October 1825; P244 CW 2/4gs; RR229.2gs)

Little's Mill, SO 858098

The mill was driven by the waters of a small spring which rises near Painswick house and enters the Wash Brook by Lower Doreys Mill. John Little owned and occupied the mill between 1809 and 1845. From 1809 Thomas Gyde was tenant but was bankrupt in 1816. John Little still held the mill in the 1840s but it is unlikely to have been a cloth mill. The mill has been demolished.
(Baker's map; RR229.23gs; GJ, 12 February 1816; tithe terrier)

Upper Doreys Mill

Thomas Wood owned and occupied this mill in 1820. According to Hyett, the pond burst soon after 1838 and the mill was closed. In 1853 the mill was occupied by John Pearce but the weaving shops were said to be vacant. The mill was partly pulled down before 1860.
(D445/E8; Colleen Haine, 'The Cloth Trade along the Painswick Stream', IV, *GSIA Journal*, 1984; F.A. Hyett)

SHEEPSCOMBE STREAM, A TRIBUTARY OF THE PAINSWICK STREAM

Sheepscombe/Wight's Mill, SO 894104

This was a corn mill in 1717, and was owned and converted to a cloth mill by 1806 by the brothers John and Edward Wight, who continued in the cloth trade at the mill for many years. Adjoining the L-shaped mill were a dyehouse, drying house and teazle house. The mill was driven by a 17ft-diameter overshot waterwheel. A curate of Sheepscombe described the mill as it was before 1817 when the 'manufactory [was] little better than a miniature of hell'. The mill had steam power by 1822. In 1826 the mill was described as the 'lately erected … good and substantial mill for the manufacturing of woollen cloths' on the site of a water corn mill, with two steam engines (6hp and 24hp), lofts and workshops. There were eight handlooms in the mill in 1838. In 1840 the mill was put up for sale. The mill has been demolished.
(RR229.23gs; D1241 Box 16/5; Elizabeth Skinner, *Sheepscombe: One Thousand Years in this Gloucestershire Valley*, 2005; Colleen Haine, 'The Cloth Trade along the Painswick Stream: Pt IV', *GSIA Journal*, 1984)

Gardner's/Flock Mill, SO 897105

Through the eighteenth century the mill was owned by members of the Gardner family. While they are known to have been a clothing family there is no proof that this was a fulling mill in the early eighteenth century. In 1743 it was a flock mill. In 1800 Edward Parker was owner and Joseph Gardner tenant. In 1804 Edward Parker's stock in trade was offered for sale, including jennies, wool, yarn and cloth. In 1837

J. & R. Wight were tenants but by the tithe survey it was a corn mill. In 1840 Flock Mill was recorded as 'taken down'. Although unused and largely demolished in the early 1900s, a portion was incorporated into a residence.
(D1347 acc. 454; Baker's map; Directory 1839–40; Colleen Haine, 'The Cloth Trade along the Painswick Stream: Pt IV', *GSIA Journal*, 1984)

Hore's/Ebworth/Back Mill, SO 898106
John Hore, clothier, was buried in Painswick churchyard in 1761 but there is no evidence to connect him with this mill. A corn mill at the site is recorded in 1796. From 1818 to 1821 John Fletcher was owner and the tenant was J. King, a corn mill being recorded. From 1825–33 Edwin Hoare, clothier, was tenant. Hoare was a manufacturer of Saxony broadcloths. In 1838 Mrs Welch owned the mill and John Wight (of Sheepscombe Mill) was tenant, but by 1839/40 the mill was void. It has been demolished.
(RR229.23gs; Colleen Haine, 'The Cloth Trade along the Painswick Stream: Pt IV', *GSIA Journal*, 1984)

MILLS NOT DRIVEN BY WATER POWER IN THE PAINSWICK VALLEY

Samuel Wood's Cloth Factory, New Street, SO 867097
This was worked by Wood during the boom period of the 1820s.

Bank House, New Street, SO 867097
This was worked by another member of the Wood family.

Zachariah Powell's Factory, Edge Street, SO 864097
This was worked in 1820 by Powell.
(This information in F.A. Hyett, *Glimpses of the History of Painswick*, p.79)

WORKSHOPS AND DYEHOUSES IN THE PAINSWICK VALLEY

Workshops near Skinner's Mill, SO 864092
These were occupied by Daniel Spring in 1820.

Dyehouses near Skinner's Mill, SO 864091
These were worked by C. Gyde in 1820 and by John Dowell in the 1830s.

Dyehouse at Brookhouse, SO 871095
See Brookhouse Mill.
(This information in F.A. Hyett, *Glimpses of the History of Painswick*)

SLAD VALLEY

Cuttles Mill, SO 847050 (for Dyehouse see p.219)
The mill leat was fed by the Slad brook and benefited from the construction of the canal, surplus water being fed towards it by a tumble weir. Formerly a fulling mill, it was used for shear grinding at the end of the eighteenth century by the Coles family. The mill and dyehouse was in the hands of Peter Watts in 1780. The garden, separated from the house by the canal, was 'a mythological paradise'.

(P.H. Fisher, *Notes and Recollections of Stroud*, 1871, repr. 1986, pp.128–9; Ian Mackintosh, 'Into the Picture', Stroudwater Textile Trust; the painting of this site is in the Museum in the Park, Stroud)

Badbrook Mill, SO 852058

Benjamin Barnfield was making cloth here in 1822. In 1830 George Fluck, woollen manufacturer, occupied the mill and in 1844 Matthew Grist was making cloth here.

It later became a flour mill occupied by Butt & Skurray and afterwards by Lewis Silk Gardner, who was the owner in 1898 when the mill was destroyed by fire.
(Directories 1822–3, 1830, 1842, 1844; Miller, 26 June 1898; 3 October 1898)

Little Mill/The Mill behind the Church, SO 854055

In 1671 Badbrook House and some unspecified mills were leased by Richard Hawker, fuller, to Thomas Lysons, clothier. In 1677 the mills comprised 'all those new erected buildings and mills', two fulling mills and one gig mill. In the following year the mills were sold to Robert Hawker of Rodborough, dyer. In 1728 Hawker mortgaged Little Mill, which at this date comprised a fulling mill with one whole stock and one half stock, and a gig mill, to Anthony Pettat, narrow weaver. The mill was sold to William Cole in 1729. In 1733 the property comprised a fulling mill containing one and a half stocks, one gig mill, one corn mill, one cloth press with a newly erected dyehouse. In 1739 Cole conveyed the property to John Fowler. Samuel Butt of Badbrook, dyer, was here in 1763. By 1810 the cloth mill and workshops were conveyed by Thomas Holbrow, dyer, to John, his son. Samuel Weddall, cloth manufacturer, was at Little Mill in 1820, but by 1822 Nathaniel Driver was making cloth here. The Little Mill or 'the mill behind the Church', comprising a fulling and gig mill, was occupied by Joseph and John William Partridge, in 1839, together with press shop, wool loft and counting house, formerly occupied by Stephen Carter. By 1856 it was a saw mill worked by William Ridler. By 1863 James Ockford, miller, occupied the mill. It has been demolished.
(D1512; D1159 acc. 1753; D873/T97; Directories 1820, 1822–23, 1856, 1863)

New Mills, SO 859056

In 1589 Richard Watts enfeoffed a messuage 'and mills called ye newe mills in Strowde' to Richard Stephens, Edward Trattman, James Stephens and John Trottman.

William Helme reported that 'deeds of the estate call the Place New Mills in the time of Charles II and state its then being used as a fulling mill'. The mills, together with the integral New Mills Court, were rebuilt by Thomas Bayliss in 1776 in Palladian style (see p.50). Nathaniel Winchcombe who held the mills in 1798 surrendered them to Thomas Bayliss. In 1820 Robert & William Helme were making cloth there. The firm continued as William & Mashiter Helme during the 1830s and were here in 1839. The power for the mills was both water and steam in 1834. In 1863 Charles Howard was making cloth at New Mills. By the 1870s John Libby was occupying the premises. The firm became known as Libby Edmunds & Co., and cloth continued to be manufactured there until 1898, when the mills were closed. New Mills Court and a small part of the mills remain, although much changed.
(D547a/T4/G; D547a/T65; PRO E315/394; PP 1834, XX, p.269; S. Rudder, *A New History of Gloucestershire*; P.H. Fisher, *Notes and Recollections of Stroud*, p.151; Directories 1820, 1822–23, 1830, 1839–40, 1849, 1863; PP 1840, XXIV, p.398)

Piggas/Peghouse/Woodlands Mill, SO 863058

In 1639 Giles Davis died seised of two messuages called Piggas and all the water, grain and fulling mills there which he had lately purchased from Samuel Hopson. In the early eighteenth century John Cripps purchased Piggas Mills. John and Mary Cripps leased the mills to Daniel Fowler of Minchinhampton. The fulling and gig mills were conveyed to Joseph Fowler on Daniel's death. William Hodges occupied the mills from the late eighteenth century into the second decade of the nineteenth. Weaving was being done on the premises and there were four looms there in 1802.

In the early nineteenth century there was a rope walk in part of the mill premises which was run as a separate concern by tenants.

In the 1820s Nathaniel Driver was making cloth here. He was responsible for rebuilding part of the mills and stoves in 1823. After Driver went bankrupt, Harman & Co., and later Barnetts, Hoare & Co., occupied the premises. In 1830 they leased the mill to Nathaniel Samuel Marling. The mill contained a waterwheel, five stocks, five gigs, a steam engine, press shop, dyehouse and a drying house over the boiler. In 1834 Marling told the factory inspector that there was a 30hp steam engine and that the waterwheel generated 10hp. The 'latter used only at present, power regular used solely for purpose of fulling'. In 1846 Barnetts, Hoare & Co. conveyed the mill to Eli James who entered a paper-making partnership with Joseph Bence Palser, of Quenington. This was dissolved in 1851. Richard Barton was using all or part of the premises in 1863 for silk throwing. The rope walk was still in existence in the 1860s for Frederick James, twine spinner, was at Woodlands in 1865. J. & A. Tannahill & Co. were making cloth at the mill in 1870, but by the 1880s Northcott Cartwright & Co. occupied it. They had forty-two looms in the mill in 1889 and 2,500 spindles in 1891. The firm had ceased production by 1906. Later, Humphreys & Co., woollen manufacturers, occupied the premises until the early twentieth century. It has been demolished.
(PRO E315/394; IPM (Index Library) III, p.43; D185/VI/2; D1159; PP 1802–3, VII, p.16; LG, 21 March 1851; Directories 1822–23, 1863, 1865, 1870, 1879, 1906; D873b; D873 E2; PP 1834, XX, pp.279–80; Kelly's Textile Directory 1889; Worrall's Directory 1891)

Wade's Mill, SO 865059
This mill was used by Alfred Haycraft for dyewood grinding in 1870.
(Directory 1870)

Hazel Mill, SO 867063
This mill was surrendered by Nathaniel Winchcombe to Thomas Bayliss in 1798. Thomas Hughes held the mill between 1801 and 1819, although in 1808 it was offered for sale and Robert Hughes was said to be the occupier. In 1814 Henry Wyatt is recorded as making cloth at Hazel Mill. In 1820 John Mills held the mill which was occupied by George Wyatt. At the tithe survey Nathaniel Samuel Marling owned and occupied the premises. Between the 1850s and 1870s Alfred Haycraft was using the mill for grinding dyewood. It is a three-storey stone building with dormer windows.
(*VCH Gloucestershire*, II, p.77; D1940; GJ, 9 May 1808; Directories 1814, 1858–9, 1879; RR229.23gs)

Vatch/New/Upper Vatch/Hermitage Mill, SO 872066
There were four distinct mills at this location. The Clissold (clothier) family owned Vatch Mill for much of the seventeenth and eighteenth centuries, so Vatch Mill was probably a fulling mill. By the late eighteenth century there was also a paper mill sometimes called Vatch Mill but, more accurately, Hermitage Mill – it was at the site of the present Woodside Cottages. In 1776 Thomas Bayliss agreed to the diversion of a spring to the pond at Vatch Mill. In 1820 Mr Mason owned the fulling mill. By 1833 Nathaniel Samuel Marling was in occupation at Vatch Mill. He told the factory inspectors that the mill had burned down and was rebuilt in 1827 and that the mill contained three steam engines which produced 66hp and two waterwheels of 12hp each. In 1838 Marling installed six power looms in the mill in addition to the

five handlooms which he had there. By 1844 William Fluck was at Vatch Mill. In the 1860s and 1870s Robert Hastings was making cloth there. The mill was closed sometime before the 1890s.

Two sets of buildings, both known as Vatch Mill, were run as a single unit. One is now covered by Vatch House and its gardens. A small stream enters the Slad brook below a trackway and the (main) Vatch Mill was driven by this small stream, being called New Mills on the tithe map. This was a highly developed, steam-powered industrial site (see p.94) – the more extraordinary considering its location. It has been demolished, but lay by the junction of the two streams; the pond remains.

Upstream of the main mill lay a small mill called Upper Vatch Mill. This was owned and occupied by Nathaniel Samuel Marling in the 1830s and run together with Vatch Mill. Marling told the factory inspector that the mill was rebuilt in 1830. It was water-powered and used for fulling, the upper floors being (hand) weaving shops. At the tithe survey, while William Fluck owned and occupied Vatch Mill, Nathaniel Samuel Marling owned and occupied Upper Vatch Mill. A converted workshop survives at the site.
(PRO E315/394; G.C.L. Hawker MSS; RR229.15(2)gs; Sun Fire Insurance Policy 3962423-4, 18 March 1778; D1159; PP 1834, XX, p.279; PP 1840, XXIV, pp.396–7; Directories 1844, 1849, 1863, 1865, 1870, 1874)

Wyatt's/New Mill, SO 873073
This does not appear in Baker's 1820 survey of Painswick, nor in valuations of 1822 and 1838. At the tithe survey it was called New Mill and was owned and occupied by Nathaniel Samuel Marling. Presumably George Wyatt occupied the premises shortly before or afterwards. The mill was still standing in 1895.

Lower Steanbridge/The Weaving/The Slad Mill, SO 875076
This was probably the mill owned by Richard Webb in 1608 for he was said to hold the watercourse from 'Stenebridge' which ran to his mill. Henry Townsend, clothier, probably owned the mill in 1691. William Townsend was at Steanbridge in 1721 and at his death in 1779 Henry Townsend owned a fulling mill with two stocks, two grist mills, a dyehouse and Steanbridge House. His brother, Theyer Townsend, was the heir. When he died in 1801 he left the property to his brother Charles, who died two years later. The Steanbridge estate went to Robert Lawrence, a cousin, who following Townsend's wish took the name Townsend. Robert was an Anglican priest and, on moving to the living of Bishop's Cleve, commissioned murals depicting the Steanbridge estate for the upper room there. The mills were advertised to be let in 1815, a 12ft overshot waterwheel being mentioned. In 1820 Dr Townsend owned the mill and Rack Mead and it was occupied by Benjamin Wood. At the tithe survey, the mill was occupied by Nathaniel Samuel Marling. Joseph Wright is recorded as woollen manufacturer of Slad Mill in 1842 and Thomas Crissold the corn miller. In 1846 Edwin Hoare 'late of the Slad Mill … clothier' was bankrupt. Mathew Lister was making cloth at Slad Mill prior to a petition for bankruptcy being filed against him in 1846. In 1853 Nathaniel Samuel Marling leased: 'All that fulling mill called … Steanbridge Lower Mill', which contained three stocks and two large waterwheels, to Horatio Collier, blanket manufacturer who was bankrupt two years later. The mill, probably of seventeenth century date, is depicted on the Bishop's Cleve mural and was standing in 1895 but has since been demolished (see p.42).
(PRO E315/394; Baker's map; LG, 20 February 1846, 4 December 1855; Directories 1842, 1849; D1159; Patricia Hopf, *The Turbulent History of a Cotswold Valley*, Nonsuch, 2006, pp.59–65)

Steanbridge/The Jenny Mill, SO 879078
This was worked by John Pegler, clothier, between 1763 and 1765, then by James Woodfield, whose stock in trade was sold in 1774. It was acquired by Theyer Townsend in 1781 from Pegler's executors. The mill was advertised to be let, together with Lower Steanbridge Mill in 1815, and was described as being

'employed wholly for machinery'. At the tithe survey Nathaniel Samuel Marling owned the mill which was occupied by William Lay. Later in the nineteenth century the mill was run by N. Partridge. The mill appears to be of early eighteenth-century date and is now converted into a dwelling (see p.77).
(F.A. Hyett, *Glimpses of the History of Painswick*, p.78; Patricia Hopf, *The Turbulent History of a Cotswold Valley*, pp.59–65, 164–5)

DYEHOUSES AND WORKSHOPS IN THE SLAD VALLEY

Window's Dyeworks, SO 851054
This was owned by 'Bustle' Window in the nineteenth century. The dyeworks were owned and occupied by John Holbrow at the tithe survey. It has been demolished.
(P.H. Fisher, *Notes and Recollections of Stroud*, p.93)

Warmer's Dyeworks, SO 864059
William Glover held the dyehouse in 1820. At the tithe survey, Mr Warmer occupied it.

Press Shop, SO 874077
This existed in 1820. At the tithe survey it was owned by G. Driver and occupied by James Hamands.

TOADSMOOR VALLEY

Gussage Mill, SO 876025
Roger Fowler left 'Gusshis Mill' to his wife in 1540. His son Thomas, clothier, occupied it in 1559. In 1653 Thomas Freame leased the tuck mill to Walter Sewell, clothier. At the tithe survey, James Taylor owned the mill and William Anthill occupied it. In 1863 Dangerfield & Foot were using the premises for silk throwing, and by 1879 R. Grist was making mill puff here. It was later a saw mill.
(*VCH Gloucestershire, II*; Directories 1863, 1879)

Toadsmoor Mill, SO 879031
In the 1740s, the mill was apparently worked by the Hancox family who made cloth there. It was later a corn mill, possibly occupied by Thomas Creed who was bankrupt in 1854. The corn mill was advertised for sale in 1863, containing an iron waterwheel with a fall of 13ft. It has been demolished.
(LG, 13 October 1854; Stroud Museum, 1962, 146)

Snow's/Rowbridge/Toadsmoor/Selwyn's (Upper & Lower) Mill, SO 880035
There were two mills at this location by the late seventeenth century, with a mill pond between them. In 1549 John Snow, tucker, was assessed at 30s. In 1690 William Hayward desired to be admitted to Snow's Mill. When he was assessed for the poor rates in 1728, the Upper Mill was valued at three times that of the Lower Mill, the latter having been built between 1690 and 1728. In 1820 Joseph Weare was making cloth at the mill but by the 1830s Aaron Evans held the Upper and Lower mills with Amos Jones, clothier. In 1856 Charles Freeman, described as a tweed manufacturer, was at Toadsmoor Mills. Charles Stanbridge & Co., woollen manufacturers, were at the mills in 1865. Freeman, described as a flock manufacturer in 1864, purchased Upper Mill and Lower Mill as well as Middle Mill. Both he and Stanbridge & Co. were at Toadsmoor Mills in 1867. In 1879 Wollaton & Co. were making mill puff and flock at the mills. The firm was in financial difficulties by 1881. Later, William Selwyn & Co. made flock there.

This site is complex. The remaining mill is a stone building lying at right angles to the valley, the windows being elliptically headed. This was the lower of the two main mills and the wheels were fed by the square pond which lay between this and the upper mill (see p.133). The upper mill, whose foundations can still be traced in places, lay immediately to the north of the pond. This was fed by a small leat which led to the east side of the mill, from a long narrow pond in the field north of the lane leading to Mackhouse wood. The tithe map shows an external wheel.
(M.A. Rudd, *Historical Records of Bisley with Lypiatt*, pp.356–8; D1241 Box 19 bdl 1 (3), (4); D1241 Box 30 bdl 14; Directories 1820, 1839–40, 1856, 1867, 1879 & PP 1840, XXIV, p.398; reproduction of 1868 watercolour by Alfred Newland Smith in Museum in the Park, Stroud)

Middle/Shortwood Mill, SO 881035

A small mill, probably run in conjunction with Upper and Lower Mills; it was driven by a spring running from Bussage towards the Toadsmoor Valley and situated near the junction of the road from Old Bussage to Toadsmoor. It is mentioned in 1864 and was run by the Selwyns, in conjunction with Toadsmoor Lower and Upper Mills in the 1890s. Water from the pond was taken to the overshot wheel by means of an iron pipe. The mill has been demolished.
(D1241 Box 19 bdl 1 (4); Hammond Papers Box 8)

Wiselands/Millgrove/Rowbridge Upper Mill, SO 879039

This mill was rented by the Sewell family in the sixteenth century. In 1740 Ann Yeats conveyed to Richard Rogers, clothier of Toadsmoor, a fulling mill called Millgrove; by 1763 it was a fulling mill or grist mill and by the late eighteenth century was in the hands of Richard Rogers. It was definitely a fulling mill in Rogers' day and, as the Sewells were clothiers, probably was in the sixteenth century. George Rogers owned the adjoining house which was called Picking House. The mill passed through many hands at the end of the eighteenth century, including William Grimes and William Restall, both clothiers. A former dyehouse was also mentioned in the 1790s. In 1799 the mill was conveyed to John Trotman and William Tayloe of Chalford, clothiers, and then to William Toghill, clothier, of Chalford. By 1801 'all that fulling mill or grist mill … now fallen into decay' was leased in 1827 to Aaron Evans. Nothing remains of the mill. The mill was fed by a very large pond which occupied almost the entire field below Toadsmoor Lake.
(D1241 Box 19 bdl 1; M.A. Rudd, *Historical Records of Bisley with Lypiatt*, p.358)

Cricketty Mill, SO 901054

The mill lies on the small stream which flows from Bisley to Toadsmoor. Richard Butler gave the mill to his grandson Thomas in the later seventeenth century. In 1828 Robert Smyth Owen and others conveyed 'all that new erected washhouse' and the 'clothing mill at Nashend on Well Close' to Henry Daubeny of Bath. It is a small two-storey stone building adjoining a small mill house; now a residence.
(Juliet Shipman, *Roockwoods*, privately printed, n.d., pp.55–62)

NAILSWORTH VALLEY

Hawker's Mill, SO 835045

This site is usually known as Hawker's Dyeworks but it was a fulling mill earlier. In 1659 Arthur Browning of Oldbury on Severn, clothworker, conveyed a messuage and fulling mill, containing two stocks, to Daniel Fowler of King's Stanley. In 1743 the property containing two stocks, one gig mill and one knapping mill was conveyed to Richard Hawker, dyer, of Dudbridge. By 1772 cloth finishing had become important, for three knapping mills with 'all those my dyehouses' were left by Hawker to his son John. Under Hawker family ownership, the business increasingly specialised in dyeing. Rudge commented 'often 42 pieces are dyed in one day'. When John Hawker's son Richard died in 1848 the devisees of his will sold the estate to Samuel Stephens Marling. Christopher Smith occupied part of the premises. At some stage in the nineteenth century dyeing became the sole function carried out at the site, and in 1873 Samuel Stephens Marling leased 'several dyehouse buildings including the new vat house [and] stoves' to Adolphus Charles Smith of Dudbridge, dyer. Arrangements were made for Smith to have a water supply. There was a 10hp engine for pumping water.

In 1880 the lease was renewed and various improvements were noted, in particular a steam heating stove. On the floor beneath the racks lay an iron pipe which took water from Selsley Hill to the old knapping mill. The site has been completely altered but some old buildings remain incorporated into Dudbridge Garage. (D873b; D873 E8; Rudge, I, p.38; R.L. Rose, 'An Industrial History of Dudbridge', *GSIA Newsletter*, 8 August 1966, pp.25–40)

Lightpill Mill, SO 839039

In his will dated 1661, Jasper Estcourt of Lightpill left his 'warpeing barr and frame, iron way beame and skales, Tuckers sheares and the press in the Mill' to his son Richard.

In 1782 the property comprised 'one Dyeing House and one Water Grist Mill and one Fulling Mill containing Two Stocks and a Gigg Mill'. By 1806 the grist mill had disappeared and the 30ft by 15ft New Mill had been built, supplied with water from a spring-fed pond. After 1805 the mill was leased to Shillito Stather who bought the premises in 1818. By 1811 both carding and spinning were carried out in the mill. Stather spent a considerable amount of money on improving and expanding the mills and in 1828 the property consisted of the Mansion, three mills adjoining called South Mill, Middle Mill and North Mill; the dyehouse, scouring house, teazle house, and the Little Mill. By this date, however, contraction of the woollen industry had led to a fragmentation of businesses on this large industrial site. During the 1830s one tenant made pins in South Mill. In 1839 Brown & Tucker were making cloth, Daniel Foote Taylor pins and Shillito Stather was dyeing on the premises. In the early 1840s J.F. Marling was making cloth here. South Mill appears to have been the main building. Middle Mill was rebuilt in 1850–54 and contained two waterwheels. William Barnard was tenant of the mills in the early 1850s and it was he who was responsible for the rebuilding of Middle Mill on the same scale as South Mill. He installed a steam engine in Middle Mill. A lease of 1854, renewing Barnard's tenancy, states that North and Middle Mills contained six pairs of stocks and two broad gigs. The occupants of the mill from the mid-1850s were Roberts Jowlings & Co., in which Barnard was a partner. By 1871 the single-storey loom sheds to the south-east of Middle Mill had been built. There were eighty looms in the mills in 1889. At some period North Mill was rebuilt and by 1907 Little Mill was absorbed into new loom sheds (p.99). In this year the firm went bankrupt. The mill was owned by Erinoids in the early to mid-twentieth century. Some buildings have now gone but others are conserved and the complex is now an industrial estate.
(R.L. Rose in appendix to K. Hudson, *The Industrial Archaeology of Southern England*; Directories 1839–40, 1842, 1844; PP 1840, XXIV, pp.396–7; Kelly's Textile Directory; John Morgan, *The Story of Erinoid*, Coleford, 2008)

Frigg's Mill, SO 839036

In 1633 Thomas and William Frigg leased a messuage called Beysomes, with the fulling mills and grist mill under one roof, to John Clutterbuck. An indenture of 1697 records that John Webb owned the property then comprising two fulling mills, one grist mill and one gig mill. By 1737 John Wicks of Rodborough possessed the mills. Cloth manufacture ceased here before the nineteenth century when

the mill was known as Lightpill Flour Mill. It continued as a flour mill until the end of the nineteenth century. A fine gabled clothier's house survives.
(D67/Z/57, 67; LG, 16 July 1847; Miller, 6 June 1892)

Unnamed Mill, SO 839032
This was owned by Nathaniel Peach at the tithe survey but was empty. It may have been run by him as a small cloth mill earlier in the nineteenth century. The mill pond is fed by a small spring. The mill has been demolished.

Rooksmoor Mill, SO 842031
In 1729 Thomas Small leased the fulling mill to Edward Peach, clothier. Nathaniel Peach worked it from 1748 to c.1773. Samuel Peach owned the mill in 1805, leasing it to Paul Wathen. In 1820 Joseph Haigh was making cloth at the mill. He continued there until the late 1830s. In 1856 John Wise was said to be at Rooksmoor, but by 1858–59 Grist & Niblet were using the premises for flock manufacture. Flock continued to be made here until 1963. The mill has been demolished.
(D1241; Box 30 bdl 13; VCH History Online, Rodborough; Directories 1820, 1856, 1858–59, 1863)

Woodchester Mill, SO 844028
This was probably a fulling mill by the early sixteenth century when a gig mill was recorded here. In 1605 Henry Dudbridge sold a 'ffullynge myll and … grystmyll' lately occupied by Henry Dudbridge to Sir George Huntley of Frocester. In the following year, the mills, containing two stocks, were leased to Dudbridge. The mill passed to Henry's second wife and then to his son, Stephen. In 1731 Holliday Dudbridge owned the mill and in c.1731, having erected a new mill by the older mill, was in a watercourse dispute with Onesiphorous Paul of Southfield Mill. In 1744 Holliday Dudbridge leased the two fulling mills, dyehouse and cloth stove to Samuel Paul of Rodborough. In 1748 they comprised 'two stocks and gigg mill under one roof [and] … all that new erected mill or mill house containing one-half stock … and also the dyehouse, sheer shops or prefs house'. By 1770 there were two dyehouses, one either side of the road leading from the main Nailsworth road to North Woodchester. In 1803 and 1804 the main mill was taken down and rebuilt but the small mill containing half a stock remained. By 1820 one of the dyehouses remained and was sublet to Thomas Stephens. The mills were run by Joseph and O.P. Wathen.

In 1838 the mills were run by Wathen & Cork using two waterwheels and a 40hp steam engine. By 1852 John & Edward Wise occupied the premises and in 1868 the mills were purchased by Edward Wise. Wise was still there in 1890 but by the end of the decade cloth manufacture had ceased. The mill was taken over by Stroud Piano Works in 1911. A severe fire destroyed the main mill in the 1930s and it was not rebuilt. The fine mill chimney has been incorporated into twentieth-century buildings. A long two-storey stone range of workshops, with living accommodation beneath, was converted into cottages but has now been demolished.
(D67/Z/50; D340a/T168; D44/A 1, 6, 11, 13, 14, 17, 18, 20, 22, 23, 24–5, 31, 33, 34, 35–6, 37–8, 43–4, 54, 61, 64; D149/375/954; D44/3I; PP 1834, XX, pp.285–6; Kelly's Textile Directory; Woodchester History 1972, App.17)

Southfield Mill, SO 843025
In 1717 Elizabeth Paul, widow, leased a messuage, the close adjacent called Rack Leaze with the cloth racks, a presshouse, dyehouse, 'cole house', madder house and loft in Woodchester, to Samuel Dudbridge, the mill excepted. The watercourse dispute between the owners of Southfield and Woodchester mills in the early 1730s referred to Southfield as having been 'built abt 40 years ago'. The mill was owned

by Sir Onesiphorous Paul and between 1730 and 1740 he invented the knapping mill here. Southfield House and the two fulling mills, dyehouses and rack stoves were advertised to be let in 1808. By 1818 it was owned by William Peach Cooper and worked by Thomas Sykes. In the 1830s Woodwark & Bird made cloth at the mill. They had twenty-four handlooms in the mill, six of which were not in use. Bird continued at the mill on his own from the 1840s to the late 1860s, and from this time Southfield was worked together with Churches and New Mill. In the early 1870s the firm was known as Bird & Bubb and by 1879 as Bubb & Co. C.T. Fox & Co. made cloth here during the 1890s. By 1906 cloth manufacture had ceased. Stick manufacture was carried out in the 1920s. The mill burned down and a new red-brick factory erected in its place, but the fine L-shaped, gabled clothier's house remains.
(T. Rudge, I, p.355; D44/31; GJ, 22 August 1808; PP 1840, XXIV, pp.396–7; Directories 1844, 1863, 1865, 1867, 1874, 1879, 1897)

New Mill (on a small tributary near Southfield Mill), SO 841025

At the tithe survey, Woodwark & Bird also occupied these premises which, although described as 'shops, buildings, workshops', were water-powered. It was worked in conjunction with Southfield and Churches mills. The mill has been converted to a house.

Churches Mill, SO 843023

In 1637 John Churches enfeoffed to his grandson John a fulling mill containing two stocks and a grist mill. By 1716 they comprised a fulling mill with two stocks, a gig mill and a grist mill under one roof. Thomas Churches died in 1762 and was buried in Woodchester church. During the nineteenth century the mill was run together with Southfield Mill by Woodwark & Bird, later by Bird, then by Bird & Bubb, Bubb & Co., and finally by C.T. Fox & Co. The mill is a two-storey building, the ground floor being of stone and the upper floor of brick. It has been converted to housing.
(D892/T100/1; D873 T67)

Frogmarsh/Bowbridge Mill, SO 841018

In 1658 Clutterbuck Deane of Minchinhampton, clothier, was seised of a fulling mill containing two stocks and gig mill with a dyehouse adjacent in Woodchester. After Richard Deane's death Thomas Shurmur bought the mills and other buildings and, in 1706, leased them to Edward Yeate of Malmsbury, clothier. The mill descended to a member of the Lovesey family by marriage. Thomas Shurmur is recorded as making cloth in Woodchester in 1784.

In 1804 William Lovesey conveyed the premises (then occupied by Thomas Cooper, clothier) to William Humphries Bennett. In 1806 the mills passed into the hands of John Knowles. In 1814 John Chalke & Co. were making cloth here. They continued at the mill during the 1820s but John Darke was also said to be making cloth at Frogmarsh in 1820. In the 1830s Francis & Flint occupied the premises for cloth manufacture. By 1846 William Haycock owned the mill but Francis & Flint are again recorded as making cloth here in 1858. By 1867 pins were being manufactured here by Perkins, Critchley & Marmont. The interest of this site lies not so much in the mill itself as in the ancillary buildings. The oldest part of the site comprises the gabled, seventeenth-century clothier's house which became absorbed into the industrial buildings. A seventeenth-century, two-storey stone-built wing and a nineteenth-century, three-storey building comprise the current 'main' mill (see p.80). The finest surviving wool stove in the county is on the opposite side of the lane, converted into a house (see p.71).
(D1347 acc. 1603; Woodchester; *200th Anniversary: A History of Playne of Longfords Mills*, p. 24; Directories 1784, 1814, 1820, 1830, 1839, 1858–9, 1867; D677/T4; N. Paterson & S. Mills, 'Cloth, Pins and Leather – An examination of Frogmarsh Mill', *GSIA Journal*, 1997)

Small unnamed mill on a stream near Frogmarsh Mill, SO 840017

This was worked as a cloth mill by William Marling in 1818 and may have been worked by J.W. Darke in 1820. By 1836 it was being run in conjunction with Frogmarsh Mill, being a wool shop, owned and occupied by Abraham Marsh Flint and Samuel Francis.
(D 677/T6;T1/201)

Merretts/Haycock's Mill, SO 843014

Daniel Webb of Merretts Mill is recorded in 1703. The mill was worked as a cloth mill by Mary Dudley & Son in 1769. By 1804, housing two stocks and a gig, the mill was owned and occupied by Thomas Haycock whose business was here until the 1830s. In the later 1830s William Martin was making cloth at the mill which was owned by William Haycock. In 1840 William Hunt was the tenant. The premises were described as fulling and gig mills and a corn mill with a shear shop. By 1863 the mill was used for flock and shoddy manufacture and was occupied by William Grist. The mill was destroyed by fire in 1894 but was rebuilt. By 1911 the Grists were making flock and bedding wool at the mill. A late nineteenth-century brick mill lies parallel to the mill pond. This has a fine tall chimney alongside, dated 1887. An earlier mill lay below the mill dam but the site is much altered.
(Hawker MSS; Directories 1814, 1820, 1830, 1839–40, 1863; D677/T4; Ian Mackintosh, 'Shoddy Work, Stroud. Well Done!', *Warp & Weft*, 2011)

Dyehouse/Philpott's Mill, SO 844012

This site, comprising a mill and dyehouse, sometimes separately tenanted, sometimes in single occupation, was associated with the Cambridge family in the seventeenth century. Two Yeates cousins, both 'woaders', were here in 1699. Yeates Senior had died in 1682 leaving 'goods and tooles and concernes at the Dyehouse', valued at £816, indicating a dyer of some wealth. Samuel Yeats 'of the Dyehouse' died in 1722, leaving his interest in it to his son, Samuel. Thomas Deverell, then a tenant, bought the site in 1740, selling it to Richard Cockle in 1777. The two parts of the site were then united in ownership under Cockle who had previously purchased the dyehouse, although they had been worked as a single enterprise for some years. In 1804 the mill contained two stocks and a gig mill. Besides the dyehouse there was a 'New Mill for machinery [of] five storeys'. Cockle entered a partnership with a member of the Paul family in 1805 but by 1812 he was bankrupt. Thomas Foxwell of Middle Mills, Berkeley, bought Dyehouse Mill in two stages in 1815 and 1818 and, by 1824, his son Philip was working the mill. In 1828 there was a 100-spindle mule powered by a 14hp steam engine and a loom shop. By 1831 Philip Foxwell was bankrupt. The mill was let to a succession of tenants over the following thirty years. By 1830 Woodwark & Hunt were making cloth here. In 1838 there were eight power looms in the mill. John & Thomas Hunt continued here alone at the end of this decade. In 1849 two firms are recorded at Dyehouse Mill: John & Thomas Hunt and Weeks & Woolright, who may have worked the dyehouse. In 1856 William H. Austie was making cloth here, but by 1858 Grist & Tabram were using the premises as a flock mill.

In 1878 Samuel Newman established a small engineering business on the first floor of the mill, with a foundry alongside. In 1896 it merged with Hender, Stevenson & Co. of Nailsworth, becoming Newman Hender Co., making valves. After a number of acquisitions and amalgamations, the site was closed. Critchley became the new owners. Shortly after 2000 Renishaw plc took over the site and developed it as a manufacturing and technology centre. Dyehouse Mill is in two sections, part stone and part brick. There is a date stone at the site of 1623, with a lion, and one of 1820, in the brick section.
(D1347, acc. 1347, D6700 3/17; A Brief History of the Dyehouse Works, Bath Road, Woodchester, Typescript (Renishaw); Directories 1814, 1820, 1830, 1839–40, 1849, 1856, 1858–59; LG, 1 July 1862;

There are two photographs of old postcards showing the mill at the Nat. Buildings Record; PP 1840, XXIV, pp.396–7; MSS notes by Ian Mackintosh)

Inchbrook/Playne's Little Mill, SO 844009

In 1714 Richard Cambridge married Mary Ridler and the settlement included three fulling mills and a gig mill with a half part of a dyehouse and some lands, all lately in the possession of Richard Cambridge, grandfather of the bridegroom. In 1758 Mary Cambridge, widow, sold the fulling mills and knapping mills to John Wade of Gloucester. They were occupied by William Hill. By 1811 they were solely corn mills and, in 1817, were sold to Peter Playne for £1,975. Thomas Williams, clothier, of Inchbrook Mills was declared bankrupt in 1832. By 1839 this was an adjunct to Dunkirk Mills for Playne & Smith. During their occupation the mill was known as Playne's Little Mill. Grist & Tabram had taken over the premises for flock making by 1867. Inchbrook Mill has had various uses in the twentieth century and has recently been well restored by Renishaw, the current owners. The mill consists of a single two-storey stone range of two sections. The west section appears to be older, possibly late seventeenth century.
(D1347, acc. 1347, Playne; LG, 3 July 1832; Directories 1839–40, 1867; PP 1840, XXIV, p.399)

Pitts Mill, SO 842008

This mill was occupied by James Norton in the early nineteenth century. In 1820 James Norton, his wife, and William Biggs Norton leased Pitts Mill consisting of two stocks and a gig mill to Nathaniel Samuel Marling who already occupied the premises. After cloth manufacture ceased in c.1841, Pitts Mill was employed in pin manufacture and also chemicals. In the 1950s it was a flock and shoddy mill. The mill is built of stone and consists of a single two-storey range. The mill house forms the north wing. It is now Oaklands Farm.
(D873b; Directory 1814)

Freames Mill, SO 839007

In 1709 John Fream of Avening and Mary, his wife, leased two fulling mills, and a gig mill adjacent, to Thomas Yates of Minchinhampton for a year. In 1814 Stephen Blackwell & Son were said to be woollen manufacturers at Inchbrook and they may have been at this site. By 1820, however, the mill belonged to James Norton and in this year he leased it together with Pitts Mill to Nathaniel Samuel Marling. The property comprised a fulling mill with two stocks and a gig with dyehouse, presshouse and stove. Marling agreed to install two new stocks in the mill within two years and, before the twenty-one years of the lease was up, to install two new waterwheels as well. This later became a flock mill. A watercolour by James Norton c.1808 shows a double-gabled building, with a two- or three-storey building alongside, at the foot of the dam.
(D873 E 1; Directory 1814)

Dunkirk/New/Cooper's Mills, SO 845005

In the 1630s there was a fulling mill, called New Mills, at the site. In 1741 New Mills was owned by Samuel Yeates who was also a blue dyer. In 1785 the mill was bought by Richard Middlemore, dealer and chapman, who was bankrupt in 1794. In 1795 the mill and clothier's house were bought by John Cooper and, in 1798, he built a new mill (Cooper's Mill) containing two waterwheels. In 1801 the mill was renamed Dunkirk Mill. In 1804 another new mill was built. The new mill was five storeys high and contained five stocks, a gig mill, and shear shop. The old mill was converted into a dyehouse and scouring house. A wool-drying stove, picking house and presshouse were erected somewhat away from the main mill to reduce the fire risk. In 1816 the mills were for sale and were bought by William and Peter Playne,

William occupying the shear shops and cottages. In 1818 Peter Playne built the first of his new ranges adjoining Cooper's Mill, also rebuilding the upper storeys of the 1798 block. A 14hp Boulton & Watt engine was installed in 1820. In 1822 Peter Playne became sole owner, William becoming the owner of Longfords Mill. A new building to house handlooms was added in 1829 as well as a wool warehouse. By 1830 the firm was known as Playne & Smith; the dates of their mill being 'part very old, no date; part built 1800, 1818, 1819, 1820 and 1821, [we] believe it has been a fulling mill more than 200 years'. In 1838 power looms were installed in the mills and in 1840 there were two, in addition to the seventy-one handlooms (four of which were not in use).

In 1844 the firm was known as Peter & Charles Playne, and in 1848 Peter Playne's three sons took over Dunkirk Mills; 28hp was still generated by water in 1848. The original New Mill was demolished in 1855 and two new blocks built, linked by a waterwheel shed.

The weaving block housed forty-four looms and finishing was done in the new block in six new fulling machines which replaced the old stocks. A 28hp compound steam engine was installed, coal being delivered via the mill siding. The 13ft-diameter waterwheel in Walkers Mill was installed at a cost of £250 in 1855. Charles Playne's son Alexander joined the firm in 1871 but, after eighteen years, he sold Dunkirk in 1889 and the mill ceased cloth production. In 1891 W. Walker & Sons, who had a large hosiery factory in Nottingham, bought Dunkirk Mills and it shortly afterwards became a stick mill.

Dunkirk Mill was reputed to have had the best water power in the Nailsworth Valley. A long, high leat took water to the wheels which were situated in pairs at right angles to one another. Two surviving cast-iron wheels are 10ft in diameter and 7ft 2in wide and, from the legend on the launder, may have been made by Daniels of Stroud. The third surviving waterwheel is 13ft in diameter and 12ft 4in wide, unnamed, but of a type made by Ferrabee at Phoenix Ironworks. The distinct building phases of this elegant stone mill can be seen from the exterior (p.140). The mill has been converted to apartments; the Stroudwater Textile Trust manages a working display of finishing machinery, operated by water power, at the north end. (LG, 5–9 April 1791; LG, 25 May 1816; PP 1834, XX, pp.260–1; PP 1840, XXIV, pp.396–7; Directories 1814, 1820, 1830, 1844, 1856, 1865, 1867, 1874, 1879, 1894; Chance & Bland, *Industrial Gloucestershire*, p.29; Hammond MSS; J.K. Major, MSS Report on Dunkirk; Ian Mackintosh, Notes on Dunkirk; Ray Wilson, 'Dunkirk Mills, Nailsworth: A New Chapter', *GSIA Journal*, 1988, pp.30–1; Ray Wilson, 'Dunkirk Mills, Nailsworth: A Progress Report', *GSIA Journal*, 1989, pp.36–8; R. Wilson, 'Dunkirk Mills, Nailsworth: Chapter Three', *GSIA Journal*, 1990, pp.60–2)

Egypt Mill, ST 849999

In 1675 George Hudson & Henry Willoughby leased a tenement, two fulling mills, one gig mill and a dyehouse in Nailsworth to Richard Webb, son of Edward Webb, clothier. A succession of Richard Webbs died in 1712, 1748 and 1749 at which point Nathaniel Webb inherited the mill. He may have been the 'Pharaoh' Webb who named the mill. In 1814 Samuel Webb leased the fulling mill with burling and scribbling shop above, the mill house with wool lofts above, dyehouse (containing two seg troughs), shear shop, dubbing shop and bleaching shop to Stephen & Edward Blackwell. In 1819 Blackwell added another gig mill and waterwheel to the mill, probably having built the eastern section of the mill. In 1836 Webb's trustees conveyed the mill to William Playne. Playne & Smith used the mill for cloth manufacture in the 1830s, running the mill in conjunction with Dunkirk Mill and flooding the meadow to enhance Dunkirk's water-power potential. Peter Playne estimated, in 1848, that 12hp was generated from the Egypt Mill waterwheels. It later became a logwood mill and then a corn mill. In 1870 it became a logwood dye factory. In the twentieth century the mill, once more, became a corn mill. In 1985 Stephen Webb (no relation to previous owners) acquired the mill and undertook its full restoration and conversion into a pub/restaurant.

The two-storey stone mill consists of a single range of two sections. The eastern section has segmental-headed windows; the older western section has horizontal-headed windows which have been partially blocked up. There are two surviving interior waterwheels; the west one a low breastshot wheel 14ft 6in in diameter and 6ft wide with wooden floats; the east waterwheel is 14ft 6in in diameter and 10ft wide (see p.95). There is a fine adjoining clothier's house with two gables on each of the four faces, dated 'RW 1698'.
(D1265; Directory 1839–40; J.K. Major, MSS Report on Egypt Mills; D3076/7; SJ, 19 February 1870)

HORSLEY VALLEY

Day's/Nailsworth Mill, ST 849996

In 1773 Jeremiah Day, clothier, bought property in Nailsworth including a house, the mill, workshops and land, from Edward Sheppard and by 1784 Daniel and Jeremiah Day were producing cloth here. A 1780 map shows a small mill without a mill pond. A new mill was built in c.1801 and in c.1814 a brick wing was added to house a beam engine and machinery. By 1820, when the firm was known as Day & Davis, two mill ponds had been constructed. Jeremiah Vick Day died insolvent in 1824. The mill was, however, enlarged and power looms installed. In 1830 Charles & Frederick Davis occupied the mill which was offered for sale in 1838 comprising clothing mill, dyehouse, workshops, steam engine house, wool stove, and teazle house. At this period the mill was called Nailsworth Mill which leads to some confusion with Birts or Nailsworth Mills on the Avening stream. In 1842 John & Edward Wise were making cloth at the mill. John Wise was joined by William Wise in the 1850s. In the 1860s the firm was known as Wise & Leonard. In 1870 Wise, Stephens & Whitby occupied the mill but, in 1871, the firm's machinery was for sale. In 1874 the mill was for sale with the 'modern power loom shed, steam engine, waterwheel and water turbine'. After the 1871 sale, Strachan & Co. occupied the premises, while Lodgemore Mill was being repaired following the fire.

T. Tabram held a temporary lease but woollen cloth production ceased in the 1870s. It became, successively, a flock and shoddy mill, a brass foundry and, by the 1890s, a corn mill. In 1890 the mill was purchased by the Davis family for their furniture business. In 1893 two of the ten bays were removed from the Fountain Street side of the mill and the road was widened. Several bays of the 1801 mill, the engine house and weaving shed, survive.
(LG, 6 December 1839; D2219/9196/box 4/1; Directories 1784, 1820, 1830, 1842, 1849, 1856, 1863, 1870; tithe map, Avening; Ian Mackintosh MSS notes)

Lock's/Johnson's Mill, ST 848994

In the 1790s the Lock family were prosperous brewers in Nailsworth and on a 1799 map the mill is designated 'Mr Lock's'. Philip Lock was making cloth here in 1820 and a map shows a mill pond, together with a wide leat, leading to what appears to be two adjoining buildings. In the c.1830 survey of the Playne family industrial estate, a grist mill was noted as being installed at Lock Mill in 1822. In the 1830s it was occupied as a cloth mill by William Hunt. Thomas & John Hunt, cloth manufacturers, occupied Lock's Mill during the 1840s. By the 1880s it was a corn mill occupied by George Rich Miller. There was a major fire in 1894 which gutted the mill. Johnson's Engineering Works was based here from 1936–97. The surviving premises were converted into housing and offices in 1998. A gabled, seventeenth-century mill owner's house is situated across the road with an industrial wing dated 1820.
(Directories 1820, 1842, 1849; Q/SR H 1799; Horsley tithe map; Ian Mackintosh MSS notes; D4644 2/101 (acc 6408))

Gigg Mill, ST 846993

In 1559 Gig Mill was granted to William Webb, clothier, and appears to have been part of Edward Webb's estate at his death in 1751. In 1786 the mill was owned by Elizabeth Castleman, the grist and fulling mills being separately tenanted. By 1796 John Remington, clothier, was the owner; he went bankrupt in 1808. In 1823 the mill was operated by James Thomas and Job Brown, clothiers. It was for sale in 1850, containing 'two waterwheels, two pair of stocks, gig, drums etc' and was then occupied by John Roberts, probably the occupier of Holcombe Mill, in 1849. By 1856 James Harris was making flock and shoddy at the mill. A fire in 1862 destroyed the roof. It was occupied by a firm of fellmongers in the early twentieth century. Since then there have been a number of tenancies. The small L-shaped building has been much altered and the stonework rendered. The Stroudwater Textile Trust demonstrates weaving machinery in part of the mill.

(Directory 1856; D1347; D2299/6527; D6700/2/2; GJ, 13 Jul 1850; SJ, 12 Apr 1862; I. Mackintosh MSS notes)

Millbottom/Great/Leather Stiffner Works/Ruskin Mill, ST 846990

A corn mill at this site was acquired in 1564 by William Webb, clothier, and in 1693 Thomas Webb, clothier, conveyed the corn mill, dyehouse and fulling mill to his son Edward, who leased the mill to Francis Clayfield, clothier. A gig mill and cloth press were added. The mill remained in the Fowler family until, in 1797, it was sold to Joseph Lock, brewer. At this date association with the woollen industry appears to have ceased. Lock was described as a clothier in 1804 and it was probably he who rebuilt the three-storey mill with a pediment in the style of Day's Mill. Three mills in Horsley, probably Millbottom, Lower Mill and, perhaps, Harley Wood Mill, were put up for auction in 1808. Two were fulling mills containing four pairs of stocks, and a gig mill 'which with the waterwheels are new and on the most approved principles'. The third mill, 'also new', was particularly adapted for shearing frames, carding machines etc.; there was also a blue and black dyehouse with vats and coppers, with a scouring washhouse, presshouse and warehouse.

John Lock sold the mill to Job Brown in 1812, Brown's 1823 will mentioning 'the Great Mill with 3 pairs of stones etc. my newly erected house nearby, my other mill Spring Mill; the carding engine, spinning jennies and billies, shearing frames in the Great Mill; the fuller's earth, teazles, dyewares etc'.

By 1826, mortgage repayments were in arrears and, in 1831, Revd George Daubeny became the owner. There were maltster tenants in the 1830s but Mr Bliss, presumably a cloth manufacturer, went bankrupt converting a stable into a stove in 1838 and Barnard & Co. took the mill at a low rent, using it for making list, the little mill being used as a lumber room. When Barnard failed in 1841 Wise of Day's Mill occupied the mill but by 1849 Millbottom had reverted to being a corn mill and was occupied by George Ford who purchased it in 1852. The old dyehouse became a malthouse. In 1856 John Ford & Son, timber merchants, and George Ford, mealman, were at Millbottom. Ford was bankrupt in 1887 and the mill, powered by both water and steam, was bought by Edward Benjamin and operated by Frank Heaven. In 1936 it was a leather stiffener works and, by 1959, used for making inks and aniline dyes. In 1973 it was converted to residential use. The mill was reconstructed as craft workshops in 1982 and renamed Ruskin Mill. An 18ft waterwheel was purchased from a mill in Devon and installed. In 1993 the central pediment was rebuilt with a cupola containing a bell cast at the mill.

(D547a/T53; D873/T104; D1297; D6700 2/2 & 31; GJ, 23 May 1808; MSS notes by Ian Mackintosh)

Lower (Horsley) Mill, ST 845986

In 1785 Thomas Howell, miller and baker, insured his stock, bakehouse and corn mill. In 1800, when the new turnpike road was built up the Horsley Valley, 'Mr Smith's Mill' was begun. (Later, the Smith family were partners in the Playne family's mills.) By 1808 it was a cloth mill, a partnership between Charles Boultbee and Richard Lowe, clothiers, expiring that year. The mill was included in the 1810 sale of

Millbottom Mill. An 1825 map of Peter Playne's mill in Horsley, dated 1825, shows a single building fed by a leat from a pond comprising the tailwater of Mr Baxter's Mill. In 1832 the assignees of Frederick Davis, clothier, a bankrupt, conveyed the lease of Lower Mill, formerly tenanted by members of the Blackwell family, to Peter Playne; the premises comprised fulling mill, burling shop, scribbling shop, 'site of the late dyeing house', shear shop, dubbing shop, drying house, coal house and Rack Leaze. In the 1830s the mill was occupied by Playne & Smith. At the tithe survey, Peter Playne owned the mill which was occupied by Richard Harris. An accident occurred in 1855, at which date Playnes were operating carding and spinning machinery here. And in that year, possibly due to a decision to concentrate all production at Dunkirk, the mill was offered for sale and described as a 'water mill, with large building formerly used as workshops and pond'. By 1865 Francis Brown 'manufacturer' was in occupation and in the late 1870s John Roberts was making flock and shoddy at the mill. The mill, only half its original length, and in 1968 owned by the Midland Fishery, is now owned by Ruskin College. It is a two-storey stone building with an attic lit by a long window.
(Sun Fire Insurance Policy 504638,1785; LG, 23–27 August 1808; LG, 23–27 October 1810; D2219/9196/2/3; Directories 1839–40, 1865, 1879; Rudkin, *History of Horsley*; D9196/Box 3/17; Ian Mackintosh, MSS Notes)

Washpool/Mr Baxter's Mill, ST 842981

Rudkin refers to 'The new mills near the Washpool … purchased by John Remington gentleman'. John Deverall of Nailsworth was a clothier who died intestate in 1798 and may have been the first person to work the mill. Remington was bankrupt in 1808. It is shown as Mr Baxter's mill on an 1825 plan.
(Rudkin, *History of Horsley*; D9196/Box 3/17)

Hartley Bridge/Boxwell Mill, ST 844983

Thomas Hanks, broadweaver, built a mill near the house he bought in 1801. In 1809 it is described as the mill at Harley Wood and it was bought, in 1810, by Richard Aldridge of Nailsworth, clothier. It was for sale in 1820 with two wool preparation machines. William Ashmead, clothier, apparently bought the site in 1819 but no mill is mentioned.
(D6700 2/31; Stroudwater Textile Trust, Nailsworth Mills, Walk 1)

NEWMARKET VALLEY (MIRY BROOK)

Price's/Old Mill, ST 847995

In the 1760s the former fulling mill had become a paper mill. This was possibly the paper mill occupied by Thomas Kench. After paper manufacture ceased this became a corn mill. A GP practice is now located here. The long workshop windows on the east side of the mill suggest that this was once a cloth mill.
(A.H. Shorter, *Paper Mills and Paper Makers in England 1495–1800*, 1957, p.166)

Hulkhay/Lower Mill, ST 843995

This was reputedly a cloth mill worked by members of the Davis family at one time. It is shown on the Horsley tithe map with no owners or occupiers.

Lot/Lowerl/Terrett's Mill, ST 841996

In 1641 Robert Wilkins leased a messuage and water fulling mill in Shortwood from Nathaniel Stephens for 8s p.a. A rent roll of 1671 states that an old presshouse lay adjacent to the mill. In 1742 the mill was

still occupied by members of the Wilkins family. In the 1790s the mill was occupied by Edward Barnfield, clothier. Carding machines and jennies were operated at the mill. Caleb Evans got into financial difficulties in the 1830s and the mill was bought by Isaac Hillier in 1839. He conveyed the mill to his stepson, H.J.H. King, a brilliant engineer, who developed a nationally known business making equipment for brewing and mining, besides power systems for the textile industry. The mill is a two-storey stone building masked, except on the west side, by additional buildings.
(D547a/T47, 55; RX167.1gs)

Nodes/Upper/Highwood/Shortwood Mill

A small fulling mill existed here from the mid-seventeenth century until 1840 (although by then it had almost certainly ceased to full cloth). In 1826 land 'used as a Rackhill', together with the mill and pond, was occupied by William Ashmead and Joseph Caraway. George Ricketts of Horsley, broadweaver, conveyed this mill to Rebecca Yerbury in 1831. Yerbury conveyed it to Isaac Hillier, bacon factor, in 1840 and, by 1866, it was a flour mill. It was demolished in the 1920s, the mill dam collapsing in 1931.
(D2219/9196/box 1/5 & 8; Stroudwater Textile Trust Nailsworth Mills, Walk 2, 2008; Betty Mills, *A Portrait of Nailsworth*, 1985, p.36)

AVENING VALLEY

Nailsworth/Birt's/Chamberlain's Mill, ST 852997

By 1635 Thomas Jayne held a house and two fulling mills from the manor of Lord Windsor and in 1686 George Small leased the '4 fulling mills, gig mill and grist mill all lately built'. In 1720 Samuel Sheppard leased the mills to John Small of Rodborough, clothier, together with the presshouse, coal house and dyehouse. John Heskins, clothier, bought, in 1796, the 'fulling mill and late grist mill now converted into a fulling mill … containing 3 stocks and a gigmill' from Edward Sheppard of Gatcombe Park. Heskins, Barnard & Bliss were making cloth at the mill in 1820. In 1830 the firm was known as Barnard, Bliss & Barnard, later as Edward Barnard & Son. They told the factory inspector the date of the mill: 'First portion very old; first employed as a cloth mill about 1730, second portion erected in 1806, third portion in 1814.' Their power was 'part water and part steam'. In 1838 there were two power looms and sixty-nine handlooms in the mill (thirty-one not in use). The firm went bankrupt in 1841. Francis & Flint were working the mill by 1849 and this firm, later called Flint & Sons, continued here until cloth manufacture ceased in 1876. They were succeeded by George Heath who made flock at the mill. In 1879 Alfred Chamberlain, leatherboard manufacturer, bought the mill and this manufacture continued until the 1960s. Little of the old mill remains but a date stone of 1814 is incorporated into a wall. The large site now comprises Nailsworth Mills Industrial Estate.
(D1265; D131/T14; D1406; D2424/19; D3102 T10; BGAS, vol. 54, 1932; SJ, 12 January 1878, 25 February 1871; Directories 1820, 1830, 1839–40, 1844, 1849; PP 1834, XX, pp.259–60; PP 1839, XLI; PP 1840, XXIV, pp.396–7)

Holcombe and Little/Spring Mill, ST 859994

This is an old mill site but from what period it was a fulling mill is unknown. In the sixteenth century it was in the hands of members of the Webb (clothier) family. In 1814 it was occupied by Clutterbuck, Davis & Brown, woollen manufacturers. In 1820 the firm was known as Joseph, Richard & William Brown. In the 1830s Edward Barnard & Son occupied the mill although it was owned by James Clutterbuck. The estate was put up for sale in 1837 comprising a mill, dyehouse, steam engine house, wool stove, teazle house, 'the round house' and 'an old building' converted into a gig house. In 1840 Barnard & Son took a new lease of the property being required to repair everything and scour it 'except the mill called the Little Mill and the wheel and gig mill'.

A survey of the Holcombe estate was made in 1840 for the purpose of ascertaining the state of repair and this showed that much needed to be done. In 1849 John Roberts was the occupier of the mill and by the early 1850s Roberts Jowlings & Co. of Lightpill Mill were making cloth at Holcombe. In 1855 the mortgage of the mills was for sale and there is a detailed description of the property which comprised a fulling mill with four pairs of stocks and four broad gigs worked by three waterwheels with spinning and warping shops over; a large air stove (see p.81); a newly erected building called the Spring Mill supplied both by a large spring of water and a culvert from the main stream, driving a 20-ft waterwheel; a scouring house with weaver's lofts over; and 'round wool stove now a workman's cottage near the entrance gate'. The whole mill was capable of making and finishing around fourteen ends of superfine broadcloth a week.

In 1858 John Wise was making cloth at the mill but by 1879 cloth manufacture had ceased here and Porter & Co. were using the premises as a flock mill. The site has been much altered but Spring Mill and the fine stone mill chimney is preserved; the former wool stove survives.
(G.C.L. Hawker MSS; D2219/919613/9; Directories 1814, 1820, 1839–40, 1849, 1856, 1858–9, 1870, 1879; Box 39gs; RX210.1gs)

Iron Mills, ST 863994

In 1673 Thomas Pinfold came into possession of the lately built Iron Mills with the two stocks. In 1734 the mill, dyehouse, and shear shop were advertised to be let. A subsequent advertisement identified an overshot waterwheel, a stock and a half stock, a gig mill, and dyehouse. 'John Webb late of the Iron Mills … clothier, died in 1756.' In 1791 the property was conveyed to John Perrin who was already a tenant of the mills. At this date the property comprised a fulling mill and gig mill, shear shops, counting house and wool lofts. In 1812 Perrin sold the mills to William Playne, although he appears to have made cloth at the mill for several years more. At the height of their prosperity, William Playne & Co. made cloth at this mill, as well as at Longfords (their main mill) and Avening Mill, the premises being called Iron House in the early 1830s. In 1817 a cloth room with drawing and marking rooms was built; in 1818 a willy room; and in 1827 a weaving shop. A high-pressure steam engine was installed in 1831. After the early 1840s, cloth manufacture at Iron Mills appears to have ceased. Very little of the mills remain. There are, however, two magnificent gabled ranges at the site, which may incorporate some of the earlier industrial buildings.
(D1265; GJ, 18 and 25 February 1734; Bigland, Minchinhampton; D4644 2/101; Directories 1814, 1839–40)

Longfords and Lake Mill, ST 867992

This is an ancient mill and was leased by members of the Elkington (clothier) family, from the mid-sixteenth century. A later mill, built in 1705, driven by two waterwheels, together with a small estate, was bought from Thomas Pinfold by Thomas Playne in 1759. In 1790 Edward Hoskins conveyed his interest in a fulling mill, grist mill and workshops to Peter & Martha Playne. Thomas Playne the elder died in 1789 and the firm was run under the name of Martha Playne & Son. In 1797 Martha Playne retired and the firm became known as William Playne & Co. In 1808 the property was described as two clothing mills and one corn mill with the workshops and dyehouse.

In 1806 Longfords Lake was constructed by building a dam 150 yards long across the main stream. The area originally flooded was 15 acres. In 1809 the Playne brothers built the 38ft-wide Lake Mill. Since they did not have complete confidence in the strength of the dam to withstand such a great force of water, the mill was built parallel to the valley, water being taken to it by leat. In 1813 the mill property was divided between William and Peter Playne. William Playne retained the original mills and workshops which were supplied by the old pond and Peter Playne took possession of Lake Mill (which contained three wheels), the dyehouses and stove. Rack Hill was divided between the two brothers. The corn mill was retained throughout the early nineteenth century. Lake Mill estate was extended over the next twenty years. A West Wing was added in 1816 and, in 1820, a small addition was made to the wing to accommodate a pair of millstones and boulting machine. A wool and rack stove were built in 1809 and 1810, a black dyehouse was built in 1812 and another small dyehouse in 1816; a scouring house was built in 1818. A sizing shop was constructed out of the 'longer building against the centre of the stove' in 1817. A wool warehouse was built in 1818. Gig mill houses, a handle drying house, with a coloured wool stove above, were built in 1829. A building was converted into a blue dyehouse for six vats in 1829 and a further adjoining dyehouse was built in 1839.

By 1829, William Playne seems to have been in control of the whole site, for Peter Playne was by then manufacturing cloth at Dunkirk, Inchbrook and Horsley mills. Some time after this, probably in the 1830s, the building history of Longfords was recorded. The blue-slated Middle Mill was constructed in 1804, a steam engine house built 'rather badly' in 1810. The 1705 mill was rebuilt on the original mill foundations in 1828. During the 1820s, three steam engines were installed. In 1838 one power loom was installed in the mills, in addition to the ninety handlooms (ten of which were out of use). New buildings and additions were made in 1857 and 1866. In 1889 there were thirty-six looms and 2,000 spindles at the mills. In 1910 it was decided that the old mill buildings were not safe for the modern machinery installed in them and new loom shops and spinning shops were built over the old pond. These were opened in 1912.

In 1920 William Playne & Co. amalgamated with the Strachans and Hunt & Winterbothams. More building took place in 1926, 1930, 1934 and 1951. New power systems were adopted in the twentieth century, including a 1904 steam engine, a *c.*1920 water turbine and two large diesel engines from *c.*1940s. Cloth manufacture ceased here in 1990. Much of the site, which was relatively intact in the 1960s, has been cleared and developed for housing.

(*VCH Gloucestershire*, X, Minchinhampton; Royal Exchange Fire Insurance Policy 121656, May 1791; D1265; *200th Anniversary: A History of Playne of Longfords Mills*, pp.24, 40, 52–3; D1347 acc. 1347; D4644 2/101 (acc. 6408); PP 1840, XXIV, pp.396–7; Kelly's Textile Directory 1889; Ray Wilson, 'Electricity Generation at Longfords Mill', *GSIA Journal*, 1992)

George's Mill, ST 879983

This mill takes its name from Robert George who owned and occupied it at the tithe survey. It was reputedly a cloth mill, and has now been demolished.

Avening/Waterley Mill, ST 883980

In 1616 John Selwyn and William Sheppard leased a water mill called Waterley to Giles Parslow. William Sheppard conveyed it to Richard Byrd in 1640/41. (The supposition that Waterley may be Avening Mill is solely based upon the proximity of deeds in the D9196 collection.) It was then a fulling mill which had been erected by William Gyde, the occupier. In 1830 John Baxter was making cloth at Avening Mill. In 1839 it was part of William Playne's property but, at the tithe survey, although William Playne was said to own the mill, it was occupied by himself, John Wise and Peter Playne Smith. By 1886, when it was offered for sale, the mill had become a corn mill. It is a four-storey stone building of early nineteenth-century date.

(D9196/8/10; D2219/9196/10/17; Directories 1830, 1839)

DYEHOUSES AND WORKSHOPS IN THE NAILSWORTH VALLEY AND ITS TRIBUTARIES

Woodchester Dyeworks, SO 844029
See Woodchester Mills (p.235).

Workshop in Woodchester, SO 841028
Situated at the upper end of the former Rack Leaze belonging to Woodchester Mills. It is a three-storey stone building with a loft lighted by a dormer window and has been converted into houses.

Shear shop and other workshops in Woodchester
These were probably near Churches Mill since in 1771 the site comprised two messuages, a shear shop, press shop and stove house which had been bought from Thomas Churches by John Hampton. (D846/III/31)

Dyehouse at Pudhill
Described in 1714 as a dyehouse near Phillpots Mill (Dyehouse Mill), it was held by Richard Cambridge. (Box 58gs)

CARRANT BROOK

Kemerton
There was a fulling mill here in the late thirteenth century, occupied by John the Walker in 1327. (IPM 123-1300, 189)

THE RIVER ISBOURNE

Hinton on the Green
The fulling and corn mill was owned by St Peter's Abbey, Gloucester, in 1266. By 1535 it was a corn mill. (*Historia et Cartularium Monasterii Sancti Petrie Gloucestriae*, III, p.60; *VCH Gloucestershire, VIII*, p.265)

Stanway Mill
Tewkesbury Abbey owned a fulling mill, in addition to three corn mills. It continued as a fulling mill until the later seventeenth century.
(*Taxatio Ecclesiastica Nicholai* IV, p.234; *Valor Ecclesiasticus*, II, p.460; *VCH Gloucestershire, VI*, p.229)

Postlip Mills, SP 008271
There were two 'tuck mills also fulling mills' called Postlip Mills on the manor in 1627. They later became paper mills.
(D1447)

Coates Mill, Winchcombe, SP 023282
Owned by Winchcombe Abbey, the fulling mill (possibly with adjacent dyehouse) was let in 1306. It was granted to Richard Horton for forty years in 1585.

(*Landboc*, I, p.259; PRO E164/39/174-5)

Throp Mill, possibly at Sudeley
This fulling mill was owned by Winchcombe Abbey in the early fourteenth century.
(*Landboc*, II, p.155)

Nohille's Mill, Winchcombe
This fulling mill was granted to Richard Horton in 1585.
(PRO E164/39/174-5)

Walk Mill, Winchcombe
This may have been the monastic fulling mill owned by Winchcombe Abbey in the late twelfth/early thirteenth century. The fulling mill was granted to Richard Horton in 1585.
(*Landboc*, I, p.63; PRO E164/39/174-5)

Blundell's Mill, Winchcombe
John Blundell received damages when the monks diverted some of the water intended for his fulling mill to their own.
(E. Dent, *Annals of Winchcombe*, p.92)

THE RIVER CHURN

Colesbourne Mill
This fulling mill was owned by Llanthony Priory at the time of the Dissolution. A tucker was on the manor in 1608.
(*Valor Ecclesiasticus*, II, p.425; *VCH Gloucestershire*, VII, p.216)

Elkstone Mill(s)
There were two fulling mills here in the sixteenth century, one a double fulling and grist mill (Lower Cockleford), 1570; the other, built in *c*.1582 lower down the stream, was demolished by 1775.
(*VCH Gloucestershire*, VII, p.189)

Hatherop Mill
Two fulling mills were recorded here in the sixteenth century.
(*VCH Gloucestershire*, VII, p.189)

Rendcombe Mill
There was a fulling mill here in the late fourteenth century.
(*VCH Gloucestershire*, VII, p.225)

Cowley Mill
A fulling mill here belonged to Pershore Abbey early in the thirteenth century.
(PRO E315/61/F26)

Perrots Brook Mill, North Cerney

Perrots Brook Mill, 'formerly a grist mill', was a fulling mill in 1799 and was conveyed to Giles Radway in 1809. In 1824 Radway built a new cloth mill south of the previous one and it was still working in 1837.
(D2525; T1/45gs; *VCH Gloucestershire, VII*, p.159)

Stratton Mill, SP 020034

In 1829 'all those two clothing mills', one of which had been converted into a grist mill, was leased to Joseph Cripps. The conversion took place sometime prior to 1780. In 1830 Joseph Cripps is described as of Stratton Mill and in 1839 there were seventy-two hands employed in the mill. The mill had one steam engine of 6hp and four waterwheels whose power totalled 11hp. Cloth manufacture at the mill ceased soon afterwards and it was converted into a store.
(PP 1839, XLII)

St Mary's/New/Cirencester, SP 032014

These mills were built on the instruction of the last Abbot of Cirencester at a cost of 700 marks, using stone from the ruinous Roman walls. In 1533–34 the mills, comprising four stocks and a gig mill, were leased to Richard Fowler of Stonehouse. In 1563, in lay ownership, they were the subject of a court case with the owner of Langley Mill, the next mill downstream. In 1570 Thomas Parry sold the mills comprising four stocks and one gig mill to Roger Dune of Cirencester. A deed of 1585 describes the mills as comprising three fulling mills and a grist mill, the latter having recently replaced a gig mill. The mills were leased to Richard Fowler and Margery his wife, of Stonehouse. In 1622 'twoe corne milles and twoe tucking milles', lately occupied by Robert Fisher and Isabell Willet, were conveyed to Richard Harding and John Davies. In 1688 'all those water corne mills' were occupied by John Selfe. The Selfe family were clothiers and it is probable that during this time cloth was fulled here. Richard Selfe, clothier of New Mills, insured his house, 'Water Corn Mill, Clothing Mill and Mill House under one roof', in 1778, the mills being under separate tenancies.
(D674b/L2, 3; D674b/T8, 21, 27; Sun Fire Insurance Policy 397687, 1778; Directory 1830; PP 1839, XXII, pp.144–7, 151)

Barton Mill or Clerk's Mill, SP 020022

This mill was owned by Cirencester Abbey. In 1570 Thomas Parry conveyed the mill to Margaret and William Partridge; William Partridge had been a tenant at St Mary's Mills previously. In 1606 William Partridge leased 'all that his two stockes or fulling milles and one water grist millne or corne millne' to Robert Seaman of Cirencester, clothier. By 1730 the fulling mills had gone and the mill contained two pairs of stones.
(D674b/T9)

Coberley Mill, SO 965156

St Peter's Abbey, Gloucester, owned a fulling mill here. Tuck Mead was recorded nearby and it may have been fulling in the later seventeenth century. In 1765 it was sold to James Bidmead. In c.1785 'the cloth mill' was valued at 18s for tithe.
(*Historia et Cartularium Monasterii Sancti Petri Gloucestriae*, III, p.60; D269B/F47; D2025)

Siddington Mill

A fulling mill was recorded here in the late thirteenth century.
(IPM, IV, pp.86, 138)

South Cerney Mill

There was a fulling mill here in 1285.
(IPM, IV, pp.86, 138)

Kempsford

Henry the Walker was drowned in a mill pond here before 1287.
(*VCH Gloucestershire, VII*, p.103)

THE RIVER COLN

Chedworth Mill

A fulling mill here is recorded in 1298.
(*VCH Gloucestershire, VII*, pp.78, 125)

Fairford Mill

A fulling mill was recorded here in 1296 and by 1307 it was a corn mill.
(*VCH Gloucestershire, VII*, p.78)

Coln St Aldwyn Mill

This was a fulling mill in 1754 but was a corn mill by 1770.
(*VCH Gloucestershire, VII*, p.51)

Quenington Mill, SP 150042

There was a fulling mill here in 1338. In 1726 'all those tuck mills' were leased to Robert Jenner of Fairford. In 1731 'a mill or tuck mill' was sold. The mill was described as 'a good fulling mill' containing one whole stock and one half stock and gig mill, 'with good conveniences for drying cloth'; a new-built dyehouse was offered to be let in 1733. By 1748 one of the tuck mills had been replaced by a paper mill. The mill has been converted into a house.
(*VCH Gloucestershire, VII*, pp.78, 125; D540/T61; GJ, 3 April 1733)

Hatherop Mill

Probably driven by a small spring entering the Coln near Quenington. This was a former Lacock Abbey-owned fulling mill which had been demolished in 1248 and rebuilt. It was recorded as a fulling mill in 1538 and was a grist mill by 1649.
(*VCH Gloucestershire, VII*, p.92)

Arlington Mill, SP 114068

In 1638 Thomas Carter of Arlington, fuller, was warned against raising the water level in his pond. In 1701 'two water grist mills under one roffe … and all that or those tuck mill or tuck mills … being under the same rooffe with the said grist mills' were leased to Richard Smart. In 1713 the manor of Bibury was divided and the grist and fulling mills became the property of Elizabeth Sackville. It is not known when cloth manufacture ceased here but Thomas Dix Cole, miller, was declared bankrupt in 1811. The buttressed stone mill probably dates from the mid-eighteenth century. The marshy land between the mill and Arlington Row is known as Rack Eye (Island).
(BGAS, LVIII, p.192; LG, 5–9 March 1811; D227; Sherborne Muniments 249; D227; *VCH Glos, VII*, p.35)

Ablington Mill, SO 104076

In 1565 Ralph Eve, fuller, sold his lease in one fulling mill lately built adjacent to his water mill to Edward Goodlake.
(*VCH Gloucestershire*, VII, p.36)

THE RIVER WINDRUSH AND ITS TRIBUTARIES

Barton Mill

The fulling mill was owned by the Knights Templar in 1185. By 1327 it was owned by the Bishop of Winchester.
(Dugdale, *Monasticon Anglicanum*, II; PRO Mins, Accts, 845-9)

Sherborne Mill

The fulling mill was owned by the Abbot of Winchcombe before 1224. It was still a fulling mill in the mid-fourteenth century.
(*Landboc*, II, p.274; *VCH Gloucestershire*, VI, 1965, p.125)

Eyford Mill

There was a woad mill here on the Ey brook in 1656.
(*VCH Gloucestershire*, VI, 1965, p.75)

Lower Mill, Bourton on the Water

Evesham Abbey owned three mills in Bourton in the twelfth century, one of which, Lower Mill, was a fulling mill. It is not clear how long this remained a fulling mill but it is likely to have reverted to corn milling by the early seventeenth century if not before.
(Dugdale, *Monasticon Anglicanum*, II, p.24; *VCH Gloucestershire*, VI, 1965, pp.43–4)

Windrush Mill, SO 193135

A tuck mill existed here in 1618. The site was probably occupied by a double corn and fulling mill.
(*VCH Gloucestershire*, VI, pp.178–9, 182)

Hound/Little Barrington Mill, SP 210131

This was given to Llanthony Priory in c.1226. It was described as a fulling mill in the fourteenth century. It included two corn mills in the early seventeenth century but, since fullers are recorded in Barrington in 1608, it may have continued as a fulling mill, too. In c.1700 it was described as a fulling and corn mill and continued as a fulling mill until the late eighteenth century in the hands of the Minchin family.
(*VCH Gloucestershire*, VI, 1965, p.23)

Donnington Mill, SP 173278

In 1580 this mill was converted from a fulling mill to a corn mill. The site is now occupied by Donnington Brewery which can still be driven by water power.
(PRO E314/15 16 Char 1. H & 1. m. 2; *VCH Gloucestershire*, VI, 1965, p.156)

STOUR

Sutton under Brailes (then in Gloucestershire)
A fulling mill here was owned by Westminster Abbey before 1362.
(D1099/T(14))

FULLING MILLS WEST OF THE SEVERN

Guns Mill, Abenhall, SO 670153
In 1701–02 Guns Mill, comprising two grist mills and a fulling mill, were mortgaged to Thomas Foley for £100. By 1743 paper was being made at the mill.
(1(1)gs)

Lydbrook Mill
In 1583 Christopher Monmouth conveyed in trust one fulling mill and water mill with lands to Thomas Smalman and William Hill.
(D33/138/267)

Two corn mills were built in 1666 by Richard Vaughan, one of which was converted to fulling prior to 1717.
(*VCH Gloucestershire, V*, 1996, p.345)

Aylburton Mill
This fulling mill was owned by Llanthony Priory in the thirteenth century and was still a fulling mill at the Dissolution. It was recorded as a tucker's mill in 1632.
(*Taxatio Ecclesiastica Nicholai* IV, p.172; P209/IN 1/1; Q/SRLgs)

Newent (near Waldens Court)
In 1640 James Morse held the Boothall, a tuck mill, garden and fishpond by copyhold. The mill was described as a tuck mill in 1733.
(D421/M81; D4647 7/1)

Flaxley
A grange adjacent to Flaxley Abbey had a fulling mill in 1291.
(*Taxatio Ecclesiastica Nicholai* IV, p.172)

Mork Mill, St Briavels
A fulling mill and dyehouse were erected shortly before 1688 when they were sold to Thomas Dale who let them to Thomas Hunt, dyer, until 1709 or later.
(*VCH Gloucestershire, St Briavels*)

English Bicknor
A fulling mill was recorded here, jointly with a corn mill, in 1301.
(IPM, IV, p.231)

A fulling and gig mill in a detached part of Ruardean within English Bicknor was held by John More (d. 1596), a Cirencester woollen draper.
(*VCH Gloucestershire, V*, 1996, p.345)

Rodmore Mill, St Briavels

This was recorded as a fulling mill in 1431 and 1478. It had become a corn mill by 1628.
(*VCH Gloucestershire, V, St Briavels*; GRO D1440)

INDEX

Apprentices 27, 111
Army 112
Bankruptcy/business failure 58, 104–6, 124
Blackwell Hall 17, 22, 27–8, 44–7, 49, 57, 59, 124
Buildings: Clothiers' houses 40, 42–3, 47–50, 60, 79, 104, 122, 144, 161, 163; Double Mill 11, 24–6, 37, 51, 79; Dyehouse 37–40, 51, 63, 66, 76, 84, 130–1, 141, 144; Factory 60, 63–4, 68, 74–5, 78, 80–1, 91, 96, 124–5, 134; Fireproof mills 78; Loom shop 111; Mill 60, 66, 114; Press shop 39, 128; Saw mill 158; Shear shop 42, 63, 68; Stove 64, 81, 84, 120, 128, 145; Weaving shed 141–2, 144
Canals: Stroudwater Navigation 71, 90, 95, 99, 162; Thames & Severn Canal 95, 99, 162
Civil War 57
Clothier 21, 42–3, 55, 58–60, 65, 67, 69, 104, 109, 111, 114
Cloth mark 27, 56
Cloth types: Beaver 45; Broadcloth 27, 51, 54, 56, 70, 72, 104–5, 107–9, 114, 124, 129–31, 133; Cassimere 67, 104–6, 108, 110, 129; Doeskin 129; Drab 44; Duffel 44; Elastic 133; Fancy 125; Felt 108; Fibro 157; Fireproof 157; India whip 144; Medina 45; Medley 36, 51; Melton 126, 143; Narrows 45, 70; New Draperies 36; Non-apparel 142–4; Nomax 157; Say 36; Scarlet 71–2, 130; Serge 36, 141; Shalloon 45; Silk 15; Spanish 36, 51, 104; Stammel 36; Stripes 103, 105–6, 108, 114, 122; Stroudwaters 1, 36, 51; Tweed 133, 141; Vicuna 134; Waterproof 140; Worsted 103–4, 140, 144, 150
Cockayne Project 54
Corn/grist mill 37–8, 40, 63, 66, 90–2, 94, 158, 161
Conservation 160, 163
Deceits 54
Depression (trade) 16, 27–8, 44, 56–60, 104, 109–10, 114, 122, 124, 130, 133, 155
Design 134, 151
Disputes 24–5, 51
Distress 12, 55–9, 103, 105–6, 134
Decline 135
Dudbridge Ironworks 158
Dyer 9–11, 43, 51, 54, 63, 69, 71, 101
Dyeworks: Bowbridge 44, 71, 106, 142, 145, 147, 155; Hawkers 26, 72, 131
East India Co. 44, 59, 103–6, 114, 122, 124
Ebley Ironworks 158
Electric lighting 141, 144
Emigration 105, 135
Engineer 69, 96
Exhibition 120, 125–6, 131–2
Export (cloth) 15–16, 37, 57
Factory clothing 127–8
Failure 63
Falsifying 37
Fullers 9–11, 14, 21
Guilds 9–10, 15, 20, 35
Innovation 74, 81, 149–50, 159

Justice of the Peace 50
Labour: 125; Children 51, 103, 107–8, 122–3; Men 103, 107, 127; Weavers 9, 12, 14, 58, 60, 108, 110, 112, 134; Women 51, 65–6, 103, 107–8, 127, 134; Wool broggers 21, 55
Limited liability 143–4, 146, 148, 152
Luddism 112, 109
Machinery & Power: 60, 125–6, 150; Billy 103, 109; Carding 152, 157; Cloth press 42, 129; Condenser 128; Cutters 100; Fly shuttle 67, 107, 110; Fulling Mill 11, 17, 21, 38, 39–40, 75, 78, 80, 90–4; Fulling stocks 63, 66, 68–9, 73, 78; Gas lighting 101, 141, 144; Gas engine 142, 149, 158; Gig mill; 22, 27, 37, 39, 51, 56, 60, 63, 65–6, 69, 73, 100, 107, 109–10, 128; Horse power 96, 100; Jack 67; Loom (hand) 63, 67–8, 101; Loom (power) 67–8, 81, 114, 128, 134, 138, 141, 144, 152; Milling 152; Racks 45, 69, 81, 100; Shears 42–3; Shearing frame 70, 101, 108–9; Spinning jenny 107, 109; Spinning mule 66, 81, 101, 114, 128, 138, 144, 152, 157; Steam engine 81, 91, 95–7, 99, 131, 140–2, 150–1, 162; Steam power 81, 96, 121–2, 144; Tentering 138; Turbine 142; Watercourses 15, 37–9, 51, 92; Water power 80, 83, 86, 108, 120–1, 144, 149; Water heel 80, 83, 91, 94–5, 100, 128, 142, 161–2; Willy 65; Wool drying 120
Magistrate 59
Management 74, 147–8
Manufacturing integration: Horizontal 149, 155; Vertical 149, 193
Markets/fairs 15–16, 21–2, 43, 54–5, 58–9, 64, 72, 104, 106, 122, 124, 133, 142, 148, 149, 151–2
Materials: Artificial fibres 143, 155, 157; Coal 99, 149; Dyestuffs 9, 11, 15, 55, 73, 127, 140, 147; Fibres: Alpaca 140, 150; Llama 140; Mohair 140, 150; Cashmere 134; Vicuna 140; Wool 8, 11, 15, 28, 45; Fuller's earth 14, 24, 26, 68, 128, 144; Mordants 9, 45; Mungo 141; Oil 9, 65; Seg 64, 68, 128; Shoddy 125–6, 141, 158; Soap 68; Sulphur 71–2; Teazles 45, 69, 128; Yarn 65
Merchant Adventurers 16, 27, 54–5, 57
Merchants of the Staple 16, 21, 28, 54
Merger 152
Mills/dyehouses: Abbey/New (Kingswood) 24, 76, 141, 175; Ablington 31; Angeston workshops 194; Arlington 39; Arundel 26, 39, 205, 219; Ashmeads 218; Avening/Waterley 245; Avening, Tortworth 174; Aylburton 13, 24, 251; Badbrook 39, 229; Bakers (Chalford) 26, 92; Bank Ho. 228; Barrack 199; Barrington 39; Barton/Clerk's 12–13, 23–4, 37, 248; Barton (Windrush) 250; Bayliss's Upper/Lodge 93, 225; Beard's/Leonard Stanley/Merretts 139, 141, 197; Bitton 13, 173; Bliss/Chalford 79, 98, 100, 112, 158–9, 216; Blue Row, Dudbridge 219; Blundell's 247; Bonds/New/Eycott's/Kinn's 26, 104, 139, 141, 196; Bourne/Grime's 162, 211; Bourton on the Water 13; Bowbridge 206, 219; Bridgend/Nashes 197; Broadbridge 93; Broadwell dyehouse 191; Brimscombe Lr &Up 26, 37, 39, 97, 99, 138–40, 148, 159, 210; Brimscombe Port/Field's/Hopton's 211; Brookhouse/Savory's 224; Brookside 214; Brownshill workshop 215; Bowbridge 26, 64, 78, 84, 94; Britannia 114, 181; Cam/Corrietts 40, 76, 86, 93, 97, 99, 124, 130–1, 138–41, 143–5, 147, 155–6, 187; Cambridge/Le

Dunstome 25, 186; Cambridge R 246; Cap 224; Capel's/Orpin's/Arundel Lower 26, 125, 149, 158, 205; Chalford unnamed 41, 216; Chalford workshop 218; Charfield/New 24, 76, 93, 96, 106, 122, 133, 142, 146, 149, 159, 161, 174; Chedworth 13, 24, 249; Cherynges/Norris's 26, 201; Church 162; Churchend 25, 74, 81, 100, 104, 124, 140–1, 195; Churches family 133–4, 139, 236, 246; Churn/Howards 189; Clayfield 163, 214; Cloud/Potter's Pond 84, 180; Coaley 186; Coaley unnamed 186; Coates 12, 246; Coberley 39, 248; Colesbourne 24, 31, 247; Coln St Aldwyn 39, 249; Coombe (Wotton) 25; Coombe dyehouse 183; Cowley 13, 247; Cricketty 40, 233; Crowlbrook 90, 124, 130, 185; Culvert/Packers 223; Cuttles 204; Damery 104, 174; Damsells/Cox's Lr 79, 93, 225; Dark/Leversage's/Winn's 26, 37, 212; Dauncey's/Blagden's/Top/Stout's Hill 193; Day's/Nailsworth 93, 240; Doge's 177; Donnington 24, 251; Doynton 94, 173; Draycot 38, 187; Dudbridge 26, 37, 39, 92, 99, 139, 140; Dudbridge Lower/Chance's 131, 202; Dudley 179; Dunkirk/New/Cooper's 76, 79–80, 93, 95, 114, 120, 134, 140, 158, 161–2, 238; Dursley 13, 37, 40, 79, 191; Dyehouse/Phillpot's 130, 237; Ebley/Deerhurst's/Maldon's 20, 26, 39, 78, 80–1, 83, 91–3, 122, 130–2, 139, 157, 161–2, 200; Eddel's/King's 226; Egypt 40, 47, 80, 93, 95, 162, 239; Elkstone 24, 247; English Bicknor 14, 252; Eyford 250; Eyles 40, 79, 191; Fairford 13, 249; Farrar's Dyehouse 219; Flaxley 13, 23, 251; Flock/Gardner's 227; Freames 125, 238; Frogmarsh 64, 71, 158–9, 236; Frogmarsh unnamed 237; Framilode 25, 194; Frigg's 234; Fromebridge 24, 195; Fromehall 26, 37, 80–1, 104, 125–7, 130–1, 139, 144–5, 148, 202; Gardiner 218; Gazard's/The Granaries 184; George's 245; Gigg 79, 162, 241; Griffin 90, 125, 207; Grindstone 177; Grove/Hermitage 221; Guns 251; Gussage 232; Hack 179; Hallidays/Stoneford/Smart's/Bidmeads 26, 78, 80, 90, 158, 217; Halmer 25, 91, 186; Ham 36, 57, 78, 93, 106, 112, 125, 130–1, 207; Hartley Bridge/Boxwell 242; Hatherop 13, 24, 31, 247, 249; Hawker's Dyeworks 202, 233; Hawkesbury 13, 176; Hazel 230; Hill/Hell 24, 37–8, 179; Hillesley/Byrettes 177; Hinton on the Green 13, 246; Hoggs 200; Holcombe & Spring/Little 81, 130, 243; Holywell dyehouse 180; Hope/Gough 78, 93, 130, 158, 209; Hore's/Ebworth/Back 228; Horton R 246–7; Hound/Little Barrington 63, 250; Howards Up & Lr 66, 190; Hulkhay/Lower 242; Huntingford 174; Iles 214; Inchbrook/Playne's Little 37, 40, 158–9, 238; Iron 162, 244; Ithell's/Bytheford 24, 38, 76, 174; Jackson's 193; Jeens/Dauncey's 97, 194; Kemerton 13, 246; Kempsford 249; King's 90, 158; King's/Lower/Paulin's 223; King's Stanley, unnamed 200; Kingswood presshop 177; Lake 76, 79, 84, 93; Langford 76, 141, 161, 175; Lenall's 222; Lightpill 67, 138, 140, 159, 234; Little/Upper/New (Painswick) 226; Little/Mill behind the Church 229; Little Barrington 13, 24; Little's 227; Loccumb Well 179; Lock /Johnson's 76, 83, 240; Lodgemore/Merretts Up & Lr Latemore 13, 26, 38, 39, 78, 92, 125–7, 129–31, 134, 139, 144–5, 147–8, 155, 157, 203; Longfords & Lake 15, 26, 39, 68, 75–6, 79–81, 84, 91–3, 106, 114, 124, 135, 138–9, 143–4, 148–9, 156, 162, 244; Lot/Lower/Terrett's 242; Loveday's/Bayliss's Lr/New 225; Lower (Bourton) 250; Lower Doreys/New 125, 226; Lower (Dursley) 70, 189; Lower Horsley/Spring 241; Lydbrook 251; Marsh/Adey's/Bottom 192; Mason's/Cook's/Reed's/Painswick 224; Meadow/Eastington 74, 139, 141, 195; Merretts 100, 130, 141, 158, 237; Middle 187; Middle, Shortwood 233; Millbottom/Great/Ruskin 162, 241; Millend 13, 25, 74, 104, 130, 141, 196; Millbottom (Horsley) 26, 96; Minchinhampton 13–14; Miserden 219; Monk 24, 38, 47, 59–60, 80, 93, 124, 130–1, 178; Moore's 180; Mork 251; Mugmore 216; Munday's/Picks 183; Nailsworth/Birt's/Chamberlain's 76, 130, 133, 159, 243; Neal's/Pounds Ground 180; Nether Coberley 13; New, Alderley 76, 177; New, Chalford 216; New, Stroud 26, 36, 50, 76, 131, 146, 159, 229; New, Dursley 37, 66, 190; New/Sury/Shewry Wotton 64, 76, 80, 83, 92–3, 100, 105, 138, 141, 159, 161, 175; New, Woodchester 236; Newcombe's/Sands' 205; Newent 251; Nibley dyehouse 185; Nibley presshops, shearshops 185; Nind 24, 80, 84, 91, 104, 138, 142, 176; Nind dyehouse 176; Nodes/Upper/Highwood/Shortwood 243; Nohille's 247; North Cerney 31; Nowell's 179; Oil, Ebley 106, 200; Oil, Berkeley 159; Old Town Factory 81, 105, 181; Oliver's/Cox's Up. 226; Ozlebrook 220; Paganhill/Little/Bourne's 220; Painswick workshops & dyehouses 228; Pallings 76; Park 93, 130; Park 179; Penley's 178, 192; Perrot's Brook 248; Piggas/Peghouse/Woodlands 36, 67, 125, 139, 229; Pitchcombe/Jenner's/Wades 222; Pitchcombe Up/Little(s) 222; Pitts 125, 238; Port 162; Postlip 246; Pounds Ground dyehouse 183; Price's/Old 242; Puck 218; Puckshole 220; Purnells 37, 184; Quenington 13, 31, 39–40, 249; Randall's/Wood 217; Rendcomb 13, 247; Rivers 190; Rock 221; Rockstowes 97, 192; Rodmore 252; Rooksmoor 130, 235; Rookwoods 219; Ruscombe 220; Ryeford/Phillips 199; St Mary's/Clark's (Chalford) 26, 39, 47, 78, 80, 82–3, 95, 97, 158, 162, 213; St Mary's, Cirencester 24, 39, 248; Salmons/Bliss's 47, 221; Samuel Wood's 228; Sevilles/Innell's 47, 158, 218; Sevillowes 97; Sheepscombe/Wight's 227; Sheppard's/Uley 193; Sherborne 13, 23, 250; Siddington 13, 249; Skinners 45; Slad valley presshop 232; Small's/Ayer's/Seagrim's 94, 222; Snitend 184; Snow's/Hayward's/Rowbridge 232; South Cerney 249; Southfield 49, 51, 83, 133, 138, 235; Stanley/Giles 25–6, 37, 39, 65, 78, 80–1, 83–4, 91, 97, 99–100, 122, 125–6, 130–1, 139, 157, 161, 163, 198; Stafford 78, 92, 96, 135, 146, 207; Stanway 13, 24, 39, 246; Steanbridge/Jenny 41, 76, 125, 231; Steanbridge Lower/Slad/Weaving 76, 93, 231; Steep/Church 181 76; Stone 173; Stoneford *see* Hallidays; Stonehouse Up & Lr 20, 24, 78, 80, 139–42, 146, 152, 197, 198; South Cerney 13, 24; Spring's 223; Strange's/Venn's 76, 130, 180; Stratford 159, 220; Stratton 83, 248; Stroudwater Dyeworks 219; Sury 76; Sutton under Brayles 13, 251; Sym Lane 182; Tayloe's/Woolings/Ballinger's/Great/Belvedere 26, 112, 161, 214; Throp 13, 247; Thrupp/Sewell's/Huckville/Wathen's/Phoenix IW 69, 158, 209; Toadsmoor 39, 122, 158, 232; Toadsmoor Up & Lr/Selwyn's 232; Tocknell's 226; Toghill's/Sevillowes with Upper, Little, Lower, Mr Hunts 69, 215; Tortworth 174; Townsend 44, 124, 189; Twissell's/Frampton's/Blankett's/Bakers 218; Uley scouring house 194; Upper Cam 70, 104, 188; Upper Dorey's 93, 125, 227; Upper Kilcot 93; Upper Pitchcombe 47, 76; Valley/Tyler's/Moreton's 48, 218; Vatch/New/Hermitage/Up.Vatch 39, 76, 80, 90, 93–4, 97, 112, 131, 133, 230; Vizards 189; Walk (Kingswood) 24; Woodchester 235; Woodchester unnamed 235; Woodchester workshops, dyehouses 246; Wortley/Broadbridge 178; Wade's 230; Walk, Kingswood 90, 175; Walk, Winchcombe 247; Wallbridge 26, 97, 139, 146, 149, 152; Wallbridge Dyeworks 219; Wallington's Dyehouse 191; Warehouse 158, 214; Warner's Dyeworks 232; Washbrook/Gyde's/Merretts 226; Washpool/Baxter's 242; Waterloo 76, 105, 181; Weyhouse 69, 205; Wheatenhurst 13, 195; Whitminster 195; Wick 183; Wick dyehouse 185; Wickwar 24, 176; Wimberley 100, 158; Winchcombe 12, 24, 39; Window's Dyeworks 232; Windrush 39, 250; Wiselands/Millgrove/Rowbridge Up. 232; Woodchester 26, 37, 39, 133; Woodlands 140; Wyatt's 93; Wortley dyehouse 179, 182; Wotton: dyehouse/workshops/shear grinding/scouring house 181–3; Wyatt's/New 231; Zachariah Powell's 225; Zachariah Powell's Factory 228

Millwright 74, 95–7, 100–1

Monasteries, dissolution, 23

Mudding 96

Names: Adey, D. 43, 178, 180; Adey, G. 96, 192, 201; Adey family 34, 43, 45, 182, 184, 188, 221; Adey, Wallington & Co. 190–1; Aldridge, R. 242; Allen, T. 43; Apperly, J. & D. 202, 216–17; Apperly, Curtis & Co. 100, 131, 142–3, 146–7, 201; Arkwright, R. 96, 148; Arundel family 35, 205; Ashmead, W. 243; Austin family 105, 174–5, 177, 181–2; Austin, H. 76, 84, 96–7, 105, 175, 177–8, 180; Austin, G. & E. 105, 182

INDEX 255

Baghot, P. 122, 200; Ballinger, C. 105, 214, 216; Ballinger family 214; Bamford, R. 209; Bancroft 72; Barnard & Co. 97, 234, 241, 243; Barnfield family 229, 243; Barton, J. 43; Basset, R.S. 142, 145–7; Baxter, J. 245; Bayliss, D. 50; Bayliss, E. 226; Bayliss, J. 226; Bayliss, T. 50, 229–30; Bayliss, W. 105, 225–6; Beard, N. 43, 197, 202; Beaufort, Duke of 114; Bence, R. 26; Bennett, E. 45; Bennett, L. 26; Bennett, N. 218; Bennett, T. 45, 200; Bennett, W. 200; Berkeley, Lord 38, 105; Berthollet, C. 70; Biddle, B. 40, 187; Bidmead, J.D. 217, 248; Bigge family 204; Bird & Co. 236; Bishop, W. 219; Blackwell, S. 238–9; Blagden family 42, 193; Bliss family 218; Bliss, J. 216; Blysse, T. 221; Bodley, G.F. 83, 130; Bond, T. 174; Bond, J. 196; Boultbee & Lowe 241; Boulton & Watt 81, 95–7, 100; Bourn, D. 65; Bowser, J. 174; Brindley, J. 100; Brooks, R. 68; Brown & Tucker 234; Brown, J. 241; Brown family 243; Brunel, I.K. 83; Busby, C.A. 83; Butler family 49, 233; Butler, J. 219; Butler, N.I. 189, 224, 226; Butt, S. 43, 229; Butt, T. 214
Cam, J.T. 188; Cambridge, J. 212; Cambridge, R. & N. 213, 238; Capel family 35, 205, 207, 221; Carmichael, G. 156; Carter, T. 249; Carruthers, T. 207; Chalke, J. & Co. 236; Chance, D. 201; Chambers, W.C. 206; Chapman, T. 54; Chew, J. 215; Chinn 100, 198; Churches family 236; Clayfield, F. 241; Clifford, M. 43; Clissold family 39, 101, 122, 146, 230; Clissold, S. 200–1, 220; Clutterbuck, E. 213; Clutterbuck, J. 205; Clutterbuck, Jas 42, 47, 49, 198; Clutterbuck, Jn 234; Clutterbuck, L. 207; Clutterbuck, R. 22, 25, 42, 49; Clutterbuck, M. 196; Clutterbuck, N. 43, 49; Clutterbuck, Rd 196; Clutterbuck, Rt 25; Clutterbuck, S. 47, 83–4, 97, 213; Clutterbuck, T. 42, 47; Clutterbuck, W. 197, 220; Clutterbuck family 25, 33, 43, 47, 60, 91, 198; Cockayne, W. 56; Cockle, R. 237; Cole, W. 229; Coles family 236; Compton, R. 216; Cook family 222; Cooke, N. 207; Cooke, R.H. & J. 203; Cooper, J. 84, 238; Cooper, R. 105; Cooper family 49, 236; Cother, J. 221; Counsell & Millman 176; Cox, B. 158; Cox, D. 105, 215–16; Cox family 225–6; Cripps, J. 97, 248; Crompton, S. 66
Dallaway, Jas 210; Dallaway, Jn 210; Dallaway, W. 58, 210; Damsell, W. 225; Dancer, J. 175; Dangerfield, N. 43; Dangerfield, S. 201; Daniell, J.C. 70; Daniels family 95; Darke, J. 236–7; Dartnell, H. 189; Dauncey, G. 194; Dauncey, Ja 186; Dauncey, Jn 45; Dauncey, P. 180; Davies, R.S. & Sons 142, 146; Davies, W. 70, 196–7; Davis, Beard & Co. 221; Davis, C. & F. 240; Davis, G. 36, 229; Davis family 35, 49, 214–15, 242; Davyes, G. 220; Day, Smith & Alder 96; Day family 240; Deane family 236; Deverell, J. 242; Driver, J. 224; Driver, N. 229–30; Du Pont 157; Dudbridge, H. 43; Dudbridge family 235; Dune, R. 248
Eddles, T. 66, 226; Ellis, J. 221; Essington family 186, 203, 206; Estcourt family 234; Evans, A. 232–3; Evans, P.C. 100, 138, 140, 148, 199, 210–11; Evans & Bishop 210; Evans, S. 199; Exell, T. 47, 112–13; Eycott, H. 196; Eyles, J. 36, 191
Fagin, F. 113; Falconer, K. 78, 83; Farrar, D. 212, 215, 219; Farrar, W. 212; Ferrabee, Jn 45, 69, 99, 209–10; Ferrabee, Ja 99, 131, 204, 211; Fisher, P.H. 73; Fletcher, E. 22; Fletcher family 47, 206, 221, 223; Flint, A.M. 237; Flower, T. 173; Fluck family 231, 229–31; Ford, L. 76, 181; Fowler, D. 49, 230, 233; Fowler, Rd 22, 24, 248; Fowler, Rt 178; Fowler, S. 197; Fowler, W. 196–7; Fowler family 210, 229, 241; Foxwell, P. 225, 237; Francis & Flint 236, 243; Freame, T. 211; Fream, J. 238; Freeman, C. 232; Freeman, W. 221; Fryer, Webb & Fryer 47; Fynemore, J. 25
Gardiner, W. 105; Gardner family, 22, 101, 220–1, 224–5, 227; Gardner & Herbert 70; Gainer & Bishop 197; Gayner family 34; Gainey, J. 43, 206; Gainey family 45; Gainsborough, T. 49; George, R. 245; Glover, W. 232; Gordon, J. 205; Gough family 209; Gower, T. 2; Green, A. 185; Griffin, J. & T. 207; Grime, W. 211, 233; Gyde, C. 45, 223, 228; Gyde, R. 211; Gyde, T. 50, 193, 227; Gyde, Th. 50, 208; Gyde, W. 245; Gyde, Bishop & Co. 155
Haden Bros 95, 97, 100; Haigh, J. 235; Hale, J.B. 84, 176; Halfpenny, W. 50; Halliday, F. 213; Halliday, H. 201; Halliday, M. 202; Halliday, W. 202; Halliday family 203, 213; Hancox family 232; Hanson & Mills 46–7; Hargreaves, S. 66; Harman & Adey 220; Harmer, J. 70; Harmer, T. 218; Harris, G.D. 198; Harris, M. 158; Harris family 49, 189, 242; Harris Stephens & Co. 97, 100; Harrod, T. 173; Hastings, R. 231; Hawker, A. 43; Hawker, G. 203; Hawker, J. 73, 198, 233; Hawker, Rd 45, 73, 229, 233; Hawker, Rt 43; Hawker, S. 204; Hawker & Richards 71, 72, 219; Haycock family 237; Haycraft, A. 230; Haywood, W. 232; Heaven, S. 66; Helme, W. 229; Heskins, J. 243; Heskins, Barnard & Bliss 243; Hewes, T. 100; Hicks, H. 74, 96–7, 104, 187, 193, 196; Hicks, J.P. 74; Hicks family 95, 174, 188, 195; Hill, T. 211; Hoare family 228, 231; Hodges, N. 43; Holbrow, J. 45, 174, 193, 198, 206, 229; Holbrow family 49; Halliday family 60, 217; Hood, W. 46; Hooper, C. 94, 124, 130, 133, 141–2, 195–7; Hooper family 78, 148; Hornblower, G.S. 96; Howard family 189–90, 204; Howard & Powell 139, 146, 204; Hint, T. 144; Huckvale family 205; Hughes family 230; Hunt, T. & J. 240; Hunt, W. 203, 240; Hunt & Barnard 130; Hunt & Winterbotham 141, 143, 146, 155, 188, 204, 245; Huntley, Bridges & Neal 176
Iles, Jn 105, 213, 214; Iles, Jos 214; Innell, C. 218; Innell, J. 105; Innell family 47; Ithell family 34
Jackson, E. 76, 105, 192–3; Jeens family 193–4; Jenner, R. 249; Jessop, W. 100; Jones, A. 232; Jones, N. & J. 94, 113, 214, 216–18, 223; Jones, T. 212, 214
King, J.H. 100, 142; King, T. 111; Knight family 224
Larton, M. 177; Leonard, Woolbright & Co. 198; Leversage, P. 212; Lewis, Jas 200; Lewis, Jn 69–70, 209; Lewis, W. 69–70, 97, 106, 211; Lewis & Dutton 175, 180–1; Libby, J. 131, 146, 229; Lippiatt, L. 158; Lloyd, D. & N. 67, 105, 188, 194; Lock, P. 240; Long, A.L. 148; Long, W. 96, 179; Long, S. 142, 146, 174–5; Loveday family 225
Mackie, J. 147–8; Maclean, D. 83, 198; Maclean, Stephens & Co. 198; Marling Industrial Felts 157; Marling, J.F. 122, 201, 234; Marling, N.S. 70, 90, 97, 125, 131, 148, 199, 203, 207–8, 211, 227, 230–1, 238; Marling, S.S. 83, 106, 121, 122, 130, 199, 201, 208, 210, 234; Marling, T. 106, 122, 124, 201, 208; Marling, W. 106, 130, 134, 152, 199, 201, 208, 237; Marling family 91, 131; Marling & Evans 139, 143, 155, 157–8, 210–11; Marklove, D. 96, 173; Maurice, Prince 57, 208; Mercer, T. 92, 175; Merrett & Chambers 206; Merryett, T. 220; Mason family 224; Miles, H. 199; Miles, T. 70; Milliken 157; Millman & Hunt 139; Millman & Long 106; Millman & Overbury 176; Minchin family 250; Misenor, T. 44; Nash, G. 42, 197; Nash & Dymock 72; Neal family 34, 180–1; Neale, C. 38; Nelme, Christopher 38; Nelme, H. 38; Newcombe, T. 205; Norris family 201; Norton family 238; Nowell family 179
Oldland, G. 70; Organ family 184; Osborne family 34, 197; Osborne, J. 43, 59–60, 178; Osborne, R. 43, 178; Overbury, J. 178; Owen, A. 84; Owen, R.S. 233; Ozanne, N. 158
Packer, D. 45, 59; Packer family 101, 224–5; Palling, T. 47; Palling, W. 58, 221, 224; Palling family 22, 223; Palmer, I. 45; Papps, J. 223; Parker, E. 227; Parry, F. 38; Parry, T. 248; Parry-Williams, T. 158; Parslow, G. 245; Partridge, Jn 45, 84, 106, 176, 206–7, 229; Partridge, Jos 84, 206; Partridge, N. 84, 206, 213, 231; Partridge, T. 206; Partridge, W. 47, 64, 66–9, 72, 229, 248; Partridge family 44; Partridge & Hancock 179; Pattison, R. 216; Palling family 221; Paul, Ob. 51, Paul, On. 37, 49, 235–6; Paul family 49, 51, 235; Peach family 49, 51, 60, 198, 235; Peach, J. 196, 211; Peach, S. 47, 211, 213–14; Pegler, J. 227, 231; Perrin, J. 244; Perrin, L. 175, 184; Perry, T. 174–5; Phelps, E. 45; Phelps, J. 44, 46, 220; Phelps, J. de F. 44, 47, 220; Phelps, W. 44, 46–7, 192; Phelps & Co. 44, 46, 189; Phillimore, J. 187, 192; Phillimore, S. 187; Phillips family 199, 203; Pinfold family 39, 221, 244; Playne, M. 39, 75, 244; Playne, P. 39, 68, 75, 84, 91, 100, 105, 122, 239, 242; Playne, T. 39, 244; Playne, W. 75, 84, 105–6, 112, 114, 124, 144, 148, 203, 238, 239, 244–5; Playne & Smith 114, 239, 242; Playne family 51, 68, 75, 91, 96, 100, 122, 143, 239; Plomer, S. 185; Plomer, T. 69, 111; Poole, R 178; Potter, J. 84; Potter, W. 84; Powell, Z. 225, 228; Power, S. 205; Poyntz, M. 177; Price, J. 101; Price, S. 70, 96; Prince of Wales 146; Prinn, J. 207; Prinn, W. 212; Purnell family 34, 185, 188; Purnell, C. 38, 178; Purnell, T. 47; Purnell, W. 187, 191

Radway, G. 76, 248; Read, W. 96; Remington family 100, 241; Renishaw 159; Rennie, J. 83; Restall, W. 233; Ricards, S. 45; Rice, M. 223; Ridler family 212–13, 216; Rimmington family 200; Ritchie & Co. 138, 208; Roach, R. 178; Roberts family 42, 241; Roberts Jowlings & Co. 13, 234, 244; Rogers, R. 233; Rupert, Prince 57, 208; Rudder, S. 59, 73
Sandford, J. 25, 198; Sandford, K. 25; Sandford, W. 196, 198; Sandys, J. 205; Sanford, I. 70; Savory family 101, 224; Seaman, R. 248; Selwyn, J. 187, 224; Selwyn, W. 200; Saville, W. 218; Selfe, R. 248; Sewell, T. 22; Sewell family 35, 209, 233; Sheppard, E. 68, 76, 93, 97, 101, 104–6, 111–12, 114, 122, 186, 193, 196; Shipway, E. 207; Shurmur, T. 236; Small, Jas 201; Small family 243; Smart family 36, 217; Smeaton, J. 95, 100; Smith, A.C. 131, 234; Smith, R.P. & G.A. 97, 204, 219; Smith & Gyde 205; Smyth, J. 31, 37, 57, 175, 182; Smyth family 185; Snow, J. 232; Sparrow, S. & S. 66; Spring, D. 223, 228; Stafford, R. 207; Stanbridge, C. 232; Stanley, J. 221–2; Stanton, A. 146; Stanton, C. 96, 135, 205–6; Stanton, W. 207; Stather, S. 234; Stephens, E. 22; Stephens, H. 104; Stephens, N. 242; Stephens, R. 25, 196; Strachan, J.G. 130–1, 139, 145, 148, 198, 203–4, 245; Strachan & Co. 152, 156; Strange family 84, 97; Strange, W. 180; Stratford, J. 220; Stumpe, W. 22; Sykes, T. 236
Tanner, T. 175; Tattersall, J. 180; Tayloe, T. 22, 212; Tayloe family 207, 214, 233; Telford, T. 100; Thomas, J. 241; Tindall family 186–7; Tippetts family 33, 190; Tippetts, T. 45, 47, 66; Tod, J.N. 145; Toghill, W. 215, 233; Trotman, J.L. 217; Townsend family 231; Trottman family 229, 233; Tubbs, Lewis 141; Tugwell, E. 45; Turner, Jn 200; Turner, T. 66; Twissell, W. 218; Tyler, W. 50
Underwood family 101, 187
Vizard family 185, 190–1
WSP Textiles Ltd 157; Walker, W. 140; Wallington, E. 182, 189; Wallington, J. 44, 68, 180, 188; Warner, T. 200–1, 220; Warner, W. 54; Wathen, N. 105, 110, 209; Wathen, P. 70, 110, 198, 235; Wathen, S. 68, 105, 209; Wathen family 51, 60, 91, 198, 208, 221, 235; Watt, J. 96; Watts family 35, 67; Watts, I. 45; Watts, N. 204; Watts, P. 72, 228; Watts, R. 36, 204, 219, 229; Watts, S. 204; Webb, B. 36, 55; Webb, D. 19, 217, 237; Webb, J. 36, 211–12, 215–16, 234, 244; Webb, S. 36, 57, 208; Webb, Ti. 93, 96; Webb Th. 204, 208, 241; Webb, W 207, 210, 241; Webb family 35, 111, 208, 231, 239, 241, 243; Weeks & Woolright 237; Wight, J. 46, 227–8; Wilson, E. 43; Winchcombe, R. 213; Winchcombe, J. 71, 208; Window, B. 232; Winn, W. 212; Winterbotham, B. 144–5, 148; Winterbotham, A. 144; Winterbotham, Strachan & Playne 144–5, 155–6; Wise, W.P. 196; Wise, J. & E. 235, 240–1, 244; Witchell, H. 42; Whitehead, W. 203; Wight, J. 224, 227; Wight, R. 224; Wise, J. 235, 245; Wither, A. 56; Wood, W. 105, 224, 228; Wood, T. 226–7; Woodwark & Bird 236; Woodwark & Hunt 237; Woolf 97; Wolfe, Maj Gen J. 58; Workman, T. 203; Workman, N. 206; Wyatt, B.D. 83; Wyatt, G. 231; Wyatt, H. 105, 230; Wyatt, P. 112; Wyatt, S. 83
Yeate, E. 236; Yeats, S. 43, 178, 237–8; Yeats family 124, 237–8
Organisation 103
Pin manufacture 158
Places: Abenhall 35; Ablington 250; Alderley 25, 80, 93, 123; Arlington 249; Avening 123; Berkeley 15, 20, 25, 35, 96, 123; Bisley 15, 17, 21, 105, 111, 114, 135; Bitton 37; Bristol 9–12, 16, 20–2, 57; Cam (River) 21, 25, 28, 31, 72, 86; Cam 33, 87, 123; Cambridge 25; Chalford 15, 25, 38, 40, 43, 59, 69, 73, 90, 92, 103, 105–6, 111, 114, 135; Cheltenham 11; Chipping Campden 8, 35; Chipping Sodbury 15, 25, 57; Cirencester 8, 11, 15, 20–1, 25, 31, 35, 44, 57; Coaley 25, 33, 37; Corpus Christi College Oxford 26, 216; Coventry 55; Cromhall 34; Damery 37; Doleman's Ham 36; Doynton 37, 38; Dudbridge 45, 71; Dursley 25, 33, 36, 44–5, 57, 65, 68, 73, 80, 106, 110, 112; Eastington 45, 74, 92, 104, 123, 135; Ebley 92; Ewelme (River) 25, 28, 31, 37, 43, 72–3, 90, 93, 135; Fairford 15, 31; Falfield 34; Frome River 17, 25, 31, 38–9, 43, 69, 73, 78, 93, 95; Gloucester 8, 10–11, 13, 15, 20, 31, 35, 56–7; Halmer 37; Horsley 36, 59–60, 120, 123; Huntingford 34; King's Stanley 31, 42, 56, 92, 123, 135; Kingswood 25, 37, 55, 57, 80, 121, 123; Leonard Stanley 31, 56; Lypiatt 31, 33; Little Avon (River) 21, 25, 28, 31, 34, 37, 43, 72, 73, 86, 90, 93, 135; Marshfield 35; Minchinhampton 14–15, 17, 21, 57, 73, 123, 135; Mitcheldean 35; Nailsworth 31, 93, 97, 124, 135; Newent 35; North Cerney 76; Northleach 8, 25, 31; North Nibley 31, 34, 43, 57, 123; Owlpen 33; Paganhill 31; Painswick 39, 44, 58, 68, 79–80, 93, 123; Purton 35; Randwick 43; Rodborough 15, 31, 76; Scotland 121–2, 125, 150–1; Slad Valley 39, 80; Somerset 12, 20–2, 45, 55–6, 65–7, 69, 93, 105–9, 111, 121, 124–5; Stinchcombe 33, 38; Stone 37; Stonehouse 42, 67–8, 92, 110, 135; Stow on the Wold 8, 21, 31; Stroud 14, 35, 43, 57–9, 68, 111, 114, 123, 133, 135; Stroudwater 15, 21, 25, 28, 33, 37, 56–7, 59, 86, 90–1, 104, 112, 123, 125, 134–5; Tetbury 21–2, 25, 35–7, 45, 57, 105; Tewkesbury 31; Tortworth 37; Tewkesbury 11, 13; Thornbury 35; Toadsmoor 80, 94; Uley 33, 45, 50, 67–8, 76, 91, 96–7, 104–6, 110, 123, 135; Westbury on Severn 35; Wheatenhurst 35; Wick 38; Wiltshire 12, 20–2, 28, 36, 54–6, 65–9, 71, 81, 93, 95, 97, 99, 105–9, 111–12, 121, 124–5; Winchcombe 11–12, 21, 35; Woodchester 15, 31, 43, 65, 68, 70, 83, 110; Woodmancote 33; Wotton-under-Edge 25, 40, 57, 68, 76, 84, 86, 91, 97, 105–6, 111, 114, 123; Yate 35; Yorkshire 12, 20, 43, 58–9, 64–7, 71, 74–5, 99–100, 103–6, 108–10, 112, 121–2, 124–5, 133, 135, 140, 150
Plague 16, 55
Poverty 110, 134
Prison 134
Processes: 20–1; Bleaching 65, 70–1, 130, 143, 146; Brushing 128; Burling 63, 128; Carding 8, 64–6, 96, 100; Cutting/cropping 103, 108, 128, 143; Drying 81, 100; Dressing 128; Dyeing 10, 12, 15, 21, 64, 70, 72–3, 81, 130, 146, 150; Finishing 64, 68, 103; Fulling 8, 11, 96, 100, 103, 128, 155; Knapping 37, 45, 57, 100, 131; Knitting 134, 140; Milling 68, 128, 155; Picking 128; Potting 70; Pressing 70; Raising 69, 103, 107; Scouring 68; Scribbling 65–6, 96, 100, 103, 107, 109, 127; Shears/shearing 15, 37, 39, 43, 63, 69–70, 96, 110; Slubbing 64, 103, 107; Spinning 64, 66, 96, 103–4, 107, 128, 133; Steaming 129; Steam heating 71; Stoving 65, 71; Washing 128, 155; Weaving 8, 10, 51, 64, 67, 103, 107, 128, 157; Wool combing 37; Wool preparation 64
Profits 130
Pychley Hunt 146
Railway 99, 142, 149–50
Ready-made clothing 133, 151
Regulation 22, 27, 54, 56–8
Religious Houses: Cirencester Abbey 22, 24; Evesham Abbey 13; Flaxley Abbey 13; Gloucester 13, 23–4, 26; Godstow Abbey 24; Kingswood Abbey 24; Knights Templar 12; Lacock Abbey 13; Leonard Stanley Priory 13; Llanthony Priory 13, 24; Tewkesbury Abbey 13; Winchcombe Abbey 12–13; Pershore Abbey 13; Westminster Abbey 13
Riots 55, 58–60, 65, 67, 109–12, 114
Royal Protection 16–17
Shearmen 110, 112
Strike 58, 105, 111–12, 114
Tariff 133, 151–2
Taxation 15, 20, 27–8, 46, 54
Technical education 151
Theft 103
Trade association 110–12, 114
Transport 124
Truck 58, 114, 135
Unemployment 108
Wages 28, 59, 108
Walking sticks 158
War 55, 59, 104, 130–31, 141–2, 144, 148–50, 156
Water 73
Winchester, Bishop of 13
Woollen industry organisation 21
Work study 155